VIRAL

Also by
MATT RIDLEY

The Red Queen:
Sex and the Evolution of Human Nature

The Origins of Virtue:
Human Instincts and the Evolution of Cooperation

Genome:
The Autobiography of a Species in 23 Chapters

Nature via Nurture:
Genes, Experience and What Makes Us Human

The Agile Gene:
How Nature Turns on Nurture

Francis Crick:
Discoverer of the Genetic Code

The Rational Optimist:
How Prosperity Evolves

The Evolution of Everything:
How New Ideas Emerge

How Innovation Works:
And Why It Flourishes in Freedom

THE SEARCH FOR THE
ORIGIN OF COVID-19

VIRAL

ALINA CHAN AND MATT RIDLEY

HARPER
An Imprint of HarperCollins*Publishers*

HarperCollins books may be purchased for educational, business,
or sales promotional use. For information, please email the Special Markets
Department at SPsales@harpercollins.com.

Originally published in Great Britain in 2021 by 4th Estate,
an imprint of HarperCollins Publishers.

Maps and diagrams by Martin Brown

FIRST U.S. EDITION

Library of Congress Cataloging-in-Publication Data has been applied for.

ISBN 978-0-06-313912-1

21 22 23 24 25   LSC   10 9 8 7 6 5 4 3 2 1

For the people who have suffered and lost during the Covid-19 pandemic, those who have cared for and provided aid to others during the pandemic, and those brave whistle-blowers, doctors, scientists, journalists and sleuths who have persisted in seeking and sharing the truth.

Contents

Prologue: The mystery   1

1. The copper mine   9
2. Viruses   33
3. The Wuhan whistleblowers   49
4. The seafood market   68
5. The pangolin papers   89
6. Bats and the virus hunters   109
7. Laboratory leaks   133
8. Gain of function   172
9. The furin cleavage site   202
10. The other eight   225
11. Popsicle Origins and the World Health Organization   244
12. Spillover   270
13. Accident   280
14. The origin of Covid-19   291

Epilogue: Truth will out   303

Timeline   316
Acknowledgements   326
Notes   327
List of Illustrations   391
Index   393

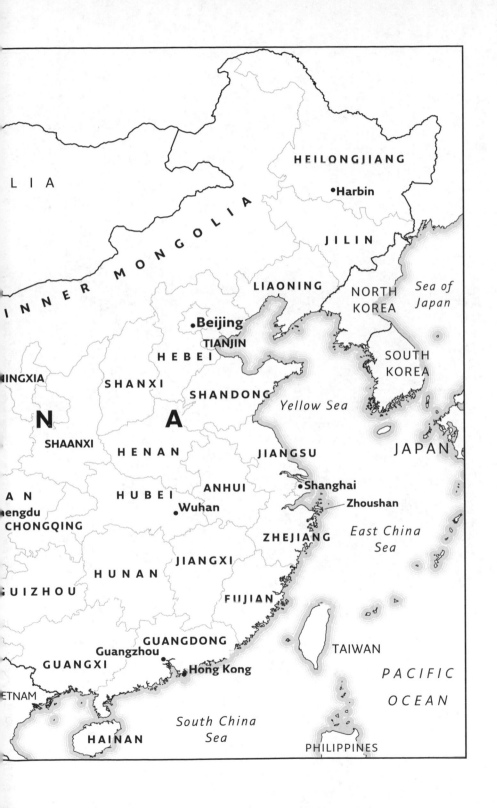

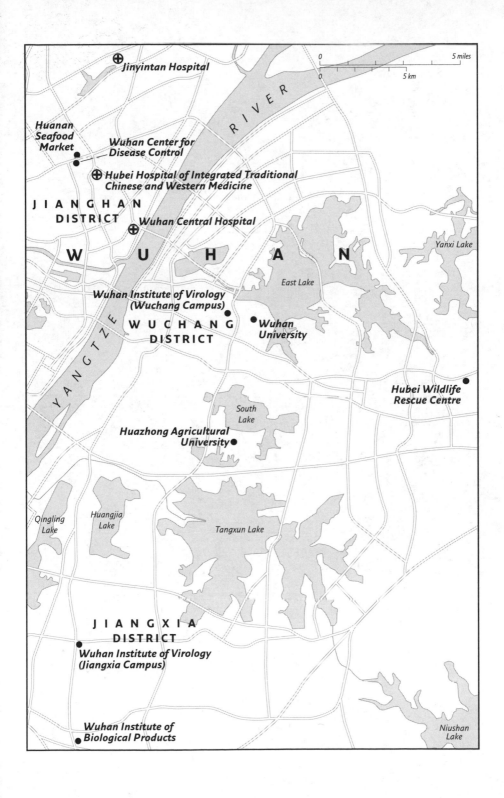

Prologue
The mystery

'The true laboratory is the mind, where behind
illusions we uncover the laws of truth.'

Sir Jagadish Chandra Bose

The authors of this book, Alina Chan, a scientist in the United
States, and Matt Ridley, a science writer in the United Kingdom,
have found ourselves inexorably drawn into the mystery of
how the Covid-19 pandemic began. Over the past year, we have
each pursued surprising revelations and rumours swirling
around the origin of the virus. These took us deep into details
of bats and SARS viruses in southern China, sick pangolins
confiscated from smugglers, and cutting-edge gain-of-function
virus research carried out in laboratories in Wuhan and elsewhere. In particular, Alina's life has been forever changed after
plunging into the question of the origin of SARS-CoV-2, the
causative agent of Covid-19, in May 2020. Her social media
has since been transformed into an open forum where prominent scientists and internet sleuths spar, publicly, on the topic.

In April 2020 Matt wrote an essay for the *Wall Street Journal*
about the virus's origin, headlined 'The Bats Behind the
Pandemic', focused on the puzzling similarity of part of the

SARS-CoV-2 genome to that of a virus from a pangolin – a small ant eater – while most of the genome resembled that of a bat virus. (A genome is the full genetic sequence of a creature.) He then read the preprint of a paper by Alina and two colleagues, Shing Hei Zhan and Ben Deverman, which came to the stark conclusion: 'Our observations suggest that by the time SARS-CoV-2 was first detected in late 2019, it was already pre-adapted to human transmission to an extent similar to late epidemic SARS-CoV.' Noticing that Dr Zhan was an expert on genomic analysis and the other two were specialists in viral vector engineering at the elite Broad Institute of MIT and Harvard, Matt judged that this preprint – a scientific paper yet to be peer-reviewed for publication – was probably a careful and reputable study. It had come to an electrifying conclusion: the virus causing the pandemic was evolving more slowly than one newly arrived in the human species from another animal normally would. This implied it was already well adapted to human beings from the moment it was first detected in Wuhan in December 2019.

On 21 May 2020, Matt wrote to Gary Rosen, an editor at the *Wall Street Journal*, proposing an article to revisit the origin question, centred on the new paper: 'There is one new paper in particular that has really thrown the issue back into the melting pot,' he wrote. 'A team of scientists finds that unlike SARS, this virus showed very little rapid adaptation/evolution in the early weeks of the epidemic, implying it was already settled into its final form. This implies that the source was almost certainly not via pangolins and probably not the wet market. It was more likely brought to the market by a person not an animal. So we don't know where it came from and that implies the source is still out there. Bad news.'

The newspaper commissioned him to write the article, so he contacted Dr Zhan and sent him a series of questions to try to understand the implications of the work for the origin of the virus. Clearly, one of the ways that a virus could have become 'well adapted for humans', in the words of the paper's title, was by having spent time in human cells or in a so-called humanised animal in the laboratory. Humanised animals, usually mice, have had their genomes altered to include a key human gene, in this case the entry receptor for certain SARS-related coronaviruses. At the time, a laboratory accident or leak still did not seem likely to Matt. He wrote to Dr Zhan: 'I'd like to focus on the issue of the lack of an apparent adaptive period of rapid genetic substitution, and its implication that the virus had been circulating in human beings for longer than expected. (I am not focusing on the lab-leak possibility, though I will probably mention it among other possibilities.)'

He received a lengthy, detailed and fascinating reply from Dr Zhan's co-author, Alina. To Matt's surprise, Alina did not pour cold water on his growing suspicions that a laboratory leak could not be ruled out. Indeed, she alerted him to a low-key announcement that week from the Chinese authorities that no animals in the Huanan seafood market in Wuhan had tested positive. This was consistent with the findings she and her colleagues reported in their paper that 'the market samples are genetically identical to human SARS-CoV-2 isolates and were therefore most likely from human sources'. On 29 May 2020, the *Wall Street Journal* published Matt's essay under the headline 'So Where Did This Virus Come From?' The article began: 'New research has deepened, rather than dispelled, the mystery surrounding the origin of the coronavirus responsible for Covid-19. Bats, wildlife markets, possibly pangolins and perhaps laboratories may all have played some role, but the simple story of an animal in a market infected by a bat that then infected several human beings no longer looks credible.'

Over the next several months, increasingly intrigued by the failure to turn up evidence of a chain of infections in a market, a village or anywhere else, Matt came to rely more and more on Alina's advice and insight as he pursued the story of the virus's origin. Eventually, he proposed that they join forces to write this book. By the time they finished writing it, because of the pandemic, they still had not met in person.

The importance of finding the origin of Covid-19

How the Covid-19 pandemic started may be the keenest mystery of our lifetime. The saga will forever punctuate the history of humanity. It has led to the deaths of millions of people, sickened hundreds of millions and dramatically changed the lives of almost every person on the planet. The impact of this invisible virus can also be measured in weddings and gatherings cancelled, jobs lost and businesses bankrupted, schools closed and parents balancing childcare and work, clinical visits missed and treatments put on hold, and innumerable people living more isolated lives than before. If we do not find out how this pandemic began, we are ill-equipped to know when, where and how the next pandemic may start.

In 2019, with more than seven billion human beings crowding the planet, crushed into dense cities, travelling frequently and far, the world was akin to a forest of dry tinder poised for a pandemic to ignite. Yet infectious diseases were in rapid retreat. The great killers of the past were under better control. Smallpox was extinct, polio endangered, typhoid rare, plague suppressed, malaria retreating, tuberculosis at a lower level than for most past centuries. Even AIDS, which emerged in the 1980s, was killing fewer people every year thanks to new treatments and public health interventions.

With each passing year the scientific tools and resources available for tracking and controlling outbreaks became ever more ingenious. A new pathogen could be detected, isolated and analysed with unprecedented speed and precision. Its genome could be sequenced, its structure elaborated and its weaknesses probed as never before. From time to time a new scare would threaten: SARS, MERS, Ebola, Zika. However, many imagined that infectious and deadly pandemics would, in the near future, be consigned to history. Few envisaged that a pandemic of Covid-19 proportions was about to begin.

There were experts who warned that this complacency was unwise, that a global pandemic, caused by Virus X, might – perhaps would – occur again. Pandemic preparedness plans and vaccine development and distribution infrastructure had to be updated. Some governments heeded this advice, focusing on the threat of a new strain of influenza, that shape-shifting menace which defied vaccine designers all too easily. The Gates Foundation and the Wellcome Trust set to work on the Coalition for Epidemic Preparedness Innovations. Surveillance of pandemic threats, by sampling wildlife for viruses with the potential to jump species, was ramped up. The United States Agency for International Development's Predict programme was established for this purpose. New antiviral drugs were tested. But none of this was enough to prevent a pandemic, as we now know.

When news broke on the penultimate day of 2019 that a pneumonia of unknown cause was sending people to hospital in Wuhan, a city of eleven million people on the Yangtze river in central China, the strongest bet was that it would soon be brought under control as SARS had been seventeen years before. It was, after all, more than a century since a respiratory disease had caused a global pandemic on the scale of millions of infections and deaths. Initial reports in January were confident that the novel coronavirus was being caught directly from animals and, like many such 'zoonoses' at first, would therefore not be so easily transmitted from person to person.

Yet in the first two weeks of January 2020, evidence mounted that the virus was spreading rapidly. By 20 January, the Chinese government was compelled to notify the world that the virus could indeed transmit between humans. Videos of overwhelmed hospitals in Wuhan began to leak online. Over the next months, scientists started to understand that pre-symptomatic individuals could transmit the virus to other people, and that the virus could spread via the air. Research and public health experts have described the SARS-CoV-2 virus as one of the trickiest pathogens to combat because of these unique characteristics that dramatically reduce the efficacy of standard public health interventions.

Identifying the source of the virus was not the most urgent priority at first, but in due course its importance would loom large, not for assigning blame, but for preventing future outbreaks. If some animal population is out there, carrying a virus exquisitely adept at infecting people, it must be found before it can trigger another pandemic. If some human practice had encouraged the spread – the farming of wildlife for food, say, or the disturbance of some natural habitat – this practice must cease. Or, if some research experiment or fieldwork project had gone awry, lessons must be learned, and laboratory practices reviewed. Searching for the origin of Covid-19 could not and cannot be some idle pastime for a few curious scientists and internet sleuths; it is a vital task for the safety of humankind and demands a rigorous, credible and evidence-based investigation by experts worldwide.

In the last year and a half, the official search for the origin of the virus has yielded no smoking gun. Tens of thousands of animals, both wild and domesticated, have reportedly been sampled across China. None has tested positive for the virus. The joint study convened by the World Health Organization (WHO) and the Chinese government resulted in more questions and fierce debates on when and how the virus emerged in Wuhan. A nagging question grew ever more urgent: why Wuhan? Out of all of the cities of the world, how did Wuhan become the original epicentre of the pandemic?

One of the most tantalising pieces of the puzzle was a medical thesis unearthed in May 2020 by an anonymous Twitter user called the Seeker, a former science teacher in India. It was around this once obscure thesis that numerous sleuths, journalists and scientists began to coalesce to trace the origin of Covid-19. The thesis carefully chronicled the story of miners in Yunnan province who had sickened with a mysterious pneumonia after working in a bat-infested mine in 2012. In the years afterwards, scientists from top laboratories, including the Wuhan Institute of Virology (WIV), home of China's most high-security virus laboratory, had repeatedly made the long journey to visit the mine to find the virus that could have infected the miners. By their accounts, they did not succeed, but in 2013 the WIV team did collect a virus that would later prove to be the closest genetic match to SARS-CoV-2. Perhaps a clue to the origin of Covid-19 lies in that distant mine in south-west China ...

1.

The copper mine

'The very cave you are afraid to enter turns out to be the source of what you are looking for.'

JOSEPH CAMPBELL

In the spring of 2012, six men were admitted to a hospital in Kunming, the capital of Yunnan province in south-west China. The chief symptoms were dry coughs, shortness of breath, high fevers, aching muscles, headaches and fatigue. All six had recently worked in the same mine in Mojiang County, clearing out bat guano, up to 150 metres deep in the bat-infested, man-made cave. The four oldest patients became critically ill and suffered respiratory failure; three eventually succumbed to the mysterious disease and died. There were signs that their immune systems had been severely damaged, allowing for opportunistic infections. This, in combination with other clinical diagnoses, suggested that an unknown viral infection was highly likely to be the cause of their affliction.

The first patient, 吕 (Lu; the full names of the patients were obscured in the thesis), aged forty-two, was admitted to the hospital on 25 April and died on 12 June. The oldest patient was admitted on 26 April and had the surname 周 (Zhou). Zhou was sixty-three years of age and died on 7 May. The two other patients also admitted on 26 April, 刘 (Liu, aged forty-six) and 李 (Li, aged thirty-two), both survived the ordeal, albeit Liu struggled in the hospital for months and was only discharged on 10 September. A fifth patient, 郭 (Guo, aged forty-five), was admitted on 27 April and died on 13 August. The last patient, 吴 (Wu, aged thirty), was admitted on 2 May. Both Li and Wu, only in their early thirties, were discharged on 28 May. The less time the patient had spent in the mine, and the younger they were, the better their prognosis and the shorter their hospital stay.

The outbreak caused alarm, and the attending physician noted afterwards that, if future cases of severe pneumonia were to be encountered in the hospital or clinic, it would be necessary to be alert to the possibility of infectious disease and take precautions against transmission in the hospital. In what sounds like an increasingly desperate attempt to diagnose and treat the cause of the sickness, which failed to respond to a barrage of antibiotics and antifungals, the doctors tested the patients for HIV, cytomegalovirus, Epstein-Barr Virus, Japanese encephalitis, haemorrhagic fever, dengue, Hepatitis B, SARS and influenza.

By the start of June, the oldest patient had died and the two youngest patients had been discharged. Senior medical experts were consulted for the remaining three critically ill cases. On 4 June, the hospital consulted Dr Xie Canmao of Sun Yat-sen University's Department of Respiratory Medicine. He thought there was a 'great possibility of fungus infection'; however, two more patients, Lu and Guo, died later in June and August respectively despite the administration of antifungal therapy. On 19 June, Dr Zhong Nanshan, also of Sun Yat-sen University, was consulted on Guo and Liu, by then the two remaining patients in the hospital, and came to a quite different conclusion: 'great possibility of virus infection'. Dr Zhong is well known in China as one of the heroes of the Severe Acute Respiratory Syndrome (SARS) epidemic of 2002–3. Born in 1936 in Nanjing, Dr Zhong trained in Beijing and at Edinburgh University Medical School, where he obtained his medical degree in 1981. He was working at the Guangzhou Institute of Respiratory Diseases in 2002 when the SARS epidemic began. It was Dr Zhong who insisted, at some risk to his own reputation, that the disease threatened a major pandemic, and he subsequently devised a treatment, based on cortisone and oxygen, that saved many lives.

For both patients Dr Zhong recommended: one, identify the type of bats at the mine; two, test the patients for SARS virus and antibodies; three, treat them with a series of antifungals and antibiotics; and four, increase airway management and apply bronchoscopy for sputum suction. Sadly, Guo could not be saved and died on 13 August. Liu survived. In May, he had been treated with antithrombotic therapy (preventing blood clots) and showed significant improvement two days later. The doctors continued the anticoagulant treatment until he was discharged in September, more than four months after he had been admitted to the hospital.

The saga of the sick miners, potentially infected by a SARS-related coronavirus in a bat-infested mine, did not go unnoticed by prominent laboratories in China. This outbreak was of such import that as well as Dr Zhong Nanshan, it drew in the Wuhan Institute of Virology (WIV), the Chengdu Military Center for Disease Control, the Beijing Institute of Pathogen Biology and even the laboratory of Dr 'George' Fu Gao, the deputy director of China's national Center for Disease Control and Prevention. (The CDC is a network of regional public health laboratories throughout China with headquarters in Beijing. Dr Gao was promoted to director in 2017.)

An abandoned copper mine

Six hours' drive south of Kunming, not far from the border with Laos, lies Mojiang County. It is designated by the Beijing government as an autonomous county, in recognition that its indigenous inhabitants, the Hani people, are a distinct ethnic group. The area is hilly, heavily wooded and sparsely populated, but with terraced fields on some of the hillsides, accessed by switchback dirt roads. The terraces have been used for growing bananas, rubber, tobacco and tea, but dense, green vegetation cloaks many of the slopes. The biggest city in the area, Pu'er, has long been famous for its tea plantations and the dark, fermented tea made from them. East of Pu'er, about twenty kilometres south of the small town of Tongguan, in the partly wooded terrain on the left bank of the Babian river, a small creek called Bengpinghe leads up into the hills. On a ridge to the south of this creek, surrounded by groves of orange trees, stands a tiny hamlet called Danaoshan. A short distance from here are the remains of an abandoned copper mine.

The location of the mine where the six miners had worked has never been officially confirmed. However, relentless digging

Tongguan Township in Mojiang County, Yunnan, September 2018.

into Chinese databases by a group of diligent sleuths unearthed a 2016 doctoral thesis, this time from the laboratory of the deputy director of the Center for Disease Control and Prevention in Beijing, Dr George Gao, which identified the mine's precise location: N 23°10'36' E 101°21'28'. As revelations about the Mojiang miners and their potential connection to SARS-CoV-2 spread on Twitter in the second half of 2020 and early 2021, a growing number of journalists took it upon themselves to visit the mine, and each faced impromptu roadblocks, official excuses and local people deterring them from getting close to it.

In October 2020, using the GPS coordinates from the thesis, the BBC's John Sudworth and colleagues attempted to drive to the mine from Kunming. According to Mr Sudworth, they were 'followed constantly for hours by as many as half a dozen unmarked cars'. As they approached from the east, the road became impassable, so they got out of their car and tried to reach the site on foot. After a long hike on rough ground, they got to the village of Danaoshan, but they were being watched. A man they met on the path refused to speak. By the time they arrived at the area the coordinates pinpointed, it was pitch black and they had to return and try again the next day. Unfortunately, their first attempt had raised the alarm and 'the authorities were more than ready' for their second visit. Mr Sudworth and his team tried again by car from a different direction but encountered a red construction truck blocking the road. They outsmarted the plain-clothes police following them by squeezing past the truck 'Houdini-like', with millimetres to spare, only to find another roadblock ahead. They set out on foot again but were 'intercepted by some very angry men, in a 4×4, clearly communicating with someone in higher authority'. Groups of men were hanging around the area, and Sudworth was warned that 'they would soon turn violent' if he did not leave. The BBC team backtracked and tried a different route but were blocked again and again. One man told them that his job was to keep them out and that they would not be able to enter Danaoshan village again. Sudworth gave up on the Mojiang mine but tried to visit a nearby bat cave, named Shitou cave, in Jinning County, where scientists had found viruses most closely matching the 2003 SARS virus. They were met by another lorry blocking the road and men in military uniforms. They were held in a field for more than an hour before being forced to leave. 'By now you get the idea. It's impossible to overstate just how large and coordinated the effort was – state-security, plain-clothes police, uniformed police, officials and local residents. When we tried to talk to anyone, they'd turn their backs,' Mr Sudworth reported in May 2021, by which time he had relocated to Taiwan.

It was clear that efforts were being coordinated to stymie journalists trying to retrace the steps of the virologists who visited the mine, and John Sudworth was only one of many thwarted reporters. A team from NBC's *Today* show were told that wild elephants were on the road so they could go no closer. In 2021, a team of undercover French journalists got close to the site and managed to speak to somebody in Danaoshan. Asked if there was a mine nearby, he replied, 'Yes, the Bengping mine is hidden over there ... It's the government that closed it. They put surveillance cameras all around the place.' In May 2021, the *Wall Street Journal* reported that its journalist did manage to get very close to the mine on a mountain bike and saw that the entrance had become overgrown with vegetation. He was detained by the Chinese police for five hours and forced to delete a photograph of the mine from his mobile phone.

The Chinese government has shown much diligence and energy in keeping people away from the mine. Yet according to its own pronouncements, it has shown less interest in conducting further investigations at the Mojiang mine. Either it has been doing such work but keeping the results secret or it has decided not to look further into what can be learned from the site. Both possibilities are concerning.

So it is to satellite images that we turn. Today the site is blanketed with greenery but images obtained from April 2012 – when the miners who fell ill were at work – show a well-used dirt road leading to a set of buildings near the top of the hill and a well-trodden path lower down the slope to what may be the entrance of a horizontal tunnel, known as an 'adit' or 'drift'. Based on research into similar sites by Brian Reed, an American engineer who has travelled extensively in rural China, including Yunnan, the buildings appear to be a U-shaped, prefab structure alongside storage tanks, typical of a mineral exploration camp. Two books unearthed by Reed reveal that the Bengping copper mining and smelting operation was originally one of the local, home-grown industrialisation ventures that were encouraged by Mao Zedong during the Great Leap Forward of the 1950s: 'In 1958, Sun Zhongxiu, deputy secretary of the county party committee, organized nearly 10,000 people to mine the Bengping copper mine.' But by 1960, owing to poor results, the mine had been abandoned. It appears there were several sites, one of which was referred to as 'Bat Cave', from which sixty tons of 6 per cent ore were extracted but not sold 'due to inconvenient transportation'. Other records show that in 1978 an attempt was made to restart the mine. Then, according to official permits, sometime around 2011, at a time of high copper prices and a boom in copper exploration, someone seems to have had the idea of reopening the mine.

In early April 2012, we know that a small group of men began to clear bat droppings from the mine, which was full of the animals, some perhaps waking from hibernation, although in this subtropical area many stay active all year. The bats were of several different kinds, but especially numerous were the

small, gregarious, insect-eating species known as horseshoe
bats. It was warm, dusty and dirty work, inhaling the dust and
noxious smell from the huge numbers of bats and their faeces.

It is not clear why the men were shovelling bat guano.
According to the 2016 doctoral thesis, before the cases of the
sickened miners, 'many people had repeatedly entered and
exited the abandoned mine, but no outbreak occurred', which
implies a regular trade was being carried out. It is probable that
the miners had been contracted either to clear the mine for
copper mining or to collect the guano to sell as organic fertiliser, or both. The bat guano trade is a lucrative one in some
parts of the world, with small amounts of nitrogen- and phos-
phorus-rich guano being used in traditional Chinese medicine,
as well as large amounts being supplied to farmers as fertiliser.
It is collected from caves mainly in Mexico, Cuba, Jamaica,
Madagascar, Indonesia and Thailand. A study of bat guano in
a Thai cave, performed in 2006 and 2007 (but not published
till 2013), found genetic material from coronaviruses in a small
percentage of the samples and warned that bat guano miners
should take preventative measures against exposure to dangerous viruses.

We do not know how many men worked in the Mojiang
mine that April and May, but six were admitted to a hospital
– not to the one in Pu'er City, less than a two-hour drive away,
nor the one in Yuxi, less than a four-hour drive away, but to the
one in the provincial capital of Kunming, almost a six-hour
drive to the north. These six men had been referred by their
local hospitals and clinics for specialised treatment at the
Kunming Medical University Hospital.

A medical thesis

We know the story of the Mojiang miners only because of the medical thesis written by 李旭 (Li Xu), a student at Kunming Medical University. Completed in May 2013 and entitled 'The Analysis of Six Patients with Severe Pneumonia Caused by Unknown Viruses', the thesis concluded that the six miners had been infected by a SARS-related coronavirus (SARSrCoV) from bats, and the author noted that it was essential to investigate the bats in the mine.

This thesis had only been discovered and shared on Twitter in May 2020 by the anonymous user called the Seeker. One of us (Alina) translated critical parts of the thesis within a day and read, with increasing distress, about each of the patients and the struggle against the unknown disease with seemingly Covid-like symptoms. We separately visited the Chinese database of academic theses and confirmed that such a medical thesis existed. Attempts by us and others to reach the supervisor and the author to further authenticate the thesis were unsuccessful.

Dr Li's 2013 medical thesis states that, after consulting Dr Zhong Nanshan, a serum immunoglobulin-M (IgM) antibody test was performed on the four living patients by the Wuhan Institute of Virology, to which the samples were sent. IgM antibodies are the body's first line of defence in response to exposure to a pathogen, so a positive IgM test means that there has been a recent exposure. The WIV test results were indeed positive, suggesting a virus infection. The thesis did not specify exactly which type of virus the WIV had found IgM antibodies for. However, the 2016 doctoral thesis from the Chinese CDC deputy director's laboratory (that had revealed the precise location of the mine) stated that the WIV had found the samples positive for SARS virus antibodies: 'The blood test results of four patients showed that: four people carried SARS virus IgG antibodies, among which two of the discharged patients with higher anti-body levels, and two hospitalised patients had lower antibody levels (Wuhan Institute of Virology).' The two discharged patients were likely the two youngest miners, Li and Wu, who had left hospital at the end of May 2012, while the two surviving patients still in the hospital were likely Guo and Liu. Curiously, the doctoral thesis described the tests as finding IgG instead of IgM antibodies. IgG antibodies are produced during the initial infection but persist for months and sometimes years as a form of long-term protection in case the body encounters a similar pathogen again. Therefore, a positive IgG test could mean that the patients had been exposed to a SARS-like virus recently or perhaps months earlier. The terms 'SARS-like' and 'SARS-related' can be used interchangeably: both refer to coronaviruses of the genus betacoronavirus in the subgenus sarbecovirus.

According to Google Maps, the WIV is 1,885 kilometres from Bengpinghe by road by the fastest route: further than New York is from Orlando or as far as London is from Rome. Yet being the leading laboratory studying SARS-like and bat-borne viruses made the WIV an obvious choice to test the patient samples. These were sent to the WIV on Dr Zhong Nanshan's instruction – to test for SARS antibodies. An immunoglobulin test result is not necessarily definitive proof of a virus having caused the disease, as opposed to some other pathogen, but both the 2013 medical thesis and the 2016 doctoral thesis certainly suggested a strong likelihood of the miners having sickened from a SARS-like virus.

Fortunately none of the patients seemed to have passed the virus on to healthcare workers or family members. The virus or viruses responsible for their illness very likely had not evolved to be as transmissible among humans as SARS-CoV-2 is. It had only managed to infect the six miners, perhaps, because they had been exposed to massive doses of the virus via the disturbed dust of bat guano within the confines of a poorly ventilated mineshaft for extended periods of time.

Virologists in the mine

No fewer than three teams of virologists visited the mine seeking the cause of the mysterious disease. According to their subsequent publications or theses, they each suspected that a virus had sickened the miners. Aside from the Chinese CDC deputy director's group, another team came from the Beijing Institute of Pathogen Biology at the Chinese Academy of Medical Sciences and was led by Dr Jin Qi. They took samples from twenty bats, nine rats and five musk shrews captured in the mine. They identified a new *Henipavirus*-like paramyxovirus in three rats and reported this discovery in a 2014 article in *Science* magazine, titled 'A New Killer Virus in China?' The article was to be the first indication received by the outside world that there had been a lethal outbreak two years earlier. The rodent paramyxovirus they discovered was similar to ones carried by fruit bats that have caused occasional lethal outbreaks in humans in Australia and Bangladesh. However, this virus was incapable of replication in monkey, human or hamster cells in the laboratory. They concluded in 2014 that it was 'more likely a curiosity'. While at the site, Dr Jin took a photograph that was later reproduced in *Science* magazine. It shows three scientists in full protective gear, gloves and masks handling specimens on a slope just outside the entrance to the mine. By lightening the background of the photograph, we were able to get a clear image of the entry to the adit, temporarily blocked by timbers and with what looked like more timbers supporting the tunnel's ceiling and walls.

The WIV team was the most persistent. It was led by Dr Shi Zhengli. Between August 2012 and July 2013, Dr Shi's group mounted at least four different expeditions to the mine, and at least three more in the subsequent two years. After all, this might prove to be the first ever recorded case of people catching a SARS-like virus directly from a bat. 'The mineshaft stunk like hell,' Dr Shi told a reporter in 2020. 'Bat guano, covered in fungus, littered the cave.'

By 2013, Dr Shi's lab had been searching caves in southern China for SARS-like viruses for close to a decade. Her group was part of the international consortium that had successfully tracked down a reservoir of SARS-like viruses – very similar to the one that had caused the 2002–3 outbreak – in a horseshoe bat of the species *Rhinolophus sinicus*, in a natural cave elsewhere in Yunnan province. Dr Shi's group continued to sample thousands of bats in the wild, took swabs and blood samples from them, analysed the specimens for virus, sequenced the virus genomes, studied the interactions between viruses and cells in the laboratory, and even altered the genomes of those viruses in a bid to understand their biology and whether they could become human pathogens. She gained fame as the 'Bat Woman' of Chinese science even before the pandemic. By 2019, Dr Shi would be the deputy director of the WIV and director of its Center for Emerging Infectious Diseases.

After sampling hundreds of bats from the Mojiang mine, Dr Shi's team discovered a single novel SARS-like coronavirus (SL-CoV) in 2013. They published a tiny part of its genomic sequence in 2016 under the name 'BtCoV/4991'. It was found in a subtly different species of horseshoe bat from the one that by then was known to harbour the progenitors of SARS: *Rhinolophus affinis*, the intermediate horseshoe bat, rather than *Rhinolophus sinicus*, the Chinese rufous horseshoe bat. 'We detected a SL-CoV-related sequence in *R. affinis*. This strain is distantly related to the previously discovered bat SL-CoVs in other *Rhinolophus* species and represents a new strain of SL-CoVs.' Years later, in November 2020, Dr Shi revealed that at least another eight SARS-like coronaviruses had been discovered in the Mojiang mine by her team in 2015.

The lack of fanfare regarding these new sarbecoviruses was puzzling. Keeping in mind the mysterious SARS-like illness that had afflicted the workers in the mine, the BtCoV/4991 sample should have been of immense scientific interest. Here was a SARS-like virus, from a horseshoe bat, found at a site where three people had died from a SARS-like disease. It was exactly what the expeditions to the mine had been seeking. Yet Shi's 2016 paper made no mention of the miners' deaths. It was left to readers to join the dots by connecting this paper with Dr Jin Qi's *Science* article about the Mojiang miners.

A sarbecovirus by any other name

Now the scene shifts to the city of Wuhan in central China in December 2019, when the Wuhan Center for Disease Control and Prevention detected a novel coronavirus in patients with atypical pneumonia. One of the first teams to obtain a sequence of the genome of the virus causing the disease was led by Dr Liu Yingle from Wuhan University's State Key Laboratory of Virology. On 2 January 2020, they took samples from two unusual pneumonia patients in Zhongnan Hospital at Wuhan University: a thirty-nine-year-old man who worked at the Huanan seafood market in Wuhan and had fallen ill on 20 December 2019, and a twenty-one-year-old woman who had had contact with Huanan seafood market staff on 22 December. By 7 January, the scientists had sequenced the virus genome, something that would have been an extraordinary achievement twenty years earlier but is now fast and routine. On 8 January they looked in databases for a match and noticed that a section of the genome of the virus they had isolated shared a 98.7 per cent identity with BtCoV/4991, the published fragment of the genome of the virus found in the Mojiang mine in 2013 by the WIV. Dr Yingle's paper was published in a prominent journal, *Emerging Microbes & Infections*, on 5 February 2020.

This ought to have been big news: a nearly 99 per cent match to the 4991 fragment was strikingly high and would have raised eyebrows, implying a possible connection to the outbreak. It got little attention, however, because two days earlier on 3 February a paper had been published in the prestigious journal *Nature* by Dr Shi's team from the WIV just across town. In that paper, Dr Shi and her colleagues had also reported assembling a full genome sequence of the new virus. They found a 79.6 per cent match between the new virus and the 2002–3 SARS epidemic virus. Only after noting this did they then mention a match between part of the SARS-CoV-2 virus and 'a short region' of the genome of another bat coronavirus called RaTG13. This fragment, they said, derived from a virus sample collected from a *Rhinolophus affinis* bat from Yunnan province. They carried out full-length sequencing on this sample and found an overall 96.2 per cent genome match to SARS-CoV-2.

This discovery caused headlines around the world. It was the first evidence that the epidemic in Wuhan had been caused by a 'new coronavirus of probable bat origin', as the title of Dr Shi's paper put it. At the time, there was speculation about the virus having come from snakes, bamboo rats or some other kind of creature sold in the Huanan seafood market – speculation encouraged by the Chinese authorities' announcement on 22 January that the virus probably came from wild animals sold at the market.

One of us (Matt) recalls noticing with bafflement that the sentence announcing the discovery of RaTG13 had no citation: 'We then found that a short region of RNA-dependent RNA polymerase (RdRp) from a bat coronavirus (BatCoV RaTG13) – which was previously detected in *Rhinolophus affinis* from Yunnan province – showed high sequence identity to 2019-nCoV.' That was it: no footnote or reference. A day was wasted fruitlessly searching for the original report of the discovery of RaTG13. Where in Yunnan was it 'previously detected' and how? The internet and scientific literature had no mention of that name 'RaTG13' before 2020. Nor did the WIV paper provide a link to any previously published genetic sequence. Dr Shi's group had, by early 2020, already described hundreds of bat coronaviruses from Yunnan province alone. None was called RaTG13.

What happened next was astonishing. On 20 February, Dr Vincent Racaniello, who is a virologist at Columbia University, published an article on his Virology Blog, speculating about how the new coronavirus had evolved naturally. In the comments beneath this essay, Dr Rossana Segreto of the University of Innsbruck in Austria noticed an anonymous post left on 1 March, saying there was a 98.7 per cent identity between part of the new virus and an entry in GenBank, the world's public gene database.

Let's meet Dr Segreto, because she is among the first of the volunteer investigators who played a vital role in disentangling the details of what happened – mostly in their spare time and at their own expense. Born and educated in Turin, it was there that Rossana Segreto earned a PhD studying the genetics of fungi that live symbiotically with orchids. She also met her German husband, a geo-ecologist, while they were both working on orchids. They moved to Norway, where she worked in Trondheim's natural history museum, using genetics to sort out the relationships among species of mosses, corals and midges. Later, they moved to Germany, where Dr Segreto commuted across the Austrian border to work at the Institute of Microbiology at the University of Innsbruck. Here her work involved the genetic manipulation of *Trichoderma* fungi, 'knocking out' genes to understand their role in parasitising other fungi with the ultimate aim of using this fungus as a form of biological control in agriculture. Seeing how easy it would be to leak an organism from a laboratory by mistake, and with an increasingly deep knowledge of genetic manipulation, Dr Segreto was unpersuaded when, in early 2020, people started labelling the possibility that the novel coronavirus had leaked from a laboratory a conspiracy theory.

The comment Dr Segreto spotted on the blog on 1 March read 'What about KP876546? 98.7% nt match.' The serial number took her to the GenBank database, where a sequence had been deposited linked to the 2016 paper from Dr Shi's group in which they announced the discovery of a virus with the name BtCoV/4991. The anonymous commenter was pointing out that one part of the SARS-CoV-2 genome was 98.7 per cent the same as a sequence from that bat virus. Dr Segreto downloaded the short sequence of 4991 and compared it with RaTG13. It was a 100 per cent match. She posted this revelation on the Virology Blog on 16 March and wrote to *Nature* to ask for clarification. She was told the authors would be contacted, but then nothing happened. So she continued to dig: 'The more I discovered, the more I had to dig. It was possessive.'

She also came across the Wuhan University paper published on 5 February, which had identified the 98.7 per cent match between the new virus and 4991. Suddenly it all made sense. The WIV's RaTG13 and 4991 were from the same sample from that mine in Mojiang, and these sequences both closely resembled the new coronavirus that was wreaking havoc in Wuhan and by then elsewhere. At this point, Dr Segreto had not yet joined the dots to find the outbreak among the miners.

On 7 March, the entry for 4991 in a Chinese database was altered to include a reference to RaTG13. This was later spotted by an anonymous Twitter user, schnufi666, who asked a close collaborator with the WIV, Dr Peter Daszak, the president of the New York-based EcoHealth Alliance, to confirm the identity of the two samples. On 9 May, Dr Daszak replied: 'The answer is already in the papers & obvious to people working in virology', which was curious given that it had been far from obvious to the Wuhan University team or Dr Segreto. It was not until July 2020 that Dr Shi would publicly state that RaTG13 and 4991 were the same thing. The *Sunday Times* in London ran a story, as did the BBC, which explicitly stated: 'The BBC has had it confirmed, from the researchers running a respected Chinese database, that it is the same virus as one that the WIV previously referred to as RaBatCov/4991. There is no explanation for the change of name.'

A charitable interpretation is that the renaming may not have been intended to disguise the origin of the virus. Samples do get renamed occasionally and the original sequencing was of just a short stretch of the virus's genome, whereas the new name now referred to a whole genome. Dr Daszak claimed in an interview that there had been a change in the naming conventions: 'The conspiracy folks are saying there's something suspicious about the change in name, but the world has changed in six years – the coding system has changed.' (But, in 2021, it emerged that the virus that had been fully sequenced was still referred to as Ra4991 in a 2019 master's thesis from the WIV.) Yet even if the naming system had changed, scientific readers would still expect an explicit reference to the prior publication of the sample in 2016. Why leave out the details of where it was found and the link to mysterious, unresolved cases of severe pneumonia? Certainly, the outcome of the renaming and lack of citation obfuscated the origin of RaTG13.

The real story of RaTG13

It was not until 18 May 2020 that the true story of RaTG13 was understood, thanks to the discovery of the 2013 Kunming medical thesis by the Seeker, an anonymous Twitter user – but no thanks to the WIV or *Nature*. The thesis had been dredged from cnki.net, the official website for master's and PhD theses in China. On 20 May, the Seeker sent it to a husband-and-wife research team, Monali Rahalkar and Rahul Bahulikar, at the Agharkar Research Institute and the Central Research Station, Pune, India. They had just that day published a preprint online pointing out the electrifying similarity between 4991 and RaTG13. By 29 May, the Seeker had found and shared the 2016 doctoral thesis from the laboratory of the Chinese CDC deputy director.

In the pages that follow, we will partly rely on sources like the Seeker. Where the scientific and intelligence establishments had, in 2020, displayed only a surprisingly shallow interest (at least publicly) in the origin of the pandemic, these online sleuths have filled the gap. The Seeker's real name is Prasenjit 'Jeet' Ray. He is a slim thirty-year-old, with shoulder-length hair, living in the city of Bhubaneswar in the Indian state of Odisha. With the exception of a brief excursion into Nepal, he has never left India. Many of those who have followed his repeated revelations and penetrating observations speculate that he might be a professional intelligence agent, or the spokesman for a team of data analysts. He has been accused of working for the CIA or Indian intelligence. The truth is more surprising and less dramatic. He is just a very clever young man who says he has prided himself on his internet research skills ever since his college days. 'I learned how to make search engines work for me,' he told us. 'It was more madness than method.' Born in Patna, he grew up in six different cities, as his father, a civil servant working in the airport and military sectors, moved around. As a child, he learned Sanskrit from a pundit who would cycle ten kilometres to teach him and he did well enough to debate in the ancient language. He went to college in Pune to study architecture, then took a certificate from the Zee Institute of Media Arts in Mumbai. Though he has worked as a science teacher and considers himself scientifically literate, he has no special expertise in medicine, biology or technology. But, time and again, we found that he combined a ruthless preference for facts over speculation with a sharp nose for new insights. Drawn to the question of where the virus came from, the Seeker posted his first findings in response to one of Alina's tweets on 8 May 2020. Ten days later, rifling through the cnki.net website and using a login that he had been given on an open forum, he stumbled on the Kunming medical thesis, with its astounding tale of six men sickening, probably from a virus caught from bats and investigated by Wuhan virologists.

The Seeker became a member of a particularly tenacious, loose confederation on Twitter, calling itself the 'Decentralized Radical Autonomous Search Team Investigating COVID-19', or Drastic, that keeps turning up vital details about the possible source of the virus. Drastic describes itself as a 'rag-tag team of internet researchers and sleuths' who 'came together on Twitter as a radical and subversive scientific collective to fight back against mainstream bamboozlement around the origins of COVID-19, and help spread the word about the dangers posed by unchecked gain-of-function research'. (We will come back to what 'gain of function' means in Chapter 8.) Drastic has no leader, but the closest it has to a coordinator goes by the pseudonym of Billy Bostickson, an indefatigable collator of information who claims to have a combination of experience with Chinese culture and expertise in biosafety and biosecurity. Prior to his involvement with Drastic, Bostickson told us that he had spent nearly a decade empowering 'activists living under repressive regimes through collaborative sharing of electronic and audio-visual materials'. Drastic had its beginnings on Twitter, one of the few places that did not censor discussion of the virus origin throughout the first year and a half of the pandemic. Facebook flagged as 'false information' much of the work of these sleuths, even when it proved to be true information, while Reddit simply deleted it. Some of the two dozen or so people listed as members of Drastic seem more reliable, others less so; some are experts in biology and bioinformatics, but others are not; some use their real names while others are anonymous. They are not the sort of sources that authors usually turn to. But these internet sleuths have found nuggets of information that have since been confirmed more formally by qualified experts and even by the WIV itself. Bostickson came up with the name for Drastic: 'We are decentralised and autonomous and a little radical.'

Reconstructing the events of early January 2020, when the SARS-CoV-2 virus genome was first sequenced, it must have been with a mixture of excitement and concern that Dr Shi found a near perfect match for the new virus genome from within her own laboratory: 4991 aka RaTG13. She later told *Scientific American* that she had frantically searched through her lab's records to see if there was a possibility of the SARS-CoV-2 virus having leaked from the lab. Every laboratory-related incident was reviewed. None of the viruses her team had sampled from bat caves were a match to SARS-CoV-2. 'That really took a load off my mind,' said Dr Shi. 'I had not slept a wink for days.'

Yet in publishing their discovery, Dr Shi and her colleagues neglected to connect the newly renamed RaTG13 to 4991, neglected to cite her own 2016 paper describing its discovery and origins, neglected to identify the mine where the bat sample had been collected, and neglected to mention that RaTG13 was from a site where three people had died of a respiratory illness of unexplained origin.

This was not the end of the revelations about how the seminal *Nature* paper by the WIV had been economical with the truth. The story of RaTG13 continued to evolve. Dr Peter Daszak of the EcoHealth Alliance, who had helped fund and had collaborated with Dr Shi's group, gave an interview in which he asserted that the sample, whatever its name, had remained neglected in a freezer in Wuhan for six years until its similarity to the virus causing Covid-19 was noticed. He repeated this in a July interview with the *Sunday Times*. Only when the match was noticed in January had the entire sequence of that virus been assembled. That sequencing event, he claimed, depleted the sample so that nothing was left for further analysis. The WIV's *Nature* paper had given the same impression. This turned out to be untrue. When the raw data underlying the RaTG13 genome sequence was uploaded to the GenBank database in May, its date stamps revealed that its various parts had been sequenced in 2017 and 2018. This discovery by internet sleuths and one of us (Alina) makes it all the more puzzling that the WIV scientists first fixated on the underwhelming 79.6 per cent similarity to a SARS virus, rather than the whopping 96.2 per cent similarity to a complete bat virus genome that was already in their database.

Once again, Dr Shi eventually confirmed this account of events deduced by sleuths. In her interview with *Science* magazine in July 2020, she conceded that her group had actually sequenced the complete genome of RaTG13 in 2018, not after the Covid-19 outbreak as originally implied by the *Nature* paper. But she failed to address the glaring question of why this had been obscured, in addition to the links to the Mojiang miners.

To summarise, an outbreak of mysterious pneumonia in a copper mine, more than 1,800 kilometres by road from Wuhan, led to patient samples being sent to Wuhan for analysis. A 2013 medical thesis concluded, after incorporating results shared by the WIV, that these miners had likely been infected by a SARS-like coronavirus from bats in the mine. An expedition by Wuhan virologists to seek the viral cause brought back hundreds of samples from bats. Their repeated visits to the mine turned up a bat-borne coronavirus in 2013, which was recognised to be a novel SARS-like coronavirus. The WIV team partly sequenced this new virus in 2017 and then fully sequenced it in 2018. When its sequence was found to closely match the sequence of the virus causing Covid-19, the Wuhan scientists published it under a new name and failed to cite their own paper detailing its discovery or to reveal that they had been studying the virus over the past few years or to mention that it had come from a mine where there had been a fatal outbreak of pneumonia.

The November addendum

Eventually, on 17 November 2020, *Nature* published an addendum to Dr Shi's February paper. The addendum acknowledged the existence of the Kunming medical thesis; it confirmed the story about the Mojiang mine, the testing of the miners' samples at the WIV, the new name assigned to the sequence from the 4991 sample, and the sequencing of its full genome by 2018 rather than 2020. It had taken nine months for the WIV scientists to come clean about the scintillating history of RaTG13. Dr Segreto, Dr Rahalkar, the Seeker and other sleuths had been correct about these points. They had been dismissed as conspiracy theorists at the time and derided by prominent virologists. Crucially, the addendum confirmed that 'we suspected that the patients had been infected by an unknown virus'.

The addendum, however, contradicted both the 2013 medical and the 2016 doctoral theses by stating that 'we tested the samples using PCR methods developed in our laboratory targeting the RNA-dependent RNA polymerases (RdRp) of Ebola virus, Nipah virus and bat SARSr-CoV Rp3, and all of the samples were negative for the presence of these viruses'. It also reported that they had tested for antibodies against the same three viruses with negative results. Remember that the medical thesis claimed that the WIV had detected IgM antibodies against viruses in the miners. The doctoral thesis said that the four living patients at the time of sampling carried IgG antibodies against SARS virus.

The addendum revealed two further startling pieces of information. Not only were the samples from four sick miners tested at the time for viruses, but they were retested after the Covid-19 pandemic began. In other words, the samples from the patients had been stored at the WIV throughout the intervening period, yet had not been shared with independent research groups outside of China for verification.

The second revelation was that far from turning up just one SARS-like, bat-borne virus in their expeditions to the Mojiang mine, the WIV team had in fact found at least nine: 'Between 2012 and 2015, our group sampled bats once or twice a year in this cave [sic] and collected a total of 1,322 samples. From these samples, we detected 293 highly diverse coronaviruses, of which 284 were designated alphacoronaviruses and 9 were designated betacoronaviruses on the basis of partial RdRp sequences. All of the nine betacoronaviruses are SARSr-CoVs.' The WIV did not disclose more information about these other eight SARSrCoVs.

The mine in Mojiang County remains, in 2021, one of the closest clues yet found to the origin of the pandemic: it is the place where the virus most closely matching SARS-CoV-2 was found – a virus that was studied in a Wuhan laboratory before the detection of the Covid-19 outbreak.

Four months after the *Nature* addendum, the WIV appeared to revise its story; it told members of the WHO-China global study of origins of SARS-CoV-2 that 'the reported illnesses associated with the miners, according to the WIV experts, were more likely explained by fungal infections acquired when removing a thick layer of guano'.

2.

Viruses

'We live in a dancing matrix of viruses; they
dart, rather like bees, from organism to
organism, from plant to insect to mammal to
me and back again, and into the sea, tugging
along pieces of this genome, strings of genes
from that, transplanting grafts of DNA,
passing around heredity as though at a great
party.'

LEWIS THOMAS

It is the year 1964. A woman on the young side of middle age,
dressed in a white lab coat, is adjusting a large machine in a
laboratory at St Thomas's Hospital in London. Originally from
Glasgow, June Almeida, née Hart, left school at sixteen, but
after training as a laboratory technician has become skilled at
operating electron microscopes while living and working in
Toronto with her Venezuelan artist husband and daughter. So
skilled that she has recently been lured back from Canada to
help with the electron microscope at this London hospital. She
has developed a technique for gathering virus particles beneath
the electron beam so that they can be photographed.

On this particular day in 1964, she is focusing the electrons on a sample sent from the Common Cold Research Unit, a British government laboratory devoted to understanding what causes runny noses. The unit's Dr David Tyrrell has identified one cold, taken originally from the nose of a boy at a Surrey boarding school, that seems to be unusual – it won't grow in the normal cell cultures, although it readily causes colds if squirted up the noses of volunteers. It is clearly a virus, passing through filters that sequester bacteria; however, it fails tests

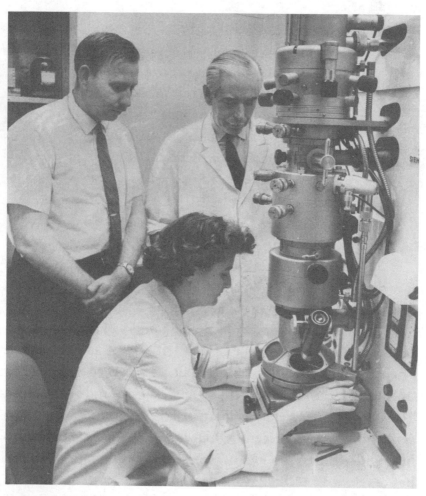

June Almeida in 1963 at the Ontario Cancer Institute in Toronto.

that single out influenza viruses, rhinoviruses, adenoviruses and other viruses. Unlike these other viruses, it is easily killed by ether, which implies that it might have an oily or fatty component.

Dr Tyrrell had sent a sample to Almeida. She skilfully managed to take a photograph of the virus, and sure enough it had a distinctive appearance: round and decorated with a ghostly crown that reminded her of the gaseous effusions from the surface of the sun known as the solar corona. This rang a bell with Almeida. She had seen similarly shaped viruses in mice and chickens. But when she submitted a paper to a journal and included the pictures, it was rejected – these are just blurred images of ordinary influenza viruses, she was told. Undeterred, Almeida stuck to her guns, gathered more data and, in 1968, she and seven colleagues wrote to the journal *Nature* and suggested that a new class of virus be recognised, the 'coronaviruses'. The sample from the Surrey boarding school was later lost so we do not know which coronavirus it was, but we now know that there are four types of coronaviruses that cause the common cold. They go by the names 229E, NL63, HKU1 and OC43.

Today the coronaviruses are divided into four groups based on their genetic differences. The rarest of those studied to date are the gamma- and delta-coronaviruses, none of which infect people. These are mostly carried by birds and pigs. The most commonly found are the alpha-coronaviruses, mostly from mammals, and two of which infect human beings: 229E and NL63. The remainder are the beta-coronaviruses, found frequently in bats and rodents, of which four have infected people before 2019: HKU1, OC43, MERS and SARS. Within the beta-coronaviruses lies a 'species' known by three different names: lineage-B, SARS-like (or SARS-related) or sarbecoviruses. SARS-CoV and SARS-CoV-2 are both sarbecoviruses.

The nano-world

Viruses can only make more copies of themselves by hijacking a host cell and leveraging the host machinery to make more viruses. This inability to replicate on their own leads most biologists to conclude that they cannot be described as living things, even though they are clearly made of protein and nucleic acids and are parasites of living organisms. Similar to living organisms, viruses are built on a genetic code that is enclosed by a protein shell and sometimes a membrane envelope. Once inside the cell of a host, viruses reveal their true nature as they churn out ribbons of genetic tape instructing the cell to turn over all its machinery to replicating and multiplying virus particles.

Viruses are all around us all the time. All are highly specialised so the vast majority cannot infect people, though they each have to infect some living organism to reproduce. In the oceans alone there may be a million trillion trillion viruses. Many viruses, called phages, actually infect and kill bacteria, so many scientists today are innovating phage therapies as an alternative to antibiotics in the face of rising bacterial resistance against frontline antibiotics.

The SARS-CoV-2 coronavirus that causes Covid-19 in human beings can also infect animals such as cats, tigers, lions, minks, ferrets, otters, raccoon dogs, dogs and non-human primates. It is very, very small. Each particle of SARS-CoV-2 weighs about thirty quintillionths of an ounce, or approximately one quadrillionth of a gram. All the SARS-CoV-2 viruses in all the millions of patients in the world at any given point in time could fit inside a single soda can.

The infectious diseases that have ravaged human populations throughout history were not all viruses. It was a bacterium, the plague, that devastated the population of much of the Old World in the 1300s. It was a virus, smallpox, that wiped out much of the population of the New World after European contact. And one of today's biggest killers, malaria, is neither: it is a protozoan. Very roughly, in terms of their size, if malaria were a cat, the plague would be a mouse and small-pox a flea.

Bacteria were known from the late 1600s, but the first virus was not discovered until the mid-1800s. Scientists studying tobacco crops found that the plants were not afflicted by a bacterium or fungus, but by an entirely new type of 'invisible contagion' that was much, much smaller than single-cell microbes such as bacteria. Unable to satisfy the criteria known as Koch's postulates, which had become the gold standard to demonstrate a microbial infection, the scientists nonetheless proved that the plant disease could be transmitted by inoculating healthy plants with sap from diseased plants. In 1898, a Dutch scientist named Martinus Beijerinck coined the word 'virus', derived from a Latin word meaning a liquid poison. Within a few years, the first human virus was identified: the yellow fever virus.

Today, although some viruses get into us through our food, such as norovirus, and some arrive via the mouthparts of insects, like yellow fever, dengue and Zika, the most common viruses that afflict people in developed countries are respiratory viruses. These get into our bodies via our noses or throats. Some two hundred kinds of common cold, most of them rhino-viruses, adenoviruses and coronaviruses, and a bunch of flu strains, swarm around the world, descending on schools with special glee. The average child catches seven colds a year. The fact that people live in dense cities, crowd into schools or bars, and travel long distances, suits such viruses just fine. It may be no accident, as we shall see, that bats carry a galaxy of viruses, for they live in larger aggregations than any other mammals bar us. In one Texan cave, at certain times of year, twenty million Mexican free-tailed bats roost together, or roughly the (human) population of the Mexico City area.

Deciphering the genomic code of viruses

As with influenza, a coronavirus genome is made of RNA, a slightly less tidy cousin of DNA. These genomes spell out instructions for building more copies of the virus in a four-letter code: A, T, G and C for DNA, but A, U, G and C for RNA. As we shall see, scientists have chosen to represent the genomic texts of RNA viruses in the language of their DNA equivalent. This means very simply that they use the letter T instead of U.

Much of the argument about where the SARS-CoV-2 coronavirus came from will turn on this textual analysis of its genome. Just as the people who study Jacobean plays can trace the authorship by close analysis of the text and can follow the pedigree of different versions of the plays by the mistakes that were made by one scribe and copied by others, so we can trace the ancestry of a virus by the mutations that happened to its text in one generation and were inherited by later viruses. If you compare three genomic texts – one taken from a bat virus, one from a pangolin virus and one from a human SARS-CoV-2 virus – and find one part of the text to be more similar in the bat and the person, while another part is more similar in the pangolin and the person, then you might be looking at a genome that was the result of recombination: a paragraph of one text replaced by the equivalent paragraph from another text. By such methods can we inch towards an understanding of where and how the text originated.

Here's an analogy. The play 'All Is True' was later renamed as 'The Famous History of the Life of King Henry the Eighth' when it was included in the first folio of William Shakespeare's works. Until 1850 it was thought to have been written by

Shakespeare alone. Then the scholar James Spedding argued that it had been partially written by John Fletcher, who was Shakespeare's successor as the principal playwright for the King's Men theatre company. He made this claim based on the appearance and style of many eleven-syllable lines that were characteristic of Fletcher. More than a century later, another scholar, Cyrus Hoy, also tried to map the play, dividing the scenes between Shakespeare and Fletcher, based on the words used. Fletcher, for example, uses 'ye' more often than Shakespeare, who prefers 'you'. Hoy's scheme mapped neatly onto Spedding's hypothesis. Meanwhile, the contemporary sources used by the two authors can also be deduced from their text. The majority came from Raphael Hollinshed's *Chronicles*, but both also consulted the authors John Foxe, John Stow and John Speed.

Our point is that the textual analysis of the genome of a virus can be equally illuminating and in a surprisingly similar way. Tiny characteristic features of the genome can indicate something critical about where the virus came from, how it evolved, or what it is related to, and yet such signs can be ambiguous and highly debated among scientists.

RNA messages are translated into another: the language of proteins. Whereas RNA has a four-letter alphabet, each letter being a different nucleotide, proteins have a twenty-letter alphabet, with each letter being a different amino acid. The RNA text is read in three-letter words known as codons, and there is redundancy: several different three-letter RNA words can translate into the same amino acid. So some changes in the RNA message do not alter the protein code. Such changes are said to be synonymous. However, changes in the RNA message that do alter the protein code are described as non-synonymous and are more likely to affect the properties of the resulting protein. As an analogy, in transcribing Shakespeare, changing 'ye' to 'you' is a synonymous change, but 'ye' to 'we' is not. Virus genomes also experience both synonymous and non-synonymous changes.

An unfortunate and dangerous combination

The four coronaviruses that cause common colds attack the upper respiratory tract. The viruses that cause Severe Acute Respiratory Syndrome (SARS) and Middle East Respiratory Syndrome (MERS) are less effective in the upper respiratory tract but instead target the lungs to cause severe respiratory disease. SARS-CoV-2 excels in both the upper and lower respiratory tract. In patients where the virus stays in the upper respiratory tract, symptoms similar to the common cold can develop, such as coughing and a sore throat. But when the virus makes it into the lungs, it can cause severe disease by ravaging the fine structure of the pockets known as alveoli, resulting in characteristically opaque shadows on scans. In especially severe cases, the infection can trigger a 'cytokine storm' of inflammation caused by an overreaction of the immune system, and in the worst case it can lead to irreversible tissue damage.

SARS-CoV-2 does not stop at the respiratory tract. It also infects other organs of the body, including the intestines, heart, blood, sperm, eyes and parts of the nervous system. The infection is systemic, sometimes damaging even the kidney, liver and spleen. Less well understood is how the virus affects the immune system – whether it can infect and kill immune cells, how it camouflages itself to avoid detection by the immune system, and how different proteins in the virus can disrupt anti-viral host responses. No other virus has been the research focus of so many scientists across such diverse disciplines and across so many countries. Scientists have generated a tremendous amount of data characterising this virus, but are still learning the tricks it has up its sleeve.

How does the virus get inside a cell? When it first encounters a susceptible human cell, let's say in the throat, the virus uses the spike proteins protruding from its surface to latch onto a receptor on the human cell surface. Specifically, the spike protein has a receptor-binding domain (RBD) that determines exactly which host receptor the virus can use for infecting cells. For SARS-CoV-2, this receptor is called Angiotensin Converting Enzyme-II (ACE2) and is present across a plethora of cell types in the human body; ACE2 has its own job to do, chiefly to modulate blood pressure. Unfortunately, this means that the cells that line arteries and veins, as well as lungs and small intestine, are covered in ACE2 receptors, offering an open invitation to the virus. ACE2 can also be found in heart, kidney, liver, testes and brain. So it is no surprise that Covid-19 has more severe implications for patients who are obese, diabetic or suffer from hypertension – medical conditions that have already weakened the organs targeted by SARS-CoV-2. Covid-19 symptoms range from common ones, such as fever, aches, cough, shortness of breath and loss of smell, to less expected ones such as headache, diarrhoea and stroke. Some Covid-19 patients who have recovered also report 'long Covid' symptoms: a lasting fatigue, difficulty breathing or engaging in rigorous physical activity, and mind fogginess that plague even individuals in their prime.

Other animals also have their own version of ACE2, which is why SARS-CoV-2 can infect a wide range of species. The first SARS virus uses ACE2 as a receptor too and was also found to infect a variety of animals: during the 2002–3 epidemic, palm civets, raccoon dogs and hog-badgers among other species were found to carry the virus. Scientists later discovered that the SARS virus could also infect ferrets and hamsters, which are used in laboratories to better understand the disease. Over the past two decades, in the study of SARS viruses in the lab, scientists have tested them on cultured cells that express ACE2 from different species and have engineered mice that express the human ACE2.

SARS-CoV-2's spike was found to bind to human ACE2 at least as strongly as the spike of the 2003 SARS virus. But that was not all. Scientists inspecting the genome of SARS-CoV-2 quickly noticed that the novel coronavirus possesses a feature called a furin cleavage site in its spike that has not been found in any other SARS-like virus, though it does appear in other distantly related coronaviruses. This short genetic sequence helps the virus to infect varied types of cells, specifically those that include a protein called furin, and is probably one of the reasons the virus is capable of causing a pandemic. Where it got the furin cleavage site from is controversial, as we shall discuss later.

After entering the cell, the virus takes off its coat, as if visiting a friend's house. But this guest is not here for idle chit-chat. The genetic material of the virus (approximately thirty thousand letters of single-stranded RNA in the case of SARS-CoV-2) gets to work commandeering the cell's machinery. The SARS-CoV-2 virus has fifteen genes spelled out by its RNA, which between them specify the recipes for twenty-nine proteins. These include sixteen non-structural proteins that are not parts of new virus particles but play a role in transforming the host cell into a virus factory. For instance, nsp12 is the RNA-dependent RNA polymerase (RdRp) that forms part of a complex that makes more copies of the virus genome; nsp3, nsp4 and nsp6 reorganise the internal structure of the host cell to be more conducive to virus replication; and nsp14 proofreads the replicated genomes to ensure that they are accurate. This is particularly critical for coronaviruses because their genomes are almost double the length of some other RNA viruses such as the influenza virus. Too many errors can render the virus incapable of infecting new hosts. The RNA genome also encodes structural proteins that form part of the final virus particles that are released from the host cell: the nucleocapsid (N) protein that wraps the RNA, the membrane (M) protein that holds the membrane together, the envelope (E) protein that encloses the core, and the spike (S) protein that sticks out from the membrane of the virus and latches onto host cells to facilitate infection. Several accessory proteins, which have a variety of jobs, are also encoded in the genome; there are eight of these in the 2003 SARS virus and nine predicted in SARS-CoV-2.

The Russian 'flu' pandemic

Of the four coronaviruses that cause versions of the common cold, two are thought to have been caught originally from bats and two from rodents. The most common of them is called OC43. It is a highly seasonal bug that causes about one in every ten common colds. There is good evidence that this one is quite a recent arrival in our noses and throats. Its genome is a 96 per cent match to a coronavirus that causes diarrhoea in young calves, called bovine coronavirus or BCoV. In 2005, Dr Marc van Ranst at Leuven University in Belgium examined parts of the genomes of the two viruses and, based on estimates of the rate of evolutionary change, concluded that they diverged in around 1890.

It may be just a coincidence, but 1889–90 saw a pandemic, the worst in the nineteenth century. It has always been assumed that it was caused by an influenza virus, but there is no direct evidence of this. Given that human coronaviruses were unknown till the 1960s, the possibility that this was a coronavirus pandemic was not considered until, in the early 2000s, Dr van Ranst, a coronavirus expert, noticed the coincidence of the date of the pandemic and the likely date of OC43's arrival in the human species. That hypothesis has now been somewhat strengthened by the Covid-19 pandemic, because of some similarities between the epidemiology and symptoms of the two pandemics. In particular, both diseases largely spared children, unlike influenza, and both affected men more severely than women. Both cause loss of taste and smell in some people.

The outbreak began in central Asia, in the independent city state of Bukhara in May 1889. By October, it had reached Krasnovodsk on the Caspian Sea, from where it took the train along the Volga eventually to Moscow and St Petersburg, the Russian capital. The symptoms included high temperatures, swollen hands, face rashes and agonising body aches. The acute phase of the illness lasted for five or six days but sometimes left the victim exhausted for weeks. That autumn, according to one estimate, the disease laid low 180,000 people in St Petersburg, out of a population of about a million. By December, military hospitals in the capital were unable to cope and several factories had shut down for the lack of workers. Sweden's turn came in November, when most of the soldiers of an artillery corps stationed on the island of Vaxholm fell ill. Within eight weeks, more than half the population of Stockholm had caught the virus. In December, the Prussian capital of Berlin was paralysed by the novel pathogen. Universities suspended lectures, government officials were said to be unable to carry out their functions and the fire brigade was disabled for lack of manpower. A contemporary report stated that schools were controversially closed throughout Germany even though the disease was not as great a threat to children as to adults. In Vienna, the schools closed early for Christmas and stayed shut till late January.

The same month Italy, France, Spain and Britain fell to the invisible invader. By Christmas, Paris's hospitals were overwhelmed. In Madrid, in early January, three hundred people a day were dying, and being buried at night so as not to spread alarm. In London, on one day in January at St Bartholomew's Hospital in the City of London, Dr Samuel West found more than a thousand people crowded into the casualty ward, most of them men. According to a modern analysis, the death rate peaked in the week ending 1 December 1889 in St Petersburg, 22 December in Germany and 5 January 1890 in Paris. By February, the disease had reached America, South Africa and India. By April, it was in Australia and China. This was the first global pandemic made possible by railways and steamships.

As fast as it came, the plague subsided and by the summer it was apparently over, though not in the southern hemisphere where it was winter. It returned north in the autumn for a second wave, killing Queen Victoria's grandson Albert Victor, among other prominent people, then again and again in the next few years with diminishing ferocity.

Do deadlier or milder viruses spread better?

The question of whether new viruses evolve to be more or less virulent over time in new host species populations has no easy answer. The sicker you get, the more viruses are being bred in your body and likely expelled into the environment and transmitted to other people. But a virus that kills its host too quickly reduces the chances of transmission in the community. A person lying on their deathbed is a dead end for the virus. Selection will therefore favour strains that keep their hosts alive and healthy long enough to spread the virus from person to person. Pumping out lots of viruses while keeping the patient active would be the perfect compromise.

A key insight came from the evolutionary biologist Dr Paul Ewald, now of the University of Louisville, beginning in the 1980s. Ewald argued that the mode of transmission influences the trade-off between virulence and contagion. Diseases spread by direct contact, and which cannot survive for long outside the body, will evolve to be low in virulence, so that the infected person remains as active as possible, interacting with a large number of people. Diseases that spread by other means – especially dirty water or insect bites – are expected to remain or become highly virulent, such as cholera, plague and malaria, because they do not pay a price for immobilising and killing the host. Diseases that spread by sex will become good at hiding in the body for a long time, so that the victim has time to move on to a new sexual partner. There is another category. Ewald argued that what he called attendant-borne illnesses – which are spread from patient to patient by nurses or other helpers – would often remain or become highly virulent, because they could spread from incapacitated and dying hosts: the sicker the patient, the more visits they get from carers.

Respiratory viruses spread through the air by coughs and sneezes can benefit from low virulence, because they have more chances of spreading from people who don't feel too ill to go to work or to parties – resulting in superspreading events.

How then would Ewald explain the high death rate from influenza in 1918? It was not spread by insects or dirty water or sex, was not durable, and yet it killed with gusto. Unlike almost every other flu outbreak, this one incapacitated many of its victims and killed a significant proportion of them. What is more, it became more deadly as time went on. The second wave in the autumn of 1918 was more lethal than the first in the spring. The episode, in which an estimated fifty million people died, cast a long shadow over the future. Every time a new variety of flu appeared, for example in 1957, 1997 or 2009, the world shuddered with fear expecting a terrible pandemic. Yet each time the flu faded into mildness, killing comparatively few people. What was so special about 1918?

Ewald hypothesised that the 'high virulence of the 1918 pandemic resulted from natural selection acting under unusual environmental conditions'. Those unusual conditions were to be found on the Western Front of the First World War. They allowed 'individuals immobilised by illness to be transported repeatedly from one cluster of susceptible hosts to another, in trenches, tents, hospitals, and trains'. A highly virulent version of the virus that rendered its victim so ill he could not move was now at no disadvantage. Imagine two soldiers who catch the flu in the front line. One has a mild case, the other a life-threatening one. The first is sent to rest in a dugout or billet somewhere, meeting a few comrades on the way. The second is taken by stretcher bearers to a crowded medical station, then on a crowded ambulance to a crowded field hospital, from where by crowded train he is sent to a nursing home back in England, say. The virus was free to hop off into wounded or convalescing troops, and those attending them, at various points along the way.

To test this hypothesis, Ewald examined the mortality in different phases of the epidemic. If the virulence was explained by the novelty of the virus in human hosts, its first outbreaks should have been the most virulent. This was not the case. The first recorded cases, in military camps in Kansas in March 1918, showed normal mortality for flu. As the virus spread through army camps and cities in the spring and summer of that year, mortality remained moderate. It was in the diary of Jefferson Kean, the deputy chief surgeon of the US Expeditionary Force, in northern France in mid-August, that reports of high death rates first emerge. He had described the flu as mild in April, May, June and July. On 17 August, he wrote, 'influenza increasing and becoming more fatal'. By September, many doctors had noticed the change. The more lethal strain spread round the world, later subsiding in both frequency and virulence during 1919. Ever since, this H1N1 flu has been not a tiger but a house cat, and so has every other strain that has come along – possibly due to the prevailing (but fading) immunity in the hundreds of millions of people who were infected by the 1918 flu and later other strains of influenza.

How SARS-CoV-2 will continue to evolve in terms of deadliness and transmissibility between humans (and other animal hosts) remains to be seen. But we do know one thing: the virus is here to stay.

3.

The Wuhan whistleblowers

'The humanities teach us the value, even for
business, of criticism and dissent. When there's
a culture of going along to get along, where
whistleblowers are discouraged, bad things
happen and businesses implode.'

MARTHA NUSSBAUM

It is still unclear when the first case of Covid-19 was diagnosed
and linked to a coronavirus. At least two western infectious
diseases experts revealed that they first learned of unexplained
pneumonia cases in Wuhan hospitals in mid December. One got
the news from a Guangzhou scientist through an online epidemiology network and the other heard about a novel coronavirus
in Wuhan that 'looks very serious' at a dinner party in the US.
One account by a lab technician, which has since been removed
from the internet, states that a doctor at the Wuhan Central
Hospital took a sample from a patient's lung on 24 December
2019 and sent it to a private firm called Vision Medicals, which
reported back two days later that it had found a genetic signature of a coronavirus. According to the Chinese authorities,
however, the first person to sound the alarm on the emerging

coronavirus was Dr Zhang Jixian, a respiratory specialist at the Hubei Provincial Hospital of Integrated Chinese and Western Medicine. In her mid-fifties in 2019, she had experienced the SARS epidemic of 2002–3 and had been investigating suspected SARS cases in Wuhan during that time. So when, on 26 December 2019, an elderly married couple from a nearby residential community were admitted to the hospital, both suffering from fever and a persistent cough, and a CT scan showed results unlike normal pneumonia, Dr Zhang was immediately concerned. She insisted that their son come to the hospital for a CT scan. The son, who did not live with his parents but had begun caring for them after they fell ill, presented with similar symptoms, and all three showed an unusual pattern of lung shadows on a CT scan.

At this point, Dr Zhang became certain that there was human-to-human transmission of what was likely a virus. On that day, 27 December 2019, the same hospital received another patient with similar symptoms – the blood tests of all four of these patients indicated a non-influenza viral infection – and Zhang reported these troubling cases of mysterious pneumonia to the head of the hospital, who two days later reported it to the local health authorities. Dr Zhang took care to ensure that the growing number of patients with similar symptoms were isolated and that the hospital staff took precautions.

Across town, on 27 December, Dr Ai Fen, the director of the emergency department of the Wuhan Central Hospital, saw her second patient with unusual pneumonia and sent samples for testing. The results came back on the afternoon of 30 December: 'SARS coronavirus'. Dr Ai circled the word 'SARS', photographed it and copied it to a doctor friend at a different hospital in Wuhan. Within an hour and a half, the photo had reached Dr Li Wenliang, an ophthalmologist at Dr Ai's hospital, who sent a message to approximately 150 of his former classmates on the Chinese social media app WeChat: 'Seven cases of SARS have been diagnosed at the Huanan Fruit and Seafood Market, quarantined in our hospital's emergency department.' He added: 'Don't circulate the message outside this group. Get your family and loved ones to take precautions.' Inevitably, the information was leaked on the internet.

Meanwhile, alerted by Dr Zhang, the Wuhan Health Commission issued the official document 'emergency notice on reporting the treatment of pneumonia of unknown causes' to hospitals, instructing them not to make anything public. Within minutes, these orders had leaked online. Later, it would be reported that the Chinese CDC director, Dr George Gao, first learned of the outbreak in Wuhan from these leaked notices on the internet; Dr Gao then called up the head of the Wuhan disease control office who informed him that the outbreak had been going on since the beginning of December, with twenty-five suspected cases. Meanwhile, lung samples from patients had been sent to at least five state labs to sequence the virus.

The WeChat message was spotted by somebody in Taiwan, who contacted Dr Marjorie Pollack, the deputy editor of ProMED-mail, the largest public global surveillance system reporting infectious diseases outbreaks. Dr Pollack, a veteran infectious diseases epidemiologist, had been the first to alert the world outside China to the first SARS outbreak in February 2003. 'We're in trouble,' she thought in a moment of déjà vu. She sent an emergency post to her eighty thousand subscribers at one minute to midnight on 30 December.

As the world learned of the outbreak on the last day of 2019, the Huanan seafood market was being shuttered by the authorities and completely disinfected – hundreds of samples were collected from animal carcasses and surfaces in the market to test for the presence of the novel pathogen. Memories of how SARS had spread from animals in live markets in 2003 were fresh, and a cluster of the pneumonia cases in Wuhan was linked to the market. Officials in protective clothing appeared in the market on the evening of 31 December and began scrubbing the stalls with disinfectant. That day the Wuhan Health Commission used its Weibo social media account to announce that twenty-seven people had contracted the viral pneumonia.

Yet the whistleblowers were penalised for alerting other healthcare workers and the public to the outbreak. On 1 January, Dr Ai was rebuked by her hospital for raising the alarm publicly. Two days later, the ophthalmologist from the same hospital, Dr Li, was interrogated by the local police and formally censured for 'spreading rumours'. He was accused of severely disturbing the social order and told in writing (he later published an image of the letter): 'We solemnly warn you: If you keep being stubborn, with such impertinence, and continue this illegal activity, you will be brought to justice – is that understood?' He signed the document, alongside his fingerprint in red ink, to indicate that he did. Later, Dr Ai would tell the Chinese magazine *Renwu* (*People*), 'If I had known what was to happen, I would not have cared about the reprimand. I would have fucking talked about it to whoever, wherever I could.'

Dr Li developed Covid-19 symptoms a week later and his death from the disease was publicly confirmed on 7 February. Dr Li was only thirty-three and left behind a pregnant wife and a young child. In response to this tragedy, there was an outpouring of grief and widespread fury at the government's management of the outbreak and Dr Li's whistleblowing.

We would learn later, from multiple accounts, that the Chinese authorities had obtained the genome sequence of the novel coronavirus by 27 December, and, according to the head of the Jinyintan Hospital, had shared it with the Wuhan Institute of Virology on the same day. In any case, the WIV obtained a genome sequence from patient samples sent to its lab. By 3 January, the Chinese CDC had shared the genome of the virus with companies developing diagnostic kits for the pathogen that was only supposedly a rumour at the time. A day after Dr Li signed the confession letter, the leading Chinese vaccine developer Sinopharm kicked into high gear manufacturing a vaccine for the novel coronavirus. It would be approximately two weeks before the Chinese authorities publicly confirmed evidence of human-to-human transmission on 20 January.

In early January 2020, the hashtag #WuhanSARS began trending on the social media site Weibo and was promptly censored. By the first day of 2020, the Wuhan police had investigated and punished eight people for 'spreading rumours' online about the disease, one of whom was Dr Li. According to Chinese Human Rights Defenders, a network of rights groups, at least 254 people would be punished in one week for the crime of spreading rumours.

As one of the countries that had been heavily impacted by the 2003 SARS outbreak, the Taiwanese health authorities were carefully monitoring Chinese social media and spotted the messages from 30 December about seven cases of atypical pneumonia, which they took to mean possible SARS. They too thought human-to-human transmission might be possible because patients were being isolated. So they alerted the World Health Organization (WHO) and requested further information. None was forthcoming, so Taiwan increased its border security and stopped flights to and from Wuhan. The Taiwanese government sent Professor Chuang Yin-ching to lead a group of experts on a fact-finding trip to Wuhan on 13 January 2020. What they saw persuaded them that person-to-person transmission was already rampant despite both China's and the WHO's stance that there was no evidence of human-to-human spread at the time. On their return, Taiwan imposed strict border controls and quarantines. Taiwan has still not been invited to join as a member of the WHO and, in recent years, has also been excluded from the World Health Assembly at China's insistence.

By 18 January, the WHO had been informed by Sir Jeremy Farrar, the director of the Wellcome Trust, a health-research charity in the UK, of the existence of data consistent with human-to-human transmission of the virus. In *Spike*, his July 2021 book on the Covid-19 pandemic, he recounts how he had been tipped off by Dr Thijs Kuiken, a leading veterinary pathologist advising the Dutch government. The *Lancet* medical journal had asked Dr Kuiken to review a research paper from the University of Hong Kong describing a family that had visited relatives in Wuhan and caught the novel coronavirus – without having visited the seafood market. One family member who had not travelled to Wuhan also fell ill. What alarmed Dr Farrar was that this study also described the lack of respiratory symptoms in some of the infected relatives and the possible spread of the virus via asymptomatic but infectious patients. He called for the *Lancet* to accelerate its peer-review procedure.

Who was the first to be infected by the novel coronavirus?

Today there are clues that the epidemic began before December 2019. Separate groups of scientists analysing genetic data from cases around the world have estimated that the most recent common ancestor of all human cases of SARS-CoV-2 was in somebody or some animal between September and November 2019.

According to government data seen by the *South China Morning Post* in March 2020, the earliest confirmed case of Covid-19 seemed to trace back to 17 November 2019. This was a fifty-five-year-old man in Hubei province (in which the city of Wuhan resides). The exact location is not disclosed, nor are details of his occupation or recent travel. The hospitals and laboratories that possessed early samples from December 2019 were instructed by the authorities to destroy them for biosafety reasons. Coincidentally, a later study by the University of Kent in Canterbury, in the UK, also came up with 17 November as its best estimate of the first human case, although with a wide margin of error – its earliest estimate for when Covid-19 might have emerged was early October. It did so by repurposing a mathematical device designed for estimating the extinction dates of animal species to the task of estimating the date of the first human case of Covid-19.

Another puzzle is that, a few weeks before, between 18 and 27 October 2019, Wuhan had hosted the Military World Games, in which nine thousand athletes from a hundred countries participated. Bizarrely, Xinhua reported that on 18 September Wuhan Customs and the Executive Committee of the Military World Games held an Emergency Response Exercise, to practise what it would do in the case of two emergencies at the city's airport: the arrival of baggage with nuclear radiation and the arrival of a passenger infected with a novel coronavirus. Later athletes from several different countries complained that they fell ill while in Wuhan with fever, coughs and diarrhoea. Some were bedridden for weeks. Yet there has been little to no effort to figure out if some of these cases could have been Covid-19. Josh Rogin of the *Washington Post*, who reported on this story, called it an 'investigative thread languishing due to neglect'.

It is unfortunate that, in the first year post-outbreak, reports of the true number of Covid-19 cases in December 2019 have been muddled and near impossible to verify. By the time Dr Zhang and Dr Ai raised the alarm, more than a hundred people could well have already contracted the disease. Many international experts have commented on the difficulty of obtaining clarity on the situation in Wuhan early in the outbreak; some have estimated several thousand Covid-19 cases in the city by mid-January 2020. Considering the nature of the Covid-19 disease – in which a majority of infected people show mild to no symptoms – the more severe cases presenting in hospitals in December 2019 would have been the tip of an iceberg of mild infection.

Nonetheless, we know from some very early scientific reports by Chinese scientists that a good portion of the index cases had no links to the Huanan seafood market in Wuhan. The first elderly couple seen by Dr Zhang on 26 December had not visited the market. Separately, a study by Wuhan and other Chinese scientists published on 24 January 2020 in the prestigious *Lancet* journal said, 'The symptom onset date of the first patient identified was Dec 1, 2019. None of his family members developed fever or any respiratory symptoms. No epidemiological link was found between the first patient and later cases.' This patient had been diagnosed with the novel coronavirus after his respiratory-tract specimen tested positive for the virus. Later, more information would emerge on this first patient, who was an elderly man with Alzheimer's disease who had not left his home. 'He lived four or five buses from the seafood market, and because he was sick he basically didn't go out,' said a senior doctor at the hospital that treated him.

On 22 June 2021, Dr Jesse Bloom of the Fred Hutchinson Cancer Research Center made a new breakthrough on the origin of Covid-19. Dr Bloom is a Howard Hughes Medical Institute Investigator who studies virus evolution and is widely recognised in the virology community. For four months, he had been tracking data from the early days of the pandemic in Wuhan and found mention of an early sequencing dataset from Wuhan University. This data was supposedly available from 30 March

2020 on the database of the US-based National Center for Biotechnology Information (NCBI). Yet, upon querying the database, Dr Bloom found that the entire project (PRJNA612766) was missing: 'No items found'. The only clues lay in a manuscript associated with the project, which described forty-five swab samples from suspected early cases of Covid-19 in Wuhan. The Wuhan authors would not reply to Dr Bloom's request for information on when the samples had been collected and why they had deleted their data from the database. The Seeker then confirmed that a project with the same serial number had been deleted from China's database of genetic sequences around the same time between 19 June and 3 July 2020.

This is when Dr Bloom pulled an ingenious move. He could not access the data from the NCBI database, but he had noticed that the files were often stored on Google and Amazon clouds. He deciphered the URL path to the actual data in the cloud. Success. Of the ninety-nine files corresponding to thirty-four early Covid-19 samples, Dr Bloom was able to obtain ninety-seven. These data were only fragments, covering the spike sequence up to the end of a section known as ORF10 in the virus genome. When Dr Rasmus Nielsen of Berkeley became aware of Dr Bloom's discovery, he described it as 'the most important data that we have received regarding the origins of Covid19 for more than a year'.

Scientists had been trying for more than a year to predict what the original SARS-CoV-2 virus had looked like at the beginning of the outbreak. Logically, the earliest virus genome should have been the most similar to the closest-related bat coronaviruses. Over time, as it passed through humans, the virus should have evolved to become more distant from what it looked like in bats or other wild animals. However, these scientific efforts had been stymied because the earliest reported virus genomes, especially those from the Huanan seafood market,

were not the most similar to bat coronaviruses. In direct oppo-
sition to the epidemiological data, none of the virus genomes
from the market looked like the parent of all subsequent virus
genomes collected elsewhere in the world.

With the deleted data in hand, Dr Bloom determined that the
Huanan seafood market sequences were not fully representative of the SARS-CoV-2 viruses circulating in Wuhan in the
earliest days of the outbreak. Yet these market sequences had
been the focus of the China-WHO study. He wondered why
such an informative and valuable dataset describing early
Covid-19 cases in Wuhan had been deleted. There was no issue
with the data or correction to the Wuhan author's manuscript.
These data, if made easily accessible to scientists worldwide,
would have pointed to earlier circulation of the virus in Wuhan
even preceding the Huanan seafood market. In view of China's
destruction of early patient samples and gag orders, Dr Bloom
observed that these suggested 'a less than wholehearted effort
to trace early spread of the epidemic'. On an optimistic note,
however, Dr Bloom pointed out that more data may exist
outside China, which scientists could examine for more clues
concerning the origin and early spread of Covid-19. He urged
a focus on gathering more data and better analysing the data
already in our possession.

On 21 July 2021, in response to Dr Bloom's preprint, the
vice minister of China's National Health Commission tried to
clarify the situation by saying that it was the journal editors
who deleted data access information in the paper, leading the
Wuhan authors to believe that they could remove the data from
the NCBI database entirely. However, in a June 2020 email to
NCBI, the authors had not brought up the journal's editorial
error, but instead cited difficulties updating their data. They
also claimed that they had already submitted an updated
version of the data to another website. Yet it was not until after

Dr Bloom had reported the deleted data that the Wuhan authors quietly uploaded their sequences onto a database maintained by one of the authors.

The first SARS-CoV-2 genomes to be made public

In Shanghai, on the afternoon of 3 January 2020, Dr Zhang Yongzhen of the Shanghai Public Health Clinical Center received a sample taken from a patient on 26 December 2019 at the Wuhan Central Hospital (where Dr Ai and Dr Li worked). After receiving the sample, he set about sequencing the genome of the virus. Within forty hours, by 2 a.m. on 5 January, Dr Zhang had succeeded in assembling a complete genome, an extraordinary feat made possible by spectacular advances in the speed of automated DNA sequencing since the 2003 SARS outbreak. Dr Zhang could see that it was a coronavirus, related to SARS but distinct. He called the head of respiratory medicine at the Wuhan Central Hospital and warned them that they were probably dealing with a pathogen more dangerous than influenza.

That same day, Dr Zhang uploaded the novel coronavirus sequence – which would come to be one of the world's most studied genome sequences – to the GenBank database, run by NCBI and based in the United States. The sequence was logged under the name Wuhan-Hu-1. However, the data was initially embargoed until 12 July; in other words not published for the world to see. An email from GenBank on 8 January sent to Dr Zhang's group clearly stated: 'You have requested that your data are to be held confidential until: July 12, 2020.' In parallel, on 7 January 2020, Dr Zhang had submitted a paper describing the novel coronavirus genome to *Nature*; and in order to publish the paper he had to obtain an accession number from

GenBank to show that he had uploaded the data supporting the manuscript.

One possible reason for the embargo, which was hardly in the best interests of scientific transparency or that of humanity as a whole, was apparently that the Chinese authorities were keen to keep the genome sequence and other information closely controlled as they developed testing kits for the new virus. Remember, at this stage, few people were expecting a global pandemic to develop. However, according to an investigation by the Associated Press, the Chinese Center for Disease Control and Prevention sold the rights to develop and distribute testing kits for the new virus to three little-known companies, all from Shanghai, for 1 million RMB each (roughly $150,000). GeneoDx Biotech, Huirui Biotechnology and BioGerm Medical Technology were relatively small companies, but with personal connections to CDC officials. The CDC conducted a secret evaluation of test kits on 10 January, and approved only those from the three obscure Shanghai firms. They were given exclusive rights to develop testing kits based on the genome of the virus, which was not released to other companies. Meanwhile hospitals were ordered to send samples to the CDC for testing and to get each positive test confirmed by a secondary test at a Beijing laboratory. Partly as a result of this time-consuming procedure, no new cases were confirmed between 5 and 17 January in the city, allowing the disease to explode in Wuhan and elsewhere.

By 10 January, *Nature* had sent Dr Zhang's manuscript out for peer review. One of the reviewers, Dr Nick Loman from the University of Birmingham, said: 'My initial response was to note the genome data wasn't available.' That same day, Dr Jeremy Farrar tweeted at both *Nature* and another prestigious journal, the *New England Journal of Medicine* (to which the Chinese CDC had submitted a similar manuscript describing a

SARS-CoV-2 genome), saying, 'If rumours of publications on the Wuhan Pneumonia situation are [*sic*] being prepared & submitted to @nature @NEJM are true & that critical public health information is not being shared immediately with @WHO – something is very wrong', following up with a second tweet: 'And if true, what are the responsibilities of the investigators, @Nature @NEJM and others? Seems clear to me.' In Dr Farrar's book on Covid-19, he revealed that 'Within minutes, the tweets attracted a private message on Twitter and a phone call from the other side of the world.' It was Dr Edward Holmes from the University of Sydney, who was a co-author on the *Nature* paper with Dr Zhang.

Within twelve hours of Dr Farrar's tweets, Dr Holmes had published the novel coronavirus genome on a site called Virological.org, run by Edinburgh University's Dr Andrew Rambaut. According to accounts by Dr Holmes and Dr Zhang, at 8 a.m. on Saturday 11 January (the time zone in Sydney, Australia is half a day ahead of the UK), Dr Holmes called Dr Zhang and urged him to publish the genome. Dr Zhang deliberated on this crucial decision and called Dr Holmes back within minutes, agreeing to share the genome, which he then sent to Dr Holmes. (Although Dr Holmes was an author on the paper, he had actually not had access to the genome up until this point.) Within an hour, Dr Holmes had uploaded the genome to Virological.org, saying that he had performed this task so quickly that he did not even check the sequence before uploading. Later, Dr Zhang connected with GenBank to release the genome from embargo on 12 January. The Chinese government also released its version of the novel coronavirus genome onto a different international virus sequence database, GISAID.

The release of the novel coronavirus genome allowed other companies in China and around the world to begin developing diagnostic tests and vaccines. Using this genome sequence, in

Boston, the now famous biotech firm Moderna would complete the design of an experimental messenger-RNA (mRNA) vaccine to fight the virus by Monday 13 January. This was a week before human transmission was even confirmed by the Chinese authorities and the WHO.

In publishing the sequence, Dr Zhang had broken the rules. On 1 January 2020, the Chinese National Health Commission had told the WIV's director general, Dr Wang Yanyi (who then notified her colleagues), that the WIV could not publish any of the novel coronavirus data on social media or in the media. On 3 January, the commission also ordered all laboratories to destroy all Covid-19 samples or transfer them to designated institutions. Sure enough, Dr Zhang's laboratory was closed for 'rectification' on 12 January.

The crackdown intensified across China. Documents leaked to the Associated Press showed 'a pattern of government secrecy and top-down control'. On 24 February 2020, the Chinese CDC issued a new policy to all of its offices requiring all scientific studies relating to coronaviruses to be approved by the higher authorities in response to important instructions from Chinese President Xi Jinping. CDC staff were not allowed to 'provide other institutions and individuals with information related to the Covid-19 epidemic on their own, including data, biological specimens, pathogens, culture, etc.' Anyone violating these regulations would be 'dealt with severely'.

In early April 2020, Chinese universities inadvertently published Chinese government directives that said all academic manuscripts on the novel coronavirus had to be reviewed at multiple levels, including by government officials, prior to submission to journals for publication. CNN reported that staff from the Chinese education ministry verified the directive but said that it was supposed to be an internal document, not meant to be published.

The first lockdown

On 23 January the city of Wuhan was locked down, preventing anyone from leaving. It was too late. A few days earlier, on 18 January, with the Chinese New Year holiday fast approaching, Wuhan's mayor Zhou Xianwang had hosted a record-breaking banquet in the Baibuting district of the city, which was attended by forty thousand families. Mayor Zhou said that he had not been given sufficient warning that there was a risk. Prior to the lockdown, millions of people had already left Hubei province to visit relatives or go on holiday. As the world would soon learn, the cat was out of the bag, the genie out of the bottle, and the virus out of China. Other countries, particularly those that had suffered from the 2003 SARS outbreak, were already on the alert for the virus appearing on their shores.

As people around the world watched the drastic measures being taken in Wuhan and other cities in Hubei – emergency hospitals erected at unimaginable speed, cases forcibly isolated in special centres, fifty-six million people locked into their homes for more than two months under harsh restrictions and enforcement – they experienced a mix of envy and horror at the unchallenged power of a totalitarian state to contain a virus outbreak. Few of us realised what was in store for the world.

On 13 January, the first case of Covid-19 outside China was confirmed in Thailand: a sixty-one-year-old Chinese woman who had arrived in Bangkok on 8 January. She was a tourist from Wuhan but had not visited the Huanan seafood market. The first case in Japan was confirmed on 15 January: a man who had returned from visiting Wuhan on 6 January; his fever had started on 3 January, but he also had not visited the Huanan market. Thailand and Japan rank among the top three air travel destinations from Wuhan.

Covid-19 cases started to bubble up around the world, all tracing back to Wuhan. The first confirmed case in the United States came on 20 January in a man who had returned to Washington State five days earlier from a visit to his family in Wuhan; he too had not visited the Huanan market. The first detected cases in France came on 24 January of a Bordeaux man and a Chinese couple visiting Paris; all had recently visited Wuhan but not the Huanan market. The first reported cases in Canada and Australia were on 25 January of men who had recently visited Wuhan. The first confirmed case in Germany came on 27 January of a man who had been in a business meeting in greater Munich with a Chinese colleague who eleven days earlier had been visited in Shanghai by her parents from Wuhan. The first confirmed case in India arrived on 30 January, that of a student who had returned home to Kerala from Wuhan University. The first confirmed case in Italy came on 30 January of a tourist who had arrived from Wuhan a week before. Britain and Sweden confirmed their first cases on 31 January, both recent arrivals from Wuhan.

Rumours about a lab leak in Wuhan

Once the outbreak became public knowledge, speculation about the origin of the virus began almost immediately. On 6 February 2020, Dr Botao Xiao and Lei Xiao, husband and wife scientists in Wuhan, published an online article, entitled 'The Possible Origins of 2019-nCoV Coronavirus'. Dr Botao Xiao was heading his own research group at the time after recently completing his postdoctoral training at Harvard Medical School. In this article, they postulated that 'somebody was entangled with the evolution of 2019-nCoV coronavirus' and that 'in addition to origins of natural recombination and intermediate host, the killer coronavirus probably originated from a

The city of Wuhan in Hubei province.

laboratory in Wuhan'. Their reasoning was that the bat source of a closely related virus had been found in a distant Chinese province, and that the Huanan seafood market in the metropolitan city of Wuhan had no exposure to bats. The husband-and-wife team pointed to two laboratories in Wuhan that had been conducting research on bat coronaviruses: the Wuhan Center for Disease Control and Prevention, which was close to the Huanan seafood market, and the WIV.

Within a five-minute walk from the market, the Wuhan CDC 'hosted animals in laboratories for research purpose, one of which was specialized in pathogens collection and identification'. Moreover, the Xiaos described how a lead researcher at this laboratory who collected bats had featured in articles and at least one broadcast video revealing that 'he was once attacked by bats and the blood of a bat shot on his skin. He knew the extreme danger of the infection so he quarantined himself for 14 days. In another accident, he quarantined himself again because bats peed on him.' Given that 'surgery was performed on the caged animals and the tissue samples were collected for DNA and RNA extraction and sequencing', relatively close to the Huanan seafood market where clusters of Covid-19 cases had been detected, it was reasonable to ask whether an accident had occurred, the two scientists argued.

The second laboratory was situated in the WIV, approximately twelve kilometres from the market. The principal investigator of this laboratory, Dr Shi Zhengli, the director of the Center for Emerging Infectious Diseases at the WIV, had 'participated in a project which generated a chimeric virus using the SARS-CoV reverse genetics system, and reported the potential for human emergence. A direct speculation was that SARS-CoV or its derivative might leak from the laboratory.' The online article by the Xiaos was quickly withdrawn. 'The speculation about the possible origins in the post was based on published papers and media, and was not supported by direct proofs,' Dr Botao Xiao wrote in an email to the *Wall Street Journal* on 26 February.

Yet we later learned that Dr Shi had experienced the same fears and doubts, except earlier in the outbreak. An interview with Dr Shi was published in *Scientific American*, on 11 March 2020, describing the moment she first heard about the novel SARS coronavirus that was causing panic in Wuhan. Patient samples had been sent to the WIV on the evening of 30 December for testing. That same night, the WIV's director general had phoned Dr Shi, instructing her to return immediately from a conference in Shanghai to investigate the atypical pneumonia cases. The *Scientific American* article reads: '"I wondered if [the municipal health authority] got it wrong," she says. "I had never expected this kind of thing to happen in Wuhan, in central China."' Her studies had shown that the southern, subtropical provinces of Guangdong, Guangxi and Yunnan have the greatest risk of coronaviruses jumping to humans from animals – particularly bats, a known reservoir. If coronaviruses were the culprit, she remembers thinking, 'Could they have come from our lab?'

4.

The seafood market

'Next time something strange and new comes anywhere in the world, let us come in as quickly as possible.'

GRO HARLEM BRUNDTLAND
(after the first SARS epidemic)

'The origin of the new coronavirus is the wildlife sold illegally in a Wuhan seafood market.' So said China's Dr George Gao, the director of the Chinese Center for Disease Control and Prevention in Beijing, on 22 January 2020. Four months later, Dr Gao was just as certain that the market was not the source of the outbreak: 'At first, we assumed the seafood market might have the virus, but now the market is more like a victim. The novel coronavirus had existed long before.' By implication therefore it was people, not animals, that brought the virus to the market. The story of how the Chinese authorities came to change their minds about the role of the Huanan seafood market may provide rich clues to what they had uncovered between January and May 2020.

The Huanan seafood market was hurriedly closed and sanitised at the end of 2019, just one day after the Wuhan Health

Commission issued its emergency notice to local hospitals about the atypical pneumonia cases. According to one employee of a local disinfection company, he received a call at 1 a.m. on 31 December to head to the Huanan market. There his team disinfected stalls selling exotic animal meat products or traditional medicine. In parallel, officials from the Wuhan CDC took hundreds of samples from dead animals in the stalls and from surfaces and sewage in the market. On 1 January, another team from the Beijing CDC arrived and continued to collect samples and remove the rest of the animals.

The swiftness with which local officials shuttered and sampled the market was a sign that, at least at the time, the authorities were convinced that the market was a probable source of the outbreak. Curiously, in mid-January 2020, a local CDC official told health experts from Taiwan and Hong Kong that they had not found wild animals at the market and these were not commonly eaten in Wuhan. The consumption of exotic wildlife for food is generally a southern Chinese speciality, though traditional Chinese medicine is more widespread. At the end of January, the Paris-based World Organisation for Animal Health (OIE) was briefed by the Chinese government, one of its members, that none of the animal samples from the market had tested positive for the virus. Over the next few months, Chinese scientists would test more domestic and farmed animals, reporting to the OIE that all samples tested negative.

The main reason for the immediate suspicion that the market was the origin of the novel SARS-CoV-2 virus was the precedent that wildlife markets and restaurants selling exotic meat had played a vital part in the first SARS virus jumping into the human species in Guangdong province in southern China more than fifteen years before ...

The first SARS epidemic

On 21 February 2003, Dr Liu Jianlun, a sixty-four-year-old medical professor from Guangzhou, the capital of Guangdong province, checked into Room 911 of the Metropole Hotel in the Kowloon district of Hong Kong to attend the wedding of a nephew. He had already been feeling ill for five days before his visit to Hong Kong, and his condition worsened so much overnight that he was rushed to the intensive care unit of Kwong Wah Hospital, where he would die within two weeks. While in hospital, he revealed that he had been treating a secret outbreak of a new disease in Guangdong and warned his carers to take precautions. Nonetheless, one doctor and five nurses at the hospital caught the infection from Dr Liu. Within days, at least seven other hotel guests staying on the same floor as Dr Liu had carried the SARS virus, unawares, to Toronto, Hanoi, Singapore and elsewhere in Hong Kong. By then the SARS virus had already separately reached Beijing.

A retrospective investigation by the Guangdong Provincial Center for Disease Control and Prevention revealed that the first person known to have fallen ill with SARS was a forty-five-year-old administrator from Foshan City, Guangdong. On 16 November 2002, he suddenly began to feel seriously unwell. He had not travelled outside the local area for two weeks, but had prepared chicken, domestic cat and snake for consumption at home. In Guangdong, domestic cat meat is a delicacy, and a 2002 Guangzhou news report claimed that ten thousand domestic cats were consumed in Guangdong each day – the demand was so large that cats were imported from nearby provinces. Before the end of 2003, domestic cats had been determined to be susceptible to infection by SARS-CoV and also capable of infecting other animals in close contact. The index case's wife also caught the virus, but his four children did not. After his aunt visited him in hospital, she, her husband and her daughter all also caught the disease.

This was not the only index case, however. Between November 2002 and mid-January 2003, SARS seems to have erupted several times across seven municipalities without human contact between the cases. On 10 December, in Heyuan, another city in Guangdong, a restaurant chef fell ill; on 21 December, a factory worker in Jiangmen City; on 26 December, another restaurant chef in Zhongshan City, all near the city of Guangzhou. This pattern left investigators wondering whether there was a missing link between these sporadic index cases or if they had each been exposed to multiple sources of the virus along the massive delta of the Pearl river.

On 3 February, Guangdong province instituted the mandatory reporting of atypical pneumonia cases to local CDCs and the provincial CDC, so that cases could be contact-traced and detailed patient data could be collected. But it was too late. By 24 February, news that more than fifty hospital staff in Guangzhou had been infected with a mysterious pneumonia reached the Global Public Health Intelligence Network. Yet it was not until three months into the epidemic that the Chinese government authorities admitted that there was a serious problem.

Back in 2003, the WHO was severely critical of the Chinese government for its lack of openness. On 14 March that year, without waiting for the Chinese government's agreement, the WHO announced a global alert, issuing a series of travel advisories. In April, the WHO's director general at the time, Gro Harlem Brundtland, gave an unprecedented dressing down to a member state, saying: 'It would have been definitely helpful if international expertise and WHO had been able to help at an earlier stage.' Despite this criticism, when the WHO sent a team of experts later in April to investigate the extent of the outbreak in Beijing, the Chinese authorities instructed doctors to hide several dozens of SARS patients in a hotel and in ambulances so that the WHO team would not find them. The patients in ambulances were driven around the city while the WHO team visited the hospital.

By this time in April, the virus had infected more than 3,500 people across 25 countries. Scientists in Vietnam had obtained preliminary evidence suggesting that individuals with mild symptoms could spread the virus, and that the incubation period could be longer than a week. An explosive cluster of 250 cases at a housing complex in Hong Kong also hinted that the virus was spreading through infection routes other than droplets from coughing or sneezing by infected individuals. After detecting that the virus was beginning to bring down younger and healthier people, the WHO commented that the SARS virus 'may have mutated into a more virulent form'.

Fortunately, the strict containment measures worked and by early July the epidemic had petered out after more than 8,000 cases worldwide and a death toll of nearly 800 people. The worst affected country outside of Asia proved to be Canada with 438 cases and 44 deaths. It had been taken by surprise by the Metropole Hotel superspreading event. Countries that were significantly affected by SARS, especially China (particularly Hong Kong), Taiwan, Singapore and Canada, learned valuable lessons from the outbreak that would stand them in good stead in 2020. One of us (Alina) was a teenager living in Singapore when the 2003 SARS outbreak hit. Scenes of quarantined individuals and patients in the intensive care unit were constantly on the news, interlaced with advice for members of the public on best practices to avoid catching the virus. So, in January 2020, when video footage of corpses piling up on the floors of hospitals in Wuhan surfaced on the internet, Alina was immediately fearful that the uncontrolled spread of the virus would soon lead to a similar calamity elsewhere in the world.

Tracking down the zoonotic culprit

Many new viral diseases stem from zoonoses, meaning they jump from another species of animal into human beings. HIV originated in chimpanzees; measles in cattle; Lassa fever in rodents; influenza in ducks and other birds; rabies, Ebola and Marburg virus in bats. Twice in the 1990s, viruses jumped from fruit bats into people: in Australia in 1994 when fruit bats infected horses which infected people with a lethal virus called Hendra; and in Malaysia where fruit bats infected pigs which infected people with a nasty virus called Nipah. In the distant past, the original zoonotic event at the start of an epidemic was almost impossible to identify. Nobody knows exactly where or how the plague that caused the Black Death made the leap from a central Asian wild rodent to a black rat and thence, via a flea, to a person. Even HIV, which is estimated to have spilled from chimpanzees to human beings sometime in the 1920s, remains challenging to trace back to its precise zoonotic source – the one from which humans first received the virus. Despite being named a novel disease in 1981, it was not until 1989 that scientists discovered a very closely related virus in wild chimps, and it was not until 1998 that a separate group of scientists leveraged new technology to sequence the virus from a 1959 human blood sample taken in Kinshasa, in what is now the Democratic Republic of Congo. Determining where and when a novel pathogen first emerged in the human population is exceedingly difficult and requires scientists to gather a large amount of data in a timely manner.

However, genomic technologies and growing scientific knowledge, particularly when applied to emerging pathogens close to world-class research hubs, have changed that and today it should be much easier. The immediate zoonotic source of the first human SARS virus was tracked down rapidly. International interest was high because of the way the outbreak had erupted in Hong Kong and spread to numerous countries and thousands of individuals within a couple of months. Tracing each transmission chain of the virus and hunting down its source in Guangdong was urgent to prevent widespread loss of life. It helped that numerous Chinese experts (several based in Hong Kong where the Metropole Hotel superspreader event had occurred) immediately began to investigate the outbreak.

In the spring of 2003, it dawned on investigators that many of the early cases of SARS had been food handlers, particularly chefs and others who prepared animals for food. In May, in the search for the zoonotic source, a team of scientists from Hong Kong visited a live animal market in Shenzhen, Guangdong, to collect animal samples from different stalls, spanning wild and domestic species, originating from different parts of southern China. All the animals appeared healthy. The scientists took samples from the noses and anuses of twenty-five animals from eight, mostly wild, species: beaver, ferret-badger, hare, muntjac deer, domestic cat, hog-badger, raccoon dog and Himalayan palm civet. Of most relevance to the story of SARS is the palm civet, a tree-climbing, fruit-eating relative of cats. An expensive dish popular with wealthy people in Guangdong is known as 'dragon-tiger-phoenix soup'. It is flavoured with chrysanthemum petals but contains the meat of both civets and snakes. From the twenty-five market animals, the single raccoon dog and a hog-badger tested positive, as did four of the six civets. SARS virus was isolated from five of the animals, and two virus isolates from the civets were completely sequenced to obtain their genomes.

In parallel, the same team of Hong Kong scientists tested dozens of stall holders for antibodies to SARS, although none had been diagnosed as SARS cases previously. Eight of twenty (40 per cent) wild animal traders and three of fifteen (20 per

Masked palm civet (*Paguma larvata*) on sale in a market in Southeast Asia.

cent) people who slaughtered animals had evidence of antibodies, while only one of twenty (5 per cent) vegetable traders had
them. As a comparison, sixty anonymised samples from
Guangdong patients hospitalised for non-respiratory diseases
were analysed. None of them tested positive for SARS antibodies. Similar studies were conducted in the same month of May
by Guangdong scientists, who detected SARS antibodies in
sixty-six of 508 (13 per cent) animal traders from three animal
markets; again, none had been diagnosed with SARS or an
atypical pneumonia previously. The highest antibody prevalence (72.7 per cent) was found in those who primarily traded
civets, with the lowest prevalence (9.2 per cent) in those who
primarily traded snakes.

When SARS broke out again in the winter of 2003–4, civets were implicated once more. This time, of the four new cases, one was a waitress and another a diner at the same restaurant in Guangzhou. The patients each had mild symptoms and did not appear to transmit the virus to others around them – even to healthcare workers. This stood in contrast to the contagiousness and virulence of the first SARS virus. The virus genome sequences from these new cases were distinct from those of the SARS virus that had proliferated earlier in 2003. In addition, unlike the first SARS virus, which had been isolated in March 2003, it was not possible to isolate the new SARS virus from any of the patient samples – signalling that the virus was not yet well adapted to infecting primate cells, the typical means by which viruses are isolated.

The waitress was diagnosed with possible SARS on 2 January 2004. She denied consuming or having been in proximity to civets. Nonetheless, when scientists immediately went to her restaurant to investigate, they found civets. In the restaurant, cages in which the animals were kept had been placed near the dining area. The scientists collected samples from all of the civets and employees at the restaurant. They also traced the restaurant civets to a live animal market in Guangzhou and sampled the animals there for SARS. All six civets sampled at the restaurant tested positive for SARS and the sequences of the civet virus matched those of the two patients. Strikingly, these 2004 SARS sequences more closely resembled the virus that had been isolated from market civets in 2003 than the viruses found in humans in the earlier outbreak. Two of thirty-nine other restaurant employees tested positive for antibodies against SARS but had not reported illness or fever in recent months. These observations strongly indicated that the new SARS cases had resulted from a more recent zoonosis from an animal source rather than an undetected continuation of the 2003 human SARS epidemic.

Based on this second emergence of SARS within months of the first epidemic, Guangdong officials implemented strong measures on 5 January 2004 to eradicate any animal in farms and markets that might transmit SARS virus. Approximately ten thousand civets were exterminated, largely by drowning the animals in vats of disinfectant, while objects that had been in contact with these animals were incinerated. Continued investigation of hundreds of the animals revealed that although there was widespread infection of SARS among civets in markets, there was none in the farms where they were reared. Oddly enough, four civets at one farm in Guangdong tested weakly positive for SARS antibodies, but this farm specialised in obtaining them from various markets to sell as pets – it was possible that the civets had been infected with SARS at these markets prior to being brought to the pet farm.

So the civets could not be the original source of the virus. The market civets sampled by the Hong Kong team in May 2003 had originated from different regions of southern China yet had caught almost identical viruses. It was more likely that they had picked up the virus during transit closer to point of sale or at the markets from any of the numerous other animal species there. The presence of SARS antibodies in animals and animal traders from several different markets in Guangdong province suggested that there was frequent and widespread circulation of SARS-like viruses, with the 2003 SARS virus being the first to demonstrate a capacity to afflict humans with severe respiratory disease and spread rapidly from human to human. Where were these SARS viruses coming from?

Between the summer of 2004 and the spring of 2005, a team of Hong Kong scientists captured 127 bats, 60 rodents and 20 monkeys from 11 locations in the wild in Hong Kong, and analysed their noses, anuses and blood for signs of coronaviruses. The anal swabs from 29 bats tested positive for coronavirus, largely those from the species known as the Chinese rufous horseshoe bat, *Rhinolophus sinicus*. The viruses that some of these bats were carrying proved to be similar to the SARS virus but were not close enough to be the immediate ancestors. Their genomes also shared a short sequence with the civet SARS viruses that was missing from the human version (excepting one of the earliest human cases from the 2003 outbreak), implying that the bats could not have caught the virus from people. More likely both the civet version and the human version were descended from a bat virus. From this moment bats took centre stage.

The MERS eruption

Between the epidemics of SARS in 2002–4 and SARS-CoV-2 in 2019, another deadly coronavirus spilled from animals into humans. In 2012, Dr Ali Mohamed Zaki, a skilled Egyptian virologist based at a private hospital in Jeddah, in Saudi Arabia, isolated a novel coronavirus, which we now know by the name of MERS (Middle Eastern Respiratory Syndrome virus). In the previous two decades, Dr Zaki had been the first to diagnose dengue in Saudi Arabia (1994) and to identify a new tick-borne flavivirus that causes severe haemorrhagic fever (1997). The patient who was admitted to his hospital in June 2012 declined rapidly, presenting acute pneumonia and renal failure, dying eleven days later. Before the patient's death, Dr Zaki took a sample from the throat and inoculated it into two types of monkey cells that are commonly used in laboratories to study viruses. Under the microscope, he observed cell death alongside 'syncytia', or large, fused cells with multiple nuclei. That told him that he was probably dealing with a virus not a bacterium. However, by then the patient had died, and others were no longer interested in investigating the mysterious new infectious agent. Undeterred, Dr Zaki conducted genetic tests for influenza, Nipah, Hendra and hantavirus (a rodent-borne pathogen), which returned negative results; but a general test for a coronavirus produced a strong positive signal. When repeated tests for SARS virus also came back negative, Dr Zaki began to realise that he was dealing with a new kind of coronavirus. Sure enough, he had discovered MERS, a nasty disease that kills more than a third of the people it infects. Dr Zaki's decision to alert the world to this novel pathogen on 15 September 2012 infuriated the Saudi authorities. He fled to his native Egypt, leaving his belongings behind in Jeddah.

Fortunately, MERS cases have been sporadic, but at the time of writing 2,574 people have had it and 886 of them have died, mostly in Saudi Arabia. There has been a single large outbreak since its discovery: in 2015, MERS spread to South Korea in an infected traveller from the Middle East, killing thirty-eight people. Once MERS was sequenced, the main clue that it had originated in bats was that a similar lineage of bat coronaviruses had been sampled in China in 2007. Some MERS-like coronaviruses have been identified in bats in Saudi Arabia and South Africa, but none of them is a recent ancestor of the MERS viruses found in humans. In the past decade, prolific research into MERS has concluded, based on strong epidemiological and genetic evidence, that camels have carried MERS viruses for decades before human cases were even identified. Multiple genetic lineages of the virus have been observed from camels in Saudi Arabia alone, with MERS antibodies detected in camel blood samples from as early as 1983 in Sudan and Somalia. Exactly how and in what form the virus got from bats to camels remains unclear.

The wildlife trade and traditional medicine

After the 2003 SARS epidemic, the Chinese government cracked down on the sale of wildlife in markets, banning the sale of wild animals altogether that year, forbidding hunting, and even outlawing the consumption of wild animals. But not all exotic – non-domesticated – animals were caught in the wild. Some were being bred and reared in farms. Within months, the rules had been relaxed for fifty-four species of wild animal reared on farms, including civets. Indeed, the business of farming exotic species was booming. By 2017, the practice of wildlife breeding had burgeoned into a $73 billion industry providing employment for more than fourteen million people, particularly in rural areas of China. In November 2019, according to the *Los Angeles Times*, the Jiangxi provincial government boasted that it had helped 1,700 people in one town to 'embark on the road to riches' by taking up the breeding of civets. 'Wildlife breeding and utilization is a rapidly developing industry in recent years,' said a director of wild animal and plant protection at the Forestry Academy in Jiangxi. 'Our province should seize the opportunity.'

As this quote and job title suggests, many of those charged with wildlife protection are also involved in its commodification. Three of the fourteen vice chairs of the China Wildlife Conservation Association had strong links to traditional Chinese medicine (TCM) companies, according to the *Los Angeles Times*, including ones that sold tiger bone wine, snake bile, seahorses and pangolin scales. There is little doubt that the exemption for farmed exotic animals also opened a loophole that was widely exploited by trappers and smugglers trafficking animals caught in the wild, including species such as the yellow breasted bunting, a small bird highly valued for its taste, sold on the black market, consumed in the millions and now driven to the brink of extinction. Many animals are sold alive in these markets and sometimes even restaurants, to show customers that they are fresh and healthy before consumption.

One reason for the boom in exotic wildlife markets is that, under Chinese President Xi Jinping, TCM has been championed. Practitioners of TCM believe that meat or medications that come from certain exotic animals, such as pangolins, bears, tigers or snakes, can be used to cure ailments. In 2019, at China's request, the WHO officially adopted TCM as a recognised form of medicine. Export markets for TCM are being developed by Chinese firms in Africa. In October 2019, weeks before the first case of Covid-19, President Xi opened a conference with these words: 'Traditional Chinese medicine is a treasure in Chinese civilisation, which carries the great wisdom of Chinese people.' In May 2020, Beijing health authorities publicly solicited submissions on a draft regulation that could criminalise anyone who slandered TCM.

So it was not just the sale of fresh meat that posed a risk of zoonosis. Although bats are eaten throughout southern China, it is mostly the much larger fruit bats that are consumed, not the small horseshoe bats, on which there is little meat. However, the smaller bats are used in TCM. For example, the 'greater' (but still small) horseshoe bat *Rhinolophus ferrumequinum* lives in the vicinity of Wuhan and its droppings are used in medications to treat eye conditions. It has been found to carry coronaviruses, although no virus closely related to SARS-CoV-2 has been detected in Wuhan or Hubei province despite thousands of bats having been sampled in the area over several years. The bat's body parts and its guano are also dried and ground into a powder to be ingested as a detox treatment or applied to the human body. Drying the dung or the body parts probably inactivates any viruses, but the handling of bats clearly carries a risk to the trader.

So it is little wonder the wildlife trade came under suspicion at once. 'Probably bats are the origin from looking at the virus itself, and it got from bats into people in the wildlife market,' said Dr Peter Daszak, president of the EcoHealth Alliance and long-time collaborator of Dr Shi at the WIV, in late January 2020. 'This is absolutely déjà vu all over again from SARS.' Dr Shi was equally certain: the outbreak was 'nature punishing the uncivilized habits and customs of humans'.

The snake theory

The Huanan seafood market was one of the largest 'wet markets' in China – so called because of their constantly wet floors due to the melting ice used to preserve meat and the regular washing down of stalls to get rid of blood and dirty water. It had nearly seven hundred stalls and an estimated ten thousand people visited every day. At the end of 2019, it was rumoured that these stalls offered a wide range of animals for sale, mostly alive or preserved, including crocodile, turtle, badger, bamboo rat and hedgehog, as well as plenty of seafood. One price list showed prices for foxes, wolf puppies, giant salamanders, snakes, rats, peacocks, porcupines and even koalas. Bats were not on sale, as far as anybody could tell, and nor were pangolins.

At first suspicion fell on snakes, which were sold alive in the market, although slaughtered at the point of sale. A study by five Chinese scientists published in the *Journal of Medical Virology* on 22 January 2020 claimed that the novel SARS-CoV-2 virus had the 'most similar codon usage bias with snake'. The argument was that viruses tend to use the amino-acid-encoding three-letter codes (codons) most frequently used by their host so that they can more effectively hijack the host protein production machinery. The new virus sported codons more commonly used in snakes than in marmots, hedgehogs, bats, birds and humans, the researchers found.

Before the pandemic, approximately nine thousand tonnes of snakes were sold in Chinese markets each year. One village in southern China, Zisiqiao, was home to a snake museum and a hundred snake farms, selling three million of the creatures a year, many of them venomous. Snakes and their venom have been a staple in TCM for thousands of years, and a few of the farmers in Zisiqiao made millions of dollars each year. These are conditions of density conducive to disease outbreaks, but cold-blooded reptiles are extremely unlikely to catch viruses from distantly related and hot-blooded mammals such as bats, let alone to pass them on to human beings. Even birds are generally too distant to share their pathogens with people, although influenza did manage to make the transition, probably via another intermediary species such as pigs. By far the greatest risk of zoonosis into humans lies with mammals. Yet when speculations emerged that the novel SARS-CoV-2 virus might have come from snakes, the Chinese authorities temporarily banned the snake industry and the lucrative business in Zisiqiao was shuttered.

The codon-usage snake origin hypothesis came to nothing. Even the snake study had acknowledged that the virus's genome was clearly related to that of bat viruses and not snake viruses. By February, the WIV's unveiling of a bat virus with 96.2 per cent genetic similarity to the human SARS-CoV-2 virus put the spotlight firmly back on bats. This seemed to bear out some dire warnings that had emanated from experts for many years. In 2007, four scientists from the University of Hong Kong, including those who had first found SARS-like viruses in bats, reflected on the SARS experience in a paper published in *Clinical Microbiology Reviews* and concluded starkly – and presciently: 'The presence of a large reservoir of SARS-CoV-like viruses in horseshoe bats, together with the culture of eating exotic mammals in southern China, is a time bomb.' Two of the authors repeated the warning in February 2019 in a paper with two other colleagues from the University of Hong Kong: 'Bat–animal and bat–human interactions, such as the presence of live bats in wildlife wet markets and restaurants in Southern China, are important for interspecies transmission of [coronaviruses] and may lead to devastating global outbreaks.'

Nonetheless, in the case of SARS, bats had not been determined to have infected people directly. They had infected palm civets or other susceptible animal hosts, which then passed the virus to people processing the civets for eating. Likewise with MERS, bats had probably infected camels which had infected people. An intermediate host, one that was likely to come into contact with bats as well as people, therefore seemed worth seeking in the case of SARS-CoV-2 as well.

The third time round

If we count the two SARS outbreaks in the winters of 2002–3 and 2003–4 in Guangdong province, the emergence of SARS-CoV-2 in Wuhan was the third time a SARS-like virus had led to a detected, reported outbreak in human beings in China. Contrast the technology available in the early 2000s with that of 2020. In 2003, scientists first had to culture the virus, then clone fragments of its genome for individual sequencing by laborious and slow machines. It took weeks to sequence a coronavirus genome. Today next-generation sequencing machines and software can spit out a full genome sequence in a few days. Yet contrast the timeline of the two epidemics. The SARS virus was isolated in March 2003, its genome sequenced in April, and animal sources in markets identified in May. At the same time in mid-2003, multiple groups of scientists were finding evidence that the animal trading community had previously undetected, widespread exposure to SARS-like viruses. When SARS emerged again in late 2003, the tracking of the spillover source was even faster. A waitress was diagnosed on 2 January 2004. Within two days, samples had been collected from all palm civets and employees at the restaurant, several of which tested positive for the virus. By 5 January 2004 the outbreak had been traced to a market. In other words, it took only a few months in the first round (expedited to a couple of weeks on the second round) to go from identifying a SARS virus to finding the proximal zoonotic source by which people had been infected. Yet today, close to two years into the outbreak, with much more superior technology and similar outbreak circumstances, we still have no idea where the first patients caught SARS-CoV-2.

In 2019, news about the novel coronavirus had leaked to the public on 30 December. The Chinese authorities knew what needed to be done. By the next day, local CDC employees had started to collect hundreds of samples from the Huanan seafood market and the market was closed. However, apart from a private meeting of the World Organisation for Animal Health (OIE) on 31 January 2020, no detailed results about the testing of Huanan seafood market samples were released during 2020. According to the *Wall Street Journal*, the OIE was told that none of the test results from animals were positive but information about the sample size and sample species was lacking. The 22 January announcement by the Chinese authorities that the virus had spilled over from illegal wildlife at the Huanan seafood market became the established wisdom throughout the world over several months. It remains the default understanding of many lay observers to this day despite the absence of any animal samples testing positive for SARS-CoV-2 from the rigorously sampled market.

Vital information about the samples tested from the market that could have helped the world to understand and, ideally, to inform public health responses to the pandemic was not shared in the timeliest manner. By December 2020, the Chinese authorities still had not confirmed whether any live animals had been tested in January and we were left piecing together scraps of leaked or half-announced information to try to ascertain whether a key hypothesis about the worst pandemic in a century was well substantiated.

During all these months, it turns out that Dr Gao's colleagues were in possession of a map, finalised around 22 January, showing exactly which stalls in the market had belonged to or been visited by people who fell ill, and which stalls had provided positive tests of environmental samples. This map was not shared with the world, although a slightly different version did reach the United States Centers for Disease Control and Prevention and the WHO, both of which also chose to keep it secret.

The Chinese CDC's map only emerged in a leak to the *South China Morning Post*, which published it in mid-December 2020, almost a year after the initial outbreak in Wuhan. It showed that thirty-three stalls in the market were linked to forty-five suspected or confirmed human cases of Covid, mostly in the western block of the market, west of Xinhua Road. In the middle of this block, there was a hotspot of positive environmental samples in one section selling wildlife and poultry and a second section selling seafood. In the news story about the map of the market, Dr Lawrence Gostin, director of the O'Neill Institute for National and Global Health Law at Georgetown University, told the *South China Morning Post* that China's 'failure to allow a full and independent investigation into the origins of the outbreak was a major failure of transparency and international cooperation'.

Eventually, when international experts convened by the WHO were granted permission by China to visit Wuhan and arrived in early 2021, they were given a detailed run-down on what had been found in the market. The China-WHO joint report published on 30 March 2021 revealed that the Chinese CDC inspectors had visited the market about thirty times from 1 January before a final clean up on 2 March 2020. In addition to animal products and frozen goods, samples had been taken from doors, stalls, transport carts, trash cans, toilets, sewage, ventilation systems, stray cats and other animal vectors such as mice. Two other nearby markets had also been sampled.

This China-WHO report revealed a different picture from that given by media reports. The market was called a seafood market for a reason: most of the stalls were selling seafood and freshwater aquatic products. Crocodiles were being sold alive. Snakes and salamanders were being slaughtered on the spot for sale. From sales records in December 2019, just ten stalls were selling meat or products from birds and mammals, including chickens, ducks, geese, pheasants and doves; and deer, badgers, rabbits, bamboo rats, porcupines and hedgehogs. According to the market authorities, all of these animals were from licensed farms and no illegal trade in wildlife was detected.

The authorities tested 457 samples from 188 animals spanning 18 species. They all proved negative for SARS-CoV-2 genetic material. This included 27 stray cats (a species that is susceptible to the virus), which were presumably living free in or around the market, as well as 52 rabbits and hares, 16 hedgehogs, ten mice, seven dogs, six muntjac deer, six badgers, six bamboo rats, a number of pigs, five chickens, three giant salamanders, two wild boar, two crocodiles, two soft-shelled turtles, two fish, one sheep and one weasel. They tested 616 animals of ten species from the suppliers to the market and found no sign of SARS-CoV-2 genetic material.

A total of 923 environmental samples were tested, meaning samples from countertops, door handles, toilets, sewage and the like. By mapping the samples that came back positive for virus genetic material to the stalls in the market, the China-WHO team were able to assess what products the vendors at those affected stalls were selling. Of the twenty-one impacted vendors, sixteen were selling 'cold-chain products' – delivered and sold in frozen form – out of eighty-seven vendors selling such products whose stalls were sampled; thirteen out of seventy-three selling aquatic products; six out of fifty-six for seafood; eight of thirty-seven for poultry; five of thirty-six for livestock; and two of eight selling vegetables. Only one out of nine vendors selling wildlife products was linked to a positive market environmental sample, and he or she had also been selling cold-chain products, aquatic products, poultry and livestock products.

Needless to say, this was a vital slug of information to emerge after so many months of speculation worldwide that the wildlife trade was bound to be the culprit. There was, after the back and forth by the Chinese CDC director and a year of waiting, no evidence that the virus had emerged in the Huanan seafood market via the wildlife trade. The epidemiological data, the genetic data and the positive environmental samples from the market were consistent with a scenario in which a sick person brought the virus into the market, where it became amplified in a poorly ventilated and crowded space. On the role of the Huanan seafood market, the China-WHO joint team stated that: 'No firm conclusion therefore about the role of the Huanan market in the origin of the outbreak, or how the infection was introduced into the market, can currently be drawn.'

5.

The pangolin papers

'In the twenty-first century we really should
not be eating species to extinction.'

JONATHAN BAILLIE

On 7 February 2020, a press release appeared on the website of
South China Agricultural University in the city of Guangzhou,
announcing that researchers had found that pangolins might be
the mysterious intermediate hosts for the new coronavirus. At
a press conference that day the university announced that two
of its professors, Dr Shen Yongyi and Dr Xiao Lihua, had found
a pangolin virus with 99 per cent genetic similarity to SARS-
CoV-2. This is the level of similarity civet coronaviruses had to
human SARS, so it was electrifying news. It seemed that the
animal that gave the SARS-CoV-2 virus to human beings might
have been found.

No such luck. A press conference is not a scientific result. In
fact, when the virus genome was finally shared in mid-February, the 99 per cent figure proved to be wrong. The virus was
only approximately 90 per cent similar to SARS-CoV-2 overall.
The 99 per cent similarity was in one short region of the spike
gene – the receptor-binding motif (RBM), a part of the RBD

that helps the virus to recognise a receptive host cell for infection. The rest of the genome was more distantly related to the human virus; the closest virus relative in February 2020 was still RaTG13. Nonetheless, here was an animal infected with a SARS-like virus and many scientists latched onto the similarity in the RBD as a clue that pangolins might have been the missing link between a bat virus ancestor and human SARS-CoV-2.

Pangolins, or scaly ant eaters, are of considerable value to practitioners of traditional Chinese medicine and hence are frequently trapped and smuggled. They are the most illegally trafficked mammal in the world. There are eight species of pangolin, four of which live in Asia and four in Africa, but all eight find their way to China in large numbers, dead, alive or skinned for their scales. One of the most trafficked is the Malayan pangolin (*Manis javanica*), also known as the Sunda pangolin or the Javan pangolin, which lives throughout Southeast Asia. It is the pangolin species of most interest in this story. The native Chinese species, smaller but similar in appearance, is almost extinct.

About the size of a small dog, a Malayan pangolin looks a bit like a long, walking brown artichoke, covered in a crazy paving of thick scales from the crown of its small head to the tip of its long tail. The scales are soft in babies but harden with age. Made of keratin, like large fingernails, these scales protect the animal against predators, especially when it rolls into a ball, a habit from which the word 'pangolin' derives in the Malayan language (*pengguling* meaning 'one who rolls up'). Their tough claws help pangolins dig into termite and ant nests, while the immensely long tongue, half as long as the body when extended and bizarrely attached near the animal's pelvis, is used to probe into the recesses of the ant nests, extracting the insects by sticking to them. Lacking teeth, pangolins swallow small stones, much like birds do, so that the muscular stomach can grind the

Sunda pangolin (*Manis javanica*) rescued from poachers and in rehabilitation.

termites and ants into soup. Malayan pangolins live mainly in forests, readily climb trees, sometimes burrow into the ground and are mostly nocturnal. They are also solitary, only occasionally meeting others of their kind, for example when breeding, which makes them poor targets for respiratory viruses. Mothers nurse their young for several months and a baby will sometimes hitch a ride on the mother's tail.

In September 2019, a group of journalists calling themselves the Global Environmental Reporting Collective put out a report on the pangolin trade called 'Trafficked to Extinction'. It uncovered a massive criminal enterprise spanning Asia and Africa, feeding pangolins almost exclusively into China. Despite a global ban on traded pangolins, which came into force at the beginning of 2017, the illegal trade apparently reached record levels in 2019. That year nearly forty tonnes of pangolin scales were seized in three operations in Singapore, equivalent to many tens of thousands of the animals. In February 2019, a raid in Sabah, Malaysia, found approximately twenty-eight tonnes of descaled, gutted and frozen pangolins alongside nearly half a tonne of pangolin scales. This was the largest pangolin raid in history.

The vast majority of these animals, alive or dead, or their scales, are imported into China and Vietnam, where they sell for up to $1,800 a kilogram on the black market. The animals are eaten as a prize delicacy while the scales are bought by TCM's pharmaceutical firms and ground up to be sold as antidotes to cancer, impotence, inflammation and poor lactation. Their medicinal value has not been substantiated by any scientific evidence: you might as well eat your own fingernails, which are made of the same stuff.

The smugglers busted

In March 2019, the Guangdong and Guangxi Anti-Smuggling Bureaus, working with more than three hundred police forces in several countries, swooped on wildlife smuggling syndicates across southern China and arrested thirty-four suspects. Along with snakes, eagles, tortoises, geckos and other endangered species that were found stuffed into bags or cages were 155 pangolins, of which 103 were still alive. The pangolins had been smuggled through Vietnam, most probably from further south. Most of the animals were held in Guangxi province, but twenty-one live pangolins were rescued in Guangdong. These were taken to the Guangdong Wildlife Rescue Center in a suburb of Guangzhou, where they were kept in two small rooms. It is these twenty-one animals that may contain some kind of clue to the origin of SARS-CoV-2.

The twenty-one pangolins were in poor health and began to die within a few days. When only five were left, on 17 April 2019, the authorities allowed a private foundation, the China Biodiversity Conservation and Green Development Foundation, to try to rescue the survivors. This was unprecedented: the foundation had been begging to care for confiscated pangolins since 2017. Yet, despite shipping in frozen ants, they were unable to save two of the animals, named 'Dahu' and 'Meidong'. Another would die later. We have been unable to discern the fate of the last two, which were taken by forestry administration officials.

When pangolin genomes were first fully sequenced in 2016, it emerged that they are unique among mammals in having a defective version of a critical immune system gene called interferon epsilon; and this seems to be true of African as well as Asian species. This gene plays a significant role in innate immunity, the first line of defence against infection on the surface of the body, especially on the skin and on mucosa-protected tissues such as lungs, intestines and reproductive tissues. The Malaysian scientists who discovered this fact concluded that pangolin 'innate immunity may be compromised, resulting in an increased susceptibility to infection, particularly in the skin and mucosa-protected organs'. This weak mucosal immunity renders pangolins 'prone to frequently fatal gastrointestinal disease, pneumonia, and skin maladies', which is one reason they so often die in captivity.

Eleven of the twenty-one pangolins seized in Guangzhou were examined after their deaths at the Guangdong Institute of Applied Biological Resources. In an October 2019 paper in the journal *Viruses*, the researchers, Dr Liu Ping, Dr Chen Wu and Dr Chen Jinping, reported that most of the dead pangolins had swollen lungs filled with 'a frothy liquid'. They took twenty-one samples from the lungs, lymph and spleens of the eleven animals, and searched for viral genomes. They found plenty. In all, there were fourteen different families of viruses detected and 85 per cent of the sequences were from two families of virus, the paramyxoviruses and herpes viruses. One of the paramyxoviruses, Sendai virus, was found in six animals.

In only two of the pangolins did they find coronaviruses, albeit possibly several different kinds. The similarity of some of the sequences to SARS-like sequences led the researchers to conclude that this might have been the cause of death in these two animals. The authors did not know whether the pangolins had died from an infection by the Sendai or SARS-like virus. However, the study observed ominously, 'Malayan pangolins could be another host with the potential of transmitting the SARS coronavirus to humans.' This paper was sent to the journal *Viruses* on 30 September 2019, and published on 24 October, shortly before the first cases of Covid-19 in Wuhan.

Curiously, the data behind this paper was only released three months after publication, on 22 January 2020, by which time the pandemic was well under way. These were the samples that the South China Agricultural University in Guangzhou reanalysed in early February, finding that they contained a coronavirus with a spike RBM close to 99 per cent the same as that of the SARS-CoV-2 virus causing sickness in Wuhan. After their press conference on 7 February, a burst of scientific preprints followed on 18 and 20 February, all purporting to present pangolin coronaviruses closely related to SARS-CoV-2. The titles of these four separate preprints told a confident and compelling story: 'Identifying SARS-CoV-2-related Coronaviruses in Malayan Pangolins' (Lam et al., *Nature*); 'Isolation of SARS-CoV-2-related Coronavirus from Malayan Pangolins' (Xiao et al., *Nature*); 'Are Pangolins the Intermediate Host of the 2019 Novel Coronavirus (SARS-CoV-2)?' (Liu et al., *PLoS Pathogens*); and 'Probable Pangolin Origin of SARS-CoV-2 Associated with the COVID-19 Outbreak' (Zhang et al., *Current Biology*).

Many people who read these papers received the impression that many different pangolins were infected with a SARS-like coronavirus closely resembling SARS-CoV-2, emphasising the likelihood that pangolins were the intermediate host that transmitted the virus to humans. Intentionally or not, the four papers left a strong impression that the problem had been close to solved.

A well-adapted virus

In February 2020, one of us (Alina) had been closely following the news of the novel coronavirus as it started appearing in countries all around the globe. Italy was devastated by the virus. Iran and the United States would soon follow. First New York was hit, then in the first week of March, news emerged that the virus had spread at a scientific conference in Boston. Within half a year, this superspreader event in Boston would eventually lead to an estimated three hundred thousand cases worldwide.

The incident led the Broad Institute, one of the world's leading biomedical institutions, to tell all employees to stay home. Alina and her colleagues wanted to find a way to contribute to the research efforts against this growing outbreak. It was unclear when the laboratories would reopen so Alina decided to take a look at what data she could analyse online. In late March, top experts who had been anticipating and watching for potentially dangerous changes in the SARS-CoV-2 genome noticed that there had been no significant change in the biology of the virus. To their relief, there were no signs that the new virus was evolving and accumulating useful mutations to infect human beings more effectively – as the SARS virus had done at the beginning of the 2003 epidemic.

This apparent difference between SARS-CoV-2 and the 2003 SARS virus piqued Alina's interest. She enlisted the help of Dr Shing Hei Zhan, a friend from graduate school at the University of British Columbia, Canada, to compare the evolution of the two SARS viruses side by side. This analysis – which would not have been possible with epidemics before the year 2000, after which genetic sequencing became readily available – revealed that the novel coronavirus more closely resembled SARS in the late phase of the 2003 epidemic after the virus had already picked up numerous advantageous adaptations for human infection and transmission. SARS-CoV-2 was very likely already well adapted to its new human hosts. The early genomes of the 2003 SARS virus diversified like a tall tree, accumulating dozens of mutations over the first three months of the epidemic and across eleven patients. In comparison, the early genomes of SARS-CoV-2 from December 2019 through to March 2020 were all highly similar and, despite the more rigorous sampling of Covid-19 cases worldwide, produced a flatter structure with fewer accumulated changes in its genome across dozens of patients.

The genomes of the 2003 SARS epidemic therefore told a story of an animal pathogen that was new to human hosts and evolved rapidly among the earliest human cases, before settling down into a much slower phase of virus evolution once it had become proficient at spreading in human populations. In sharp contrast, SARS-CoV-2 had shown no early phase of rapid evolution: it was as though a virus highly adept at infecting and transmitting between human beings had appeared out of thin air at Wuhan's Huanan seafood market in December 2019.

The puzzle for Covid-19 was that no precursor or closely related 'siblings' of the virus had been found. When Dr Zhan and Alina looked at the available SARS-CoV-2 sequences from the environmental samples collected at the market (some of which had been published in February), they noticed that these were essentially identical to the human virus isolates. Furthermore, no details had been released concerning animal samples at the Huanan seafood market. This stood in strong contrast to the market investigations for SARS in 2003 and 2004: numerous animals at multiple markets had tested positive for SARS virus back then, and the viruses isolated from the animals had proved to be siblings to the viruses isolated from human patients – in other words they were very closely related but they were sufficiently different that it could be deduced that the market animals were infected with a different variant from the one causing the human epidemic. Significant differences showed up particularly in the spike of the animal and human versions of the 2003 SARS virus, and some of the human virus mutations later proved to be functionally important, enabling the virus to better bind to the human ACE2 receptor. In the case of SARS-CoV-2, no market animals had been reported to be positive for the virus, and the market's environmental samples were more than 99.9 per cent identical to the human virus isolates, even when comparing spike sequences. This suggested human contamination of the market rather than the presence of infected animals in the virus at the market in December 2019.

This work culminated in a preprint authored by Dr Zhan, Dr Chan and Alina's supervisor at the Broad Institute, Dr Ben Deverman, on 2 May 2020, noting that: 'If intermediate animal hosts were present at the market, no evidence remains in the genetic samples available.' The paper infamously included speculation about how the novel SARS-CoV-2 might have evolved to transmit between human beings so effectively: 'There is presently little evidence to definitively support any particular scenario of SARS-CoV-2 adaptation. Did SARS-CoV-2 transmit across species into humans and circulate undetected for months prior to late 2019 while accumulating adaptive mutations? Or was SARS-CoV-2 already well adapted for humans while in bats or an intermediate species? More importantly, does this pool of human-adapted progenitor viruses still exist in animal populations? Even the possibility that a non-genetically-engineered precursor could have adapted to humans while being studied in a laboratory should be considered, regardless of how likely or unlikely.'

The authors of Zhan et al. had no inkling of the media storm that would ensue because of that single sentence about the possibility of a lab origin of SARS-CoV-2. Alina was aware that the issue had been politicised by comments made by members of both the US and Chinese governments. However, she was determined not to let politics dictate what she could or could not write in her research papers. After the preprint was posted, Alina asked the Broad Institute's communications department for advice on how to share the preprint on Twitter and discovered what a 'tweetorial' is – a Twitter thread that serves as a tutorial for a scientific paper.

Ian Birrell, writing for the *Mail on Sunday* in London, picked up on the preprint and ran a story on 17 May titled 'Landmark Study: Virus Didn't Come from Animals in Wuhan Market'. That same day, *Newsweek* and several other media outlets ran articles about the preprint, with a focus on the possible lab origin and the lack of evidence pointing to a natural zoonosis from animals to humans at the seafood market. Then, quietly, on 25 May, the Chinese CDC director, Dr Gao, announced the transformative conclusion that the virus had existed long before the market and that none of the animal samples from the market had tested positive for the virus.

When the WHO prepared a preliminary document on the origin of the outbreak in November 2020, it included a similar point to that which Drs Zhan, Chan and Deverman had made in May: 'Current findings show that the virus has been remarkably stable since it was first reported in Wuhan, with sequences well conserved in different countries, suggesting that the virus was well adapted to human transmission from the moment it was first detected. This is also corroborated by the epidemiology and transmission patterns seen since the start of the COVID-19 pandemic.' Given the furious online criticism of Alina and her colleagues for using the phrase 'well adapted' in their preprint in May, this was a welcome endorsement.

An anonymous late-night tip

On 11 May 2020, shortly after Alina had posted the Zhan et al. preprint and created a tweetorial for it, an anonymous Twitter user named lllandca posted to her thread, in Chinese, that the origins of the pangolin samples in the Xiao et al. *Nature* paper were worth looking into. Intrigued, Alina responded to lllandca seeking clarification and went through each of the pangolin papers with a fine-tooth comb. To her surprise, all the pangolin coronaviruses with a spike receptor-binding domain similar to SARS-CoV-2, described across the four pangolin papers, came from the same batch of animals at the Guangdong Wildlife Rescue Center, confiscated from smugglers in late March 2019, mostly dying from respiratory disease.

When the four papers were formally published following peer review in May 2020, Alina, together with Dr Zhan, contacted the journals *Nature* and *PLoS Pathogens* to seek the detailed genomic data behind the results, much of which had not been made available but could hold important clues as to how each viral genome was assembled. Replication is after all a key part of the scientific method. Strangely, although one of the studies, the Xiao et al. *Nature* paper, claimed to have isolated virus particles in the laboratory by infecting cultures of Vero cells (derived from monkey kidney cells), they chose instead to assemble their genome by amplifying fragments of sequences from multiple frozen pangolin samples. The very title of their paper was 'Isolation of SARS-CoV-2-related Coronavirus from Malayan Pangolins'. Virus isolates – real, complete viruses – would have provided fuller and cleaner genomes than fragments from the pangolin tissue samples.

On closer inspection, Alina and Dr Zhan found that Xiao et al. had in fact republished the same samples under different names without proper attribution to the original October 2019 Liu et al. study in the journal *Viruses*. They had also used an inconsistent definition of 'total reads' in their sample description that made it difficult to match their samples with those from the 2019 paper, thus making it challenging for other scientists to determine that the two data sets were the same. The samples described as lung02, lung07, lung08 and lung11 by Liu et al. in 2019 were the very same samples as those called M3, M2, M4 and M8 by Xiao et al. only a few months later in 2020. The high-quality profile of a single sample described by Xiao et al. was actually derived mostly from Liu et al.'s previously published lung08 and most likely a composite of more than one sample. In fact, all four pangolin papers relied on the same data from the October 2019 study to produce a virus genome with a similar spike RBD to SARS-CoV-2. It is all but impossible, based on published data and information, to determine which sequences came from which samples, or which samples came from which animals.

In November 2020, as a result of Alina's and Dr Zhan's efforts, the journal *Nature* added an 'editor's note' to the Xiao et al. paper, reading as follows: 'Readers are alerted that concerns have been raised about the identity of the pangolin samples reported in this paper and their relationship to previously published pangolin samples. Appropriate editorial action will be taken once this matter is resolved.' Despite its mild wording, this is a rare intervention by a leading journal like *Nature*. These pangolin papers were prepared by scientists, some of whom knew each other, used overlapping data sets, confused names of samples, and were not clear or accurate in their explanation of the data and sample histories. Close to a year later, no further editorial action was taken by *Nature*, however.

In the same month, the US Right to Know organisation published emails regarding the Liu et al. *PLoS Pathogens* paper of May 2020, revealing that the editor of the paper, Dr Stanley Perlman, had acknowledged that he had not checked the veracity of a sample or for the existence of data supporting the paper. 'Concerns about similarities between the two studies came to light only after both studies had been published,' wrote Dr Perlman. 'I regret that these issues were not discovered and addressed prior to publication.' Finally, in June 2021, a correction of the paper was published, noting that the similarities between the pangolin coronavirus and SARS-CoV-2 were 'not strong enough to support that pangolins are intermediate hosts of SARS-CoV-2'.

To provide an analogy, this would be similar to four different teams of clinicians publishing separate papers on the same group of patients with a rare disease without clarifying that they were all describing the same patients, or that they had already published a previous paper only a few months before describing these patients. With confused patient histories and data, and even different names provided for the same patients, the readers of these numerous papers could mistakenly conclude that there are several distinct groups of patients with the affliction and get the impression the disease is more common than it is.

When confronted by US Right to Know, the authors of the *PLoS Pathogens* pangolin paper, Liu et al., confirmed that they had not provided the novel data described in their paper and said, 'We are different research groups from *Nature* paper authors, and there is no relationship with each other ... and we don't know where the samples of the *Nature* paper from.' For context, the senior authors of both papers, in *PLoS Pathogens* and *Nature*, had co-published the data set released on 22 January 2020 and re-used it in both papers posted as preprints in mid-February 2020.

It is difficult to avoid the perception that there could have been a political or public-relations motive driving the work. We know from internal Chinese government documents, seen by the Associated Press, that the pangolin studies were sponsored by various government agencies. The virologist Dr Linfa Wang of Duke-NUS Medical School in Singapore told the Associated Press the search for the coronavirus in pangolins did not appear to be 'scientifically driven'.

A pangolin by any other name

This may seem like a storm in a scientific teacup to outsiders. Some scientists working on pangolins appear to have been republishing the same data in such a way as to generate more papers in high-profile journals. Does it affect the story of the virus origin? Yes, it does, because the world received the impression that SARS coronavirus infection in pangolins is a common phenomenon, which makes the event of a human catching the virus from a pangolin more probable. Now we know that of more than a hundred live pangolins intercepted by Chinese customs officers in March 2019, only two were certainly carrying coronaviruses at the time they were analysed in Guangdong province. They might have caught the virus from any of the other trafficked animals under duress and unsanitary trafficking conditions.

Scientists working with the China Biodiversity Conservation and Green Development Foundation, who had tried to rescue the surviving Guangdong pangolins, said: 'Pangolins are actually unlikely to be the natural reservoir or intermediate hosts due to several reasons.' They reported that more than twenty people had 'continuous unprotected physical contact with the CoV-infected animals'; one person had even cared for the sick pangolins for more than seventy days. None of them reported falling sick.

However, one of the four papers preprinted in February 2020, later published as Lam et al. in *Nature*, while rehashing the same data from the Guangdong pangolins, did also report data from sick pangolins in the neighbouring Guangxi province. These had been intercepted between August 2017 and January 2018. Samples had been collected from multiple organs such as lungs, intestines and blood. Sequencing revealed evidence of coronaviruses in samples from five animals. This virus was much less similar to SARS-CoV-2 than the one from the Guangdong pangolins, let alone the bat virus RaTG13, so it was of little relevance to the search for the origin of Covid-19, but it was nonetheless a SARS-related virus. Dr Lam and his colleagues then tested another set of samples from twelve pangolins intercepted between May and July 2018, and found that three were positive for coronavirus. Yet, to our knowledge, no sequence data has been shared concerning these coronavirus-positive samples from 2018.

Apart from smuggled pangolins, the search for pangolins in the wild with similar viruses has drawn a blank. In June 2020, Dr Peter Daszak and his collaborators reported an analysis of 334 smuggled pangolins confiscated in Malaysia between 2009 and 2019, finding no signs of coronavirus or any zoonotic virus for that matter. The team speculated that the pangolins are more likely infected on the way to market, not in the wild: 'Our samples were drawn from an "upstream" cohort of animals yet to enter or just entering the illegal trade network, whereas all others were drawn from "downstream" cohorts confiscated at their destination in China.'

Based on our understanding of pangolin biology and the scientific evidence to date, it appears highly unlikely that wild populations of these animals carry SARS-like coronaviruses. Pangolins live solitary lives in the wild, making them poor hosts for coronaviruses, which thrive on victims that regularly meet in large gatherings, like bats and people. Furthermore, the severity of the illness afflicting the smuggled pangolins, trafficked thousands of kilometres from their native habitats, suggests that if these coronavirus infections were common in wild populations, these animals would have been wiped out. To this day there remains no evidence that the SARS-CoV-2 virus passed through pangolins before emerging in humans.

To put the final nail in the coffin, in June 2021, a bombshell publication in *Scientific Reports* revealed that no pangolins or bats had been found to be traded in Wuhan's markets between May 2017 and November 2019. The Chinese scientists who authored the paper had serendipitously been cataloguing wild animal sales across Wuhan's animal markets, reaching a total of 47,381 animals across thirty-eight species. Their aim was to search for the animal origin of a tick-borne fever in the province, and they had no inkling of how valuable their data would become once Covid-19 emerged in Wuhan. The scientists were especially confident that their research was complete because vendors had 'freely disclosed a variety of protected species on sale illegally in their shops, therefore they would not benefit from specifically concealing the pangolin trade'. Although regulations in China at the time made the sale of most of these wild animals illegal, the traders and their customers were not particularly bothered, according to the scientists. Thirty-one of the species they had observed were protected under 'China's List of Terrestrial Wild Animals of Significant Ecological, Scientific, or Social Value Protected by the State', meaning that if any of these animals were found to have been taken from the wild and traded as food, the offender could face imprisonment and fines. Yet about 30 per cent of the mammals inspected had wounds indicating gun injuries or traps, suggesting illegal hunting, and the vendors had not hidden these animals from the scientists. Furthermore, the scientists had made clear that they were not from law enforcement. The vendors at the Wuhan markets had been forthcoming because they too were interested in finding the source of the tick-borne disease. Neither the scientists nor the traders had the prescience to know that bats or pangolins would later become implicated as possible sources of the 2019 novel coronavirus, so they would not have specifically hidden away bats or pangolins instead of the numerous other endangered or protected species on sale. Later, when the scientists wrote their paper, they emphasised that bats were not typically consumed in central China.

Secret pangolin research

On 15 January 2021, the US State Department released a fact sheet on 'Activity at the Wuhan Institute of Virology' highlighting, among several notable points, that 'the WIV became a focal point for international coronavirus research after the 2003 SARS outbreak and has since studied animals including mice, bats, and pangolins'. Needless to say, this announcement took many people by surprise and raised more questions than it answered. What was the animal coronavirus research that was being performed at the WIV? Were SARS viruses involved in these animal experiments, considering the presence of a world-class SARS laboratory in the institute? When had experiments involving pangolins and coronaviruses begun? Before or after the emergence of Covid-19?

As a scientist and science writer reviewing this timeline of events, we find several things unsettling about the whole pangolin episode. There was the publication of a pangolin coronavirus with such a similar spike receptor-binding domain (RBD) only a few weeks before the emergence of Covid-19; the release of the pangolin coronavirus data behind the 2019 paper on 22 January 2020, the same day that China announced the Huanan seafood market was the culprit; the posting within three days in February of four manuscripts all describing the same pangolin coronavirus; and the variety of scientific issues plaguing some of these papers' descriptions of the pangolin coronavirus genome with the SARS-CoV-2-like spike RBD.

The questions we would like to ask, if and when Chinese scientists are allowed to answer, are these. After the intercepted March 2019 pangolins had been sampled and sequenced, where was the virus data stored and who had access to this data? Could other scientists have attempted to characterise (clone and study the function of) the novel spike RBD? When were the 2017 and 2018 smuggled pangolins sampled and sequenced? Could the finding of distantly related SARS-like viruses in these animals have prompted more research interest before late 2019? Or were these samples simply sitting in freezers until Covid-19 emerged and scientists decided to test them for coronavirus?

We stress that there is no evidence that any such experiments on pangolins were attempted, let alone that they resulted in an accidental leak. But it is very much within the capabilities of some laboratories to do such a project, as we shall show in a later chapter. Indeed, it would not be an irrational experiment to conduct if you were curious about newly discovered pangolin SARS-like coronaviruses that you thought might someday cause an epidemic.

Will Covid-19 put an end to the pangolin trade?

Animal lovers rejoiced prematurely when China announced a ban on the trade of wild animals in February 2020 – the same month that the four pangolin papers were posted as preprints. It turned out that the ban applied to animals being traded for consumption but, according to the Wildlife Conservation Society, not to the trade of exotic animals for fur, medicine or research. Which meant it would be essentially pointless for an animal such as the pangolin which has been aggressively hunted and trafficked mainly for the use of its scales in traditional Chinese medicine, and for which trade was already illegal before the emergence of Covid-19.

In October 2020, the Environmental Investigation Agency, an international watchdog, revealed that eBay and Taobao (a Chinese shopping website similar to Amazon) were continuing to market pangolin products, and that more than two hundred companies were still licensed to sell these products. In spite of the Chinese government having removed pangolin scales from its list of approved TCM ingredients, the country's medical insurance system still covered pangolin-containing TCM and companies were still permitted to use the animal's scales from a national stockpile that 'never seems to run out'. This is concerning in its own right, but it has another implication. In the face of a pandemic that by the beginning of October 2020 had infected more than thirty-five million people and killed a million, these ineffectual bans seemed half-hearted if the Chinese authorities gave any credence to the hypothesis that pangolins might be the intermediate host of Covid-19.

ACTAsia, part of the Animal Care Trust non-profit organisation, criticised the Chinese ban on wildlife consumption for not being broad enough to address the substantial wildlife trade for fur, TCM, experimentation and entertainment. ACTAsia said that the new law, made shortly after China announced that Covid-19 had come from wild animals, was 'complex, with loopholes and caveats to renege on every statement of protection, surreptitiously supporting the wildlife trade as a commercial industry, as part of China's plan for five years out of poverty'. In reality, parts of the new regulation support the captive breeding of wild animals that are listed as 'livestock' and include more than forty kinds that have the potential to be intermediate hosts for human pathogens. These wildlife farms are also conducive to spillovers in scenarios in which a large number and variety of animals are held in unsanitary captivity. The farming of bears, tigers and even pangolins is still permitted, which the Environmental Investigation Agency has pointed out could be a cover for laundering illegal wild pangolins due to the severe difficulty of breeding them in captivity and the lack of supervision of such programmes. In April 2021, Reuters reported that pangolins continued to be trafficked into China, especially at lax international borders such as at the special economic zone of Mong La in Myanmar, where there was little government control.

We cannot understand why a government that officially maintains that the SARS-CoV-2 virus originated in wild animals, farmed or trafficked, would not install and rigorously enforce regulations to comprehensively eliminate these routes for the emergence of dangerous pathogens.

6.

Bats and the virus hunters

'The order Chiroptera (the "hand-wing"
creatures) encompasses 1,116 species, which
amounts to 25 percent of all the recognized
species of mammals. To say again: one in every
four species of mammal is a bat.'

DAVID QUAMMEN

SARS, SARS-CoV-2 and other SARS-like coronaviruses predominantly thrive in just one genus of bat, *Rhinolophus*, the horseshoe bats. These are a diverse group, with 106 species known so far. They are found only in the old world, having not reached the Americas, and are mostly tropical or subtropical in their range although two do live as far north as southern Britain. Noted for their manoeuvrability on their relatively short, broad wings, and their top-of-the-range, sophisticated echo-location equipment, they are among the most accomplished bats when it comes to flitting through trees and shrubs in the dead of night finding insects on the wing. Horseshoes are small or average-size bats, with large, broad, but pointed ears and rich, often reddish fur. They get their name from their most distinctive feature, a strange, fleshy sonar dish known as a nose-leaf on the end of the snout, the outer part of which is usually shaped like an upturned horseshoe. The name *Rhinolophus* translates as 'nose leaf'. The nose-leaf serves to focus the high-pitched ultrasound beams the bats send out via their nostrils, while shielding their ears from the sound. Horseshoe bats generally like to roost in caves and some species are very gregarious, gathering in roosts of thousands of animals.

China is home to many horseshoe bat species: *Rhinolophus ferrumequinum*, the greater horseshoe bat, lives in central China (and across Eurasia), so it is found near Wuhan, but not further south. In southern China, near the borders with Laos, Vietnam and Thailand, at least nine species of horseshoe bat are common, some almost indistinguishable from one another. There is the intermediate (*affinis*), the Chinese rufous (*sinicus*), the big-eared (*macrotis*), the woolly (*luctus*), the king (*rex*), the Thai (*siamensis*), Osgood's (*osgoodi*), Pearson's (*pearsonii*) and the least (*pusillus*), plus there are another ten that are either very rare in China or not necessarily recognised as being separate species. So if you walk into a cave in Yunnan in south-west

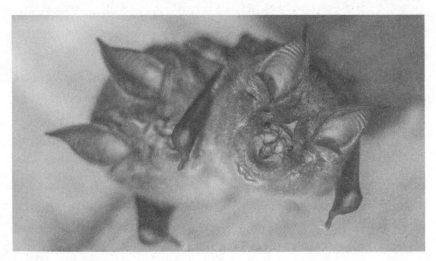

Two greater horseshoe bats (*Rhinolophus ferrumequinum*). This species occurs in Europe, Northern Africa, Central Asia and Eastern Asia.

China, shine a light on the ceiling, and catch a glimpse of a bunch of horseshoe bats huddled there, they could be several different species together. The differences in their appearance and behaviour are sometimes slight.

In the last few decades, no other group of animals has proved as prolific as bats at serving as the reservoir of new viruses that make their way to human populations. Hendra in Australia in 1994, Nipah in Malaysia in 1998, SARS in China in 2002–3, MERS in Saudi Arabia in 2012, Ebola in Sudan and Zaire in 1976, Guinea in 2013 and Congo more recently: all are deadly viruses that ultimately came from bats. There are probably several reasons that bats prove such a reservoir of zoonotic viruses. Bats are mammals, which means they are sufficiently related to other mammalian intermediate hosts and human beings that it is not a great leap for a bat-borne virus to find its way into people. In comparison, viruses that infect reptiles or fish or birds are living in very different bodies with very different cellular machinery from that in humans, making it less straightforward for a virus to cross host classes in the animal kingdom. Furthermore, bats have not been domesticated, are nocturnal, and they are usually to be found in dense roosts in remote caves. This means that although humans are sometimes exposed to bats through practices such as the guano trade, we are less likely to encounter their diseases.

Bats are highly diverse, though it is difficult to be sure how many bat species there are. 'Lumpers', who downgrade some species to subspecies, estimate around 900 species; 'splitters', who upgrade some subspecies to species, can get to nearly 1,400. That means that about one in four mammal species is a bat. Each species can have its own viruses, and many viruses evolve to become proficient at jumping between bat species and even to other animal species. For their small size, bats are surprisingly long-lived animals. A horseshoe bat has been found to live to the age of thirty and most survive for six or seven years – far longer than a mouse of the same size. It is possible the bat that was carrying the RaTG13 virus in the Mojiang mine in 2013 is still out there.

Like humans and farmed animals, bats sometimes live in dense populations. In some breeding colonies, there can be up to five hundred bat pups per square foot on the wall of a cave. At certain times of the year, one cave in Texas houses twenty million bats, a concentration of mammals paralleled only by people in cities. Another cave in Borneo is home to ten million bats. Moreover, different species roost together and share their viruses. Coronaviruses are good at recombining and shifting between host species. Also, bats are the only flying mammals. This means that, unlike mice, they travel long distances, meeting a lot of stranger bats as they do so, with the opportunity to pick up new viruses. All of this presents terrific opportunities for the spread and diversification of viruses. A tiger or a pangolin, living a solitary life and hardly ever meeting another of its species except to mate, just isn't much of a prospect for an ambitious coronavirus.

Bats have existed for an estimated fifty million years and have evolved unique immune systems (with distinct properties from the human immune system) that more often enable them to show no sign of disease when infected by various viruses. Scientists have found that bat viruses that cause severe disease in humans often trigger an excessive immune response and inflammation in human hosts but do not stimulate the same overreactive response in bats. Dr Linfa Wang of Duke-NUS Medical School in Singapore has been arguing for years that one reason bats have disproportionately more virus diseases than other mammals is that they tolerate chronic viral infections better than other mammals. Many other scientists are now persuaded that he is right.

The reason, Dr Wang thinks, is that bats run their bodies at high temperatures while flying. A flying bat's pulse might reach a thousand beats a minute and its blood temperature can exceed 100 degrees Fahrenheit. This requirement for ultra-high metabolic activity has led to faster oxygen metabolism, which can damage the DNA in the cell nuclei and sometimes cause DNA to leak into the rest of the cell. That would normally induce an immune response leading to inflammation and the production of interferons: proteins that cells make to communicate to other cells that they are under attack and to trigger a protective response. But in bats the system seems to be dampened so that this inflammation is avoided. As a consequence, viruses also do not provoke such a strong inflammatory response in bats. This may hold clues to how to treat people who are infected by viruses and it could even prolong our health span – the years of our life when we are in good health.

In 2018, Dr Peng Zhou, Dr Shi Zhengli and colleagues published the results of an experiment at the Wuhan Institute of Virology that aimed to understand how bats dampen their interferon response. Spleen cells taken from three bats of the species *Rhinolophus sinicus*, captured in the Taiyi cave in Xianning, about fifty miles south of Wuhan, were used to test the interferon response as compared with spleen cells from three mice. Sure enough, when the cells were stimulated with a molecule that is a product of DNA breakdown, the genes that produce interferons were more active in the mouse cells than in the bat cells. The scientists zeroed in on a particular mutation in the so-called STING gene found in most animals that modulates the interferon response, ensuring that their immune systems do not overreact and cause inflammation. The team concluded that 'bats are more effective in peaceful co-existence with a large number of viruses'.

Incidentally, these experiments illustrate the fact that the WIV was doing experiments on bat cells in the laboratory. In the light of this, it is interesting to read of the risks of working with bat cells, as set out in a paper by an Australian team in 2018. They pointed out that such cells can harbour latent viruses 'that can become reactivated during *in vitro* cultivation when the cells are outside the host and isolated from other components of the immune system that would otherwise control virus replication'. They caution that cell lines from bats should therefore be handled with much care. Nonetheless, different teams of scientists are beginning to create their own customised laboratory cell lines derived from wild animals in order to facilitate the collection and study of entirely novel viruses found in nature. What is in each laboratory's virology toolkit will likely be a mystery as long as scientists continue not to publish the details of the animals they have captured and used to create so-called 'immortal' cells – cells that seem to be able to grow and divide endlessly in the laboratory. It appears that several such cell lines were established at the WIV likely from bats taken from the Jinning caves (Shitou and Yanzi) and the Mojiang mine.

The virus hunters

After the 2002–3 SARS epidemic, the WHO convened a scientific mission to investigate the origins of the first SARS virus in China. The team included Dr Linfa Wang, based in Australia at the time, who had previously studied Hendra and Nipah viruses that had originated in bats, and Dr Shi Zhengli from the WIV, together with international scientific colleagues. They eventually traced SARS-like viruses to bats in southern China. After locating caves in the hillsides where the bats roosted in good numbers, the team set mist nets over the entrance to catch the bats as they emerged at dusk. The nose and anus of each one would be swabbed to provide a sample. Next morning after first light the scientists would return and collect fresh droppings that had fallen onto tarpaulins they had stretched on the floor of the cave. As one of Dr Shi's star doctoral students, Dr Ben Hu, recalled, Dr Shi often led these virus-hunting expeditions herself, trekking through the wilderness with her group, and was admired by her colleagues for her perseverance.

At first the scientists struggled to find evidence of bat infection in the caves, but when they started looking for antibodies to SARS instead of virus genetic material, they began to find strong indication of past infection in horseshoe bats. Seeking expertise in wildlife handling, Dr Wang and Dr Shi teamed up with Dr Peter Daszak, then of the Consortium for Conservation Medicine in New York. By 2005, the team had found a SARS-like virus in a faecal sample from a *Rhinolophus* bat, the genome of which proved to be approximately 92 per cent identical to the human SARS virus. The group went on to publish a review of bats and SARS, describing their discovery of several *Rhinolophus* bat species that were reservoirs of SARS-like viruses.

Over the next decade and more, many other influential papers from this research consortium describing novel viruses isolated from bats would follow. Thanks to this history of hunting down such viruses, Dr Wang earned the nickname Batman, while Dr Shi became known as the Bat Woman. While chasing the SARS-like viruses across China, they also forged friendships and collaborations with overseas scientists such as Dr Daszak and Dr Hume Field, who later became the president and the science and policy advisor respectively of the EcoHealth Alliance non-profit organisation. After their first discovery of SARS-like viruses in bats in southern China, Dr Daszak sought grants in the United States to support the work, starting a long and fruitful collaboration with Dr Shi.

Despite this early success, scientists had not found the bat virus that had given birth to the 2002–3 SARS epidemic. It remained a mystery as to how the killer virus had emerged in Guangdong. In the wake of SARS, governments around the world began to funnel money towards virologists and wildlife biologists in the hope of averting the next pandemic. The biggest of these projects, the United States Agency for International Development's (USAID) Emerging Pandemic Threats (EPT) programme began in 2009. Intended to last for five years, it would be renewed in 2014 for another five. The EPT was divided into four strands: Predict, Identify, Respond, Prevent. The first of these, Predict, aimed to estimate the spillover potential of animal viruses, based on where each virus was found and the range of species it infected. Predict was led by the University of California at Davis One Health Institute, where the veterinary scientist Dr Jonna Mazet was the director of the programme, partnering with Metabiota, a viral database firm based in San Francisco, and the EcoHealth Alliance, as well as the Wildlife Conservation Society and the Smithsonian Institution.

The EcoHealth Alliance's roots lie partly in the books of Gerald Durrell (1925–95), a naturalist who combined a career collecting animals for zoos with writing light-hearted books about his adventures. The books were so successful that he founded his own zoo and conservation charity, based on the island of Jersey in the English Channel. In 1971, Gerald Durrell's charitable legacy, the Jersey Wildlife Preservation Trust, formed an American branch, which in 1999 became the Wildlife Trust. Shortly after this, Dr Peter Daszak joined the organisation. Dr Daszak had earned his PhD studying parasitic infectious diseases at the University of East London and then followed his wife to the United States. While waiting for a work visa, he volunteered at the US Centers for Disease Control and

Prevention at the time of the Nipah outbreak, then joined the University of Georgia before in 2000 applying for a job at the Wildlife Trust. This work consisted of coordinating a project involving five universities studying diseases carried by wildlife, which became the Consortium for Conservation Medicine. The project grew to dominate the finances of the Wildlife Trust and in 2010 Dr Daszak became president of the organisation and it changed its name to the EcoHealth Alliance.

The task of preventing zoonoses was both urgent and noble, and it required expertise in wildlife as well as viruses. The new organisation secured a series of large grants, mainly through Predict, to lead work around the world on pandemic risks from new viruses found in wild animals. As the distributor of these grants to various academic institutions internationally, Dr Daszak gained considerable influence. By 2018, the EcoHealth Alliance had grown its income to almost $17 million a year, nearly all of it from government. In total, in its first decade of existence, the organisation received more than $120 million in US government grants. It takes some diligence to work out from the EcoHealth Alliance's accounts that a lot of that money came from the Pentagon. One journalist, Sam Husseini, found his emails and voicemails ignored when requesting this information, despite the EcoHealth Alliance website's own proclamation that 'A copy of the EHA Grant Management Manual is available upon request to the EHA Chief Financial Officer'. Government databases eventually revealed to Husseini that from 2013 to 2020 the EcoHealth Alliance received $39 million from the Pentagon, mainly via the US Department of Defense's Threat Reduction Agency. It received $20 million from the Department of Health and Human Services and more than $64 million from the USAID's Predict.

Predict's grants to the EcoHealth Alliance were to help it build local capabilities and test high-risk wildlife. 'After scientists collect swabs or small amounts of blood, they analyze the samples in the lab to look for evidence of disease,' the organisation explained. The money was spent partly on funding research by overseas partners, one of the most high-profile of which was the WIV. In the search for the origin of SARS, Dr Daszak and Dr Shi travelled throughout Yunnan and other parts of China collecting bats to be tested for viruses. He was a regular collaborator, co-author and funder of her work.

By 2011, Dr Shi and Dr Daszak had begun to focus on the caves of Yunnan. Slowly, the team zeroed in on the area and the species with the most SARS-like of the SARS-like viruses. Between April 2011 and September 2012, Dr Shi's team took 117 anal swabs and faecal samples from bats in the Shitou cave near Kunming, more than a thousand kilometres to the west of Guangdong where palm civets and people had become infected by SARS virus. Twenty-seven of the samples tested positive for coronavirus, and at least seven strains of SARS-like viruses could be detected. The complete genomic sequences of two strains were obtained and they were named SARS-like corona-virus (SL-CoV) RsSHC014 and Rs3367. This time the key part of the spike protein – the receptor-binding domain – was similar to that in the SARS virus; it was the first time that scientists had discovered a wild bat SARS-like virus that could use the same ACE2 receptor that the 2003 SARS virus had utilised to infect its animal hosts. Overall, the virus sequences that the WIV had found were about 95 per cent identical to the SARS viruses that had been isolated from humans and civets during the SARS epidemic. They announced the discovery triumphantly in a paper in *Nature* on 30 October 2013, entitled 'Isolation and Characterization of a Bat SARS-like Coronavirus That Uses the ACE2 Receptor'.

Not only had this expedition resulted in the identification of the closest match to the epidemic SARS virus, but the team also succeeded for the first time in 'isolating' a live virus from one of the samples. The phrase 'isolating a virus' has a specific meaning: to coax the virus into replicating in cells in the laboratory and thus produce new, infectious viruses for further study. It is a difficult task. RNA viruses are fragile and the virus particles and their genomes are usually too broken up to work with by the time they have been transported thousands of kilometres to the laboratory. Isolating a virus provides greater confidence that the genome assembled from fragments of sequences is accurate and represents a real virus. In this case, Dr Shi's team retrieved a fully viable virus from a sample taken from a Chinese rufous horseshoe bat, *Rhinolophus sinicus*. They called this virus SL-CoV-WIV1, which stands for SARS-like coronavirus Wuhan Institute of Virology 1.

Successes in pathogen surveillance

In 2017 the virus hunting at the WIV scored a significant success: identifying the source of a new epidemic and proving the worth of their work to date. However, it did not concern beta-coronaviruses in people, but alpha-coronaviruses in pigs. Between 28 October 2016 and 2 May 2017, severe diarrhoea broke out on four pig farms in Guangdong. Piglets five days old or younger had up to 90 per cent mortality if they caught this disease, known as swine acute diarrhoea syndrome (SADS). In all, 24,693 pigs died. The WIV was called in to help investigate. About a year later, the usual team from Dr Shi's laboratory, with Dr Peng Zhou as lead author, and Drs Peter Daszak, Linfa Wang and Shi Zhengli among the co-authors, announced that they had extracted more than fifteen million genetic reads from the intestines of one piglet. Among these, they found 4,225 that were from an alpha-coronavirus, which they deduced was the cause of the disease.

They then isolated thirty-three complete genomes of the coronavirus taken from pigs at all four farms and found that it was roughly 95 per cent the same as a coronavirus genome isolated from *sinicus* horseshoe bats in Hong Kong and Guangdong ten years before, known as HKU2. However, the spike protein was much less similar, 86 per cent, so they concluded that the bat virus was only a close cousin of the pig virus. They then returned to their freezers, much as they would do in 2020, and dug out the sequences from faecal swabs of bats caught at seven locations in Guangdong between 2013 and 2016. They found fifty-eight samples positive for SADS coronavirus, all from horseshoe bats, and among them were four genomes that very closely resembled that of SADS.

· In the lab, the scientists tried to isolate the novel SADS virus using a variety of different cell types from different species: Vero (monkey kidney); home-grown *R. sinicus* cells from bat kidney, lung, brain and heart; and swine cells from the intestines, kidney and testes. The reason being that it is difficult to know how to grow novel viruses in the laboratory. Even when you know the host species that the virus was collected from, the virus may not grow well in cells from that same animal species in the lab. So some scientists inoculate different types of cells, derived from different host species, with the novel virus sample. Then the scientists take what grows in that first batch of cells and inoculate another batch of cells, hoping that the fittest of the virus particles will propagate. This process can be repeated as many times as needed and is called serial passaging. However, this approach can be risky when a sample contains multiple viruses, or multiple virus samples are combined before passaging. With different viruses growing together in an entirely new host environment, unique recombinants may emerge.

In the case of the SADS virus, it was only after five passages in Vero cells that the scientists finally observed cell death due to viral infection – in other words, effective virus replication and spreading among cells. The team then tested whether the viruses could infect human cells using the receptors of SARS, MERS or alpha-coronaviruses. It turns out that the SADS virus did not use any of these but was able to cause severe disease and mortality in challenge studies, in which healthy piglets were inoculated with the virus. The conclusion was that the virus, presumably from a horseshoe bat, had somehow directly infected the pigs on the four farms once or several times, and fortunately it did not look as if it would be very dangerous for people. As Dr Shi later pointed out, 'This is the first documented spillover of a bat coronavirus that caused severe diseases in domestic animals.'

The SADS episode is one that the WIV-EcoHealth Alliance collaboration could rightly look back on as textbook: there is an outbreak of disease, the virus is identified, its natural reservoir (bats) is tracked down, and the risk from the virus to people is assessed – all made possible by years of sampling of bats in the wild, storing samples in the laboratory, and testing them against cells from different species, organs, presenting different virus entry receptors. The scientists wrote in their paper that their study demonstrated the 'value of proactive viral discovery in wildlife, and targeted surveillance in response to an emerging infectious disease event'.

Sampling human beings for novel pathogens

As well as sampling bats and other animals, the WIV-EcoHealth Alliance team sampled people. They wanted to know whether people living near bat caves in which SARS-like viruses had been found were getting infected by these viruses. In other words, what was the risk of another SARS virus outbreak stemming from the bat caves in Yunnan? In October 2015, they took blood from 139 women and 79 men living in four villages within a few kilometres of two of the caves in Yunnan in which the scientists had been catching bats: Shitou and Yanzi caves near Jinning. Most of the subjects were farmers, most kept livestock or owned pets, and one had handled a dead bat. The area had not experienced any part of the SARS epidemic. As a control, the scientists also sampled the blood of 240 donors in Wuhan. The results were intriguing. Of the 218 villagers, six had detectable levels of antibodies to the nucleocapsid protein of a SARS-like virus named Rp3 (which had been determined to be a reliable method for detecting SARS-specific antibodies). By the same assay, none of the Wuhan controls proved positive for SARS antibodies. The scientists obtained oral and faecal swabs as well as more blood samples from the six individuals but could not detect any virus genetic material.

The six positive cases were typical villagers. In the year before the human samples had been collected, only one had travelled outside the province, only one had visited Kunming, and two had not left the village. They were all among the twenty people in the study who had noticed bats flying about their villages. None remembered developing any clinical symptoms in the previous year. The scientists concluded that 'the 2.7 % seropositivity for the high risk group of residents living in close proximity to bat colonies suggests that spillover is a relatively rare event, however this depends on how long antibodies persist in people, since other individuals may have been exposed and antibodies waned'.

Over the next two years the team repeated the survey on a wider scale, throughout southern China, targeting people who regularly visited or worked around bat caves or who were involved in the wildlife trade, catching, rearing, transporting or selling animals. Of the 1,596 participants in the study, 265 had reported recent severe acute respiratory infections and/or influenza-like illness. However, only nine (of the 1,497 participants who provided samples) tested positive for SARS antibodies and none of the nine had reported experiencing any symptoms in the year preceding their interviews. Eight of them were farmers of crops with no unusual work connection to animals. The highest self-reported risk factor for having experienced these symptoms was the consumption of raw or undercooked meat from a carnivorous animal. There was no evidence of contact with bats influencing whether a person was more likely to have SARS antibodies.

Like the previous study, this one, published in September 2019, concluded that 'bat coronavirus spillover is a rare event'. Once the Covid-19 pandemic began, however, Dr Daszak (a co-author on the two studies) took a rather different line, emphasising now how common the spillovers of bat coronaviruses were, rather than how rare. In April 2020, in an interview with Vox, he said that the maths was straightforward: 'We also find tens of thousands of people in the wildlife trade, hunting and killing wildlife in China and Southeast Asia, and millions of people living in rural populations in Southeast Asia near bat caves.' Extrapolating from the Yunnan study, he argued that between one million and seven million people were getting infected every year by bat viruses.

Youth in the wild

In parallel with the work of the WIV scientists, another Wuhan laboratory also got into bat virus hunting. At the Wuhan Center for Disease Control and Prevention, a laboratory less than three hundred metres from the Huanan seafood market, Dr Tian Junhua, an expert on pathogens carried by ticks and mosquitoes, began sampling bats in Hubei province. According to a profile in the *Yangtze River Commercial Daily* in 2017, Dr Tian graduated from Huazhong Agricultural University, majoring in plant protection in 2004, before joining the Department of Disinfection and Vector Control of the Wuhan CDC. He started work on cockroaches but later shifted to a bat sampling project. By 2019, he was managing director of the division for vector biology at the Wuhan CDC (in this case, vectors refer to disease-carrying insects).

In December that year, the China Association for Science and Technology released a video, unwittingly coinciding with early cases of Covid-19, as part of a series celebrating young scientists. Called 'Youth in the Wild: Invisible Defender', it is a slickly produced seven-minute film about the work of Dr Tian. He recounted his experiences in exploring caves in search of bats, often with his wife and colleague, saying that 'the caves frequented by bats became our main battlefields', and adding 'bats usually live in caves humans can hardly reach. Only in these places can we find the most ideal virus vector samples.'

In the video Dr Tian said he had visited every corner of Hubei province over the past decade and had explored dozens of caves to study more than three hundred types of viruses. In the process, he claimed to have trapped nearly ten thousand bats. Dr Tian is a co-author with Dr Zhang Yongzhen in Shanghai on the *Nature* paper reporting the first genome sequence of SARS-CoV-2, in which Dr Tian is reported to have 'performed the epidemiological investigation and sample collection'. Although his laboratory did not itself carry out genomic sequencing, there is no doubt that bat samples were brought to the Wuhan CDC. The withdrawn paper (speculating on possible lab origins of Covid-19) by two Wuhan scientists, Dr Botao Xiao and Lei Xiao, pointed out that a 2013 publication by the Wuhan CDC had described surgeries on bats to obtain tissue and organ samples for downstream pathogen detection.

Dr Tian's extensive and diligent work apparently did not turn up SARS-like sequences in Hubei, of which Wuhan is the capital. No SARS-like virus utilising the ACE2 receptor has been found in the central province despite the thousands of bats that have been sampled there over the past decade. In fact, in 2019, when Dr Shi Zhengli and Dr Ben Hu, alongside other WIV scientists, published a paper reviewing their years of bat virus sampling across China, they stated that the 2003 human SARS virus may have originated from southern China, and that SARS-like viruses 'clustered according to their geographical location of sampling, indicating that geographical range over-lap between hosts is likely to play an important role in shaping the evolution of these viruses'. In other words, geography limits the overlap of viruses that encourages the diversification of SARS-like viruses. China is after all a very large country, with regions sometimes separated by expansive mountain ranges. Bats can travel long distances, but unlike birds they rarely undertake long seasonal migrations, preferring instead to hiber-nate during winter in colder climates. In July 2020, Dr Shi told *Science* magazine that her group had 'done bat virus surveil-lance in Hubei Province for many years, but have not found that bats in Wuhan or even the wider Hubei Province carry any coronaviruses that are closely related to SARS-CoV-2. I don't think the spillover from bats to humans occurred in Wuhan or in Hubei Province.'

For all of the WIV's hard work in collecting SARS-like viruses from bats in Yunnan, none yet found could be the immediate ancestor of the 2002–3 SARS virus. In 2015, the researchers got a step closer, finding and isolating a virus they named WIV16 in the Shitou cave that was the closest match yet (96 per cent identical) to SARS in humans and civets. But, strangely, in one gene, ORF8, it was not the closest relative of SARS. It began to look as if the 2003 SARS virus was derived from 'a complicated recombination and genetic evolution among different bat SL-CoVs', as the WIV scientists put it.

The two Wuhan labs were not the only ones to sample horseshoe bats for coronaviruses in the 2010s, although they were by far the most active. From 2012 to 2015, a study led by Dr Zhang Yongzhen sampled 1,067 bats from twenty-one species at five sites throughout China, finding SARS-like coronaviruses in five species of horseshoe bat, with the most SARS-like ones being found in Guizhou, the province between Guangxi and Yunnan.

In eastern China, near the city of Zhoushan, in the province of Zhejiang, a team of biologists from the Third Military Medical University in Chongqing and the Research Institute for Medicine of Nanjing Command visited a mountain cave on four occasions between July 2015 and February 2017, catching 334 Chinese rufous horseshoe (*R. sinicus*) bats, a quarter of which tested positive for coronaviruses. They sequenced two SARS-like viruses, named ZXC21 and ZC45, which proved to be not very closely related to SARS, but would later prove to be more closely related to SARS-CoV-2. Indeed, with the exception of the 4991 fragment from Mojiang, these were the first SARS-CoV-2-like viruses published in the scientific literature before the emergence of Covid-19.

In the laboratory, they managed to isolate one of these SARS-like viruses, ZC45, by injecting fluids from virus-infected ground-up bat intestines into the brains of suckling rats. This demonstrated that such viruses were capable of infecting other mammals. Furthermore, despite the route by which the virus had been introduced into the bodies of the suckling rats, the highest viral loads were later found in the lungs rather than the brain or other organs. The same team had already shown that least horseshoe bats, *Rhinolophus pusillus*, and greater horseshoe bats, *R. ferrumequinum*, captured in buildings in south-east China, also sometimes carried SARS-like coronaviruses.

In a 2017 interview with Xinhua News Agency, Dr Tian of the Wuhan CDC recounted how he once forgot to wear protective clothing and 'bat urine dripped from the top of his head like raindrops'. He returned home, took the initiative to isolate himself from his wife and children and quarantined himself 'for half a month'. On several occasions he was spattered with bat blood on his skin while trapping the bats. Dr Tian's account shows that, for all their precautions, virus-hunting scientists do run a risk of becoming a patient zero of a new epidemic. Photographs and accounts of people involved in the WIV-EcoHealth Alliance expeditions catching and handling bats without wearing full protective clothing have popped up online. In a Chinese television film of 2017 to showcase the work of Dr Shi in caves in Yunnan and labs in Wuhan, members of her team are shown handling bats with their bare hands, collecting bat faeces in caves while wearing shorts and short-sleeved shirts, working among flying bats without face coverings, and handling the animals indoors while not wearing masks. One researcher, Cui Jie, describes being bitten by a bat's fangs through his gloves – it was 'like being jabbed with a needle'.

In his book *Spillover*, David Quammen describes squirming deep into a narrow cave near Guilin in Guangdong province with Aleksei Chmura, Guangjian Zhu and Yang Jian, as part of an EcoHealth Alliance bat virus survey to catch bats in mist nets. 'At this moment I became conscious of a dreary human concern: Though we were searching for SARS-like coronavirus in these animals, and sharing their air in a closely confined space, none of us was wearing a mask. Not even a surgical mask, let alone an N95. Um, why is that? I asked Aleksei. "I guess it's like not wearing a seat belt," he said.'

Tourists and cavers visit caves all over the world, wearing no special protective clothing, so it is harsh to pick on scientists for sometimes doing the same. At least scientists are aware of the risks and do often take precautions. But then they also run extra risks by catching and handling bats. The possibility that one of these researchers caught an infection from a bat, possibly mild and barely noticed, is not high, but nor is it zero. Dr Shi's group at the WIV has handled several thousands of bats in less than a decade, throughout the caves and mines of southern China, deliberately going to known virus hotspots and harvesting viruses from the bats in them before taking them back to a large city. Dr Tian from the Wuhan Center for Disease Control and Prevention, across town, has handled nearly ten thousand bats.

Bats in the lab

Did the WIV ever keep bats in the laboratory in Wuhan? In April 2020, Dr Peter Daszak was clear that the answer was no: 'The researchers don't keep the bats, nor do they kill them. All bats are released back to their cave site after sampling. It's a conservation measure and is much safer in terms of disease spread than killing them or trying to keep them in a lab.' In December he repeated the claim: 'This piece describes work I'm the lead on and labs I've collaborated w/ for 15 years. They DO NOT have live or dead bats in them.'

Yet in 2009 a colleague of Dr Shi's gave an interview to *Science Times* and said: 'The research team captured a few bats from the wild to be used as experimental animals. They need to be fed every day. This Spring Festival, the students went home for a holiday, and Dr Shi quietly took on the task of raising bats.' And, as open-source intelligence analyst Charles Small discovered, in 2018, the WIV lodged a patent application for a new design of bat cages, including details of how the bats would be fed and encouraged to breed. The animals, it said, are 'captured as needed, and … freed after taking [the] required sample or are temporarily raised [for] a period of time'. The patent was granted in January 2019. Of course, it is possible that the cages might have been used by the WIV not in Wuhan but at a site nearer to the caves where bats were captured. Yet it is worth noting that, by November 2019, the WIV had filed another eyebrow-raising patent for a device to quickly bind and disinfect finger wounds.

In June 2021, Billy Bostickson and a fellow Drastic member named Jesse found a video produced by the Chinese Academy of Sciences in 2018 to mark the inauguration of the highest-security laboratory at the WIV. Among other revelations, the video clearly showed a brief clip of bats clinging to wire mesh in a metal container not unlike those shown in the patent. Every indication from the context is that this is in the laboratory in Wuhan. A researcher is then shown feeding a mealworm to a bat, clearly inside a laboratory. In October 2020, the WIV had filed a patent on bat breeding, describing the capture of fifty bats for domestication purposes. When questioned about allegations that the WIV might have 'bat rooms', Dr Daszak replied on Twitter: 'We didn't ask them if they had bats. I wouldn't be surprised if, like many other virology labs, they were trying to set up a bat colony. I know it's happening in labs here and in other countries.'

The Global Virome Project

In September 2019, Predict came to an end after two budget cycles costing $207 million. Having collected more than 140,000 animal samples, identified 1,000 new viruses, trained 5,000 people in virus surveillance across thirty African and

Asian countries, and funded sixty laboratories worldwide, the project's supporters were dismayed. The head of the Emerging Pandemic Threats programme, Dennis Carroll, blamed the decision on the 'ascension of risk-averse bureaucrats'. But much of the work continued under new headings, with the US Department of Defense funding some. Also, in 2018, an even more ambitious global project called the Global Virome Project (GVP) had been launched, an international effort to identify all of earth's viruses with epidemic potential within a decade. 'Predict showed us that we are ready to do this on a much larger scale,' Dr Jonna Mazet, implementation director of the GVP leadership board, told an interviewer in March that year. The aim was to catalogue viruses, sequence their genomes and detail their characteristics so that humanity could be one step ahead: 'We have to stop chasing the last virus that just attacked our community, and instead get prepared in advance,' said Dr Mazet.

China was to play a leading role in the GVP. 'China will help lead a project to identify unknown viruses from wildlife to better prepare humans for major epidemics – if not global pandemics,' Dr George Gao of the Chinese CDC told the *Lancet* in 2019. Dr Daszak told the *Guardian* that under the GVP, 'we are about to start initial work in China and Thailand by studying bats, rodents, primates and water birds there'. China's national chapter of this project, called the China National Global Virome Initiative, was duly launched but has not yet reported on its progress.

The GVP received sharp criticism from fellow scientists. Three prominent virologists, Drs Edward Holmes, Andrew Rambaut and Kristian Andersen, published an article in *Nature* in mid-2018 in which they expressed strong doubts about whether the approach of virus hunting would achieve foreknowledge: 'Advocates of prediction also argue that it will be possible to anticipate how likely a virus is to emerge in people on the basis of its sequence, and by using knowledge of how it interacts with cells (obtained, for instance, by studying the virus in human cell cultures). This is misguided.' They argued that no matter how many viruses were found, predicting which one out of the estimated 1.6 million animal viruses – a number derived by extrapolating from the prevalence of viruses known in animals – might jump into people would remain impractical. 'We urge those working on infectious disease to focus funds and efforts on a much simpler and more cost-effective way to mitigate outbreaks – proactive, real-time surveillance of human populations.'

Dr Daszak was undeterred. On 21 October 2018, after he spoke at a conference about more than four years of productive collaboration with Dr Shi, he could be forgiven a moment of pride. The species of bat that carried the progenitor virus of SARS had been identified, so had the region of China where the virus came from, and so had the mechanism by which it had probably evolved through recombination within a population of similar viruses co-infecting bats. The origin of the SADS coronavirus that afflicted pigs had also been quickly tracked down. As he reflected on the effort expended and the detective work done in reaching a persuasive conclusion, he tweeted about the success of their partnership: '5,370 bats sampled and released, 2 papers in Nature, one in Cell, 15 others published and more on the way.' By late 2019, the WIV's records reflected similarly huge numbers of samples – in total, descriptions of more than 22,000 samples and specimens stored on a database, at least 15,000 of them relating to bats.

Equally impressive was the laboratory work downstream of the virus hunting to sequence genomes, synthesise viruses, manipulate their genomes and test their virulence. A year after his conference speech, around the time the pandemic was starting in Wuhan, on 21 November 2019, Dr Daszak had an exchange on Twitter with the virologist Dr Andrew Rambaut who again voiced the concern widely shared among scientists that all this monitoring of bats in caves might not have improved our ability to prevent pandemics. 'The more we look the more new viruses we find,' wrote Dr Rambaut. 'The problem is that we have no way of knowing which may be important or which may emerge. There is basically nothing we can do with that information to prevent or mitigate epidemics.' Dr Daszak shot back: 'Not true – we've made great progress with bat SARS-related CoVs, ID'ing >50 novel strains, sequencing spike protein genes, ID'ing ones that bind to human cells, using recombinant viruses/humanised mice to see SARS-like signs, and showing some don't respond to [antibodies], vaccines ...' Adding that 'it's proof-of-concept in a v. important viral family with pandemic potential'.

By July 2020, Dr Daszak had told the *Economist* that approximately sixteen thousand bats had been sampled and around a hundred new SARS-like viruses discovered.

7.

Laboratory leaks

> 'A neat and orderly laboratory is unlikely. It is, after all, so much a place of false starts and multiple attempts.'
>
> ISAAC ASIMOV

In their quest to understand and defeat dangerous viruses, scientists must bring them into laboratories where they test how these viruses infect cell cultures and experimental animals. This is risky, so elaborate precautions to contain the viruses and protect the scientists are essential. These precautions involve cumbersome equipment and exacting procedures, so it is not surprising that scientists as human beings occasionally get sloppy or make mistakes. As a result, viruses can and do leak from laboratories. This includes coronaviruses: the first SARS virus escaped several times from laboratories and infected people: in Singapore in August 2003, in Taiwan in December 2003, and multiple times in Beijing in 2004.

There are four levels of biosecurity in laboratories, the safest being BSL-4, which is reserved for the most lethal viruses such as Ebola, Crimean-Congo haemorrhagic fever, Lassa fever and Marburg virus. Here, scientists work in pressurised suits, so that air goes out of any leaks, not into them, and they breathe through air tubes. Extra precautions are undertaken to ensure that the laboratory is absolutely airtight. On leaving a BSL-4 lab a scientist takes a decontamination shower while still suited, then removes the suit and moves through an air lock to another facility where he or she takes another shower before exiting to where his or her normal clothing has been stored. All this requires meticulous training and adherence to best practices.

SARS and MERS require BSL-3 conditions. At BSL-3, scientists wear all-encompassing gowns, hoods, goggles and gloves over their normal clothing, but breathe through masks, and research is conducted within negative-pressure safety cabinets. Rooms and furniture are designed without sharp edges that might puncture gloves, and with smooth seams and corners to allow deep cleaning. There are multiple layers of security and disinfection on entry and exit of people and materials.

In the Singapore case, the twenty-seven-year-old doctoral student who caught SARS was actually studying an entirely different virus, West Nile, at the Environmental Health Institute laboratory in BSL-3 conditions. It just so happened that someone else in the same lab had conducted a live SARS experiment two days before. The lab had been working with SARS virus since April 2003 to develop diagnostics for the virus. The genetic signature of the virus that infected the student matched the strain that was being studied in the laboratory, and the specimen that the student had worked with that day was positive for both West Nile and SARS virus. 'Cross-contamination of West Nile virus samples with the SARS virus in the laboratory was the source of infection,' concluded the WHO-convened international investigation. Exactly how this cross-contamination occurred remains unknown and the student and technician involved in the West Nile experiment were unable to recall any accident that could have led to it.

Thankfully, Singapore had installed active surveillance for SARS after the outbreak earlier that year and quickly detected this mild case of laboratory-acquired SARS. No one was infected by the researcher although eight household contacts, two community contacts, 32 hospital contacts and 42 work contacts were identified, with 25 of them placed under quarantine. Biosafety investigators made several recommendations to resolve inappropriate laboratory practices, including biosafety-training and record-keeping issues. The WHO biosafety expert who led the investigation recommended that all virus samples in the laboratory be destroyed.

In the Taiwan case, a forty-four-year-old researcher, known only as Lieutenant-Colonel Chan, was testing herbal remedies against coronaviruses at the Institute of Preventive Medicine in Taipei's National Defense University. It was later deduced that he had most likely been infected on 6 December 2003 while cleaning waste in the BSL-4 laboratory, without wearing basic protective gear such as a lab gown and gloves. Chan flew to Singapore the next day for a conference. After returning to Taiwan on 10 December, he developed a fever. Realising what was wrong, Chan isolated himself at home, and refused to get medical attention because he did not want to 'bring shame to his lab and the country'. It was not until 16 December that Chan called an ambulance and was admitted to a hospital on the insistence of his father, who threatened to kill himself otherwise. By then, Chan had already developed more SARS symptoms, including a cough and pneumonia.

In the case of the 2003 SARS virus, patients were known to infect others only after developing symptoms. It was a close shave. If the Singapore conference had taken place only a few days later, or if Chan had decided to extend his trip in Singapore, he would have become symptomatic and infectious while at an international meeting or during international transit. Still, when news got out that Chan had been in Singapore before developing symptoms, seventy people in Singapore were quarantined. The approach in Taiwan was slightly different. A total of thirty-four contacts there were placed under 'self-initiated health monitoring'. The health minister at the time said that because Chan was not expected to have been infectious during his trip, the government did not individually notify the passengers on the return flight he had taken. Only five non-Taiwanese people who had sat near Chan on the plane were traced and told to watch for symptoms. Yet two of Chan's colleagues had flown to the United States after Chan returned to Taiwan. The two were examined after their return on 19 December, but their travel itinerary in America was not disclosed. Fortunately, Chan infected nobody. Nonetheless, all Taiwanese laboratories working with SARS virus were told to cease their research and to disinfect. International experts from America and Japan were called in to advise.

The Singapore and Taiwan lab leaks of SARS were isolated incidents with no transmission of the virus from the scientists to others in workplaces, homes, hospitals or during travel. Not so in the case of the Beijing lab leaks of SARS in early 2004. As retrospectively documented by Drastic's Gilles Demaneuf, an engineer and data scientist based in Auckland, New Zealand, approximately a thousand people were quarantined after two researchers, a twenty-six-year-old female medical graduate student and a thirty-one-year-old male postdoctoral academic, caught SARS, apparently in separate incidents while working in a laboratory that handled SARS virus samples at the Chinese Institute of Virology, part of the Chinese CDC. The student, Ms Song, had first returned to her hometown in Anhui province on 23 March after working in the lab for two weeks, and only developed SARS symptoms on 25 March. Song then took a train back to Beijing on 27 March and visited a hospital where viral pneumonia was diagnosed on 29 March. She was hospitalised on 4 April. Tragically, Song's mother, who began to care for Ms Song on 31 March, also developed symptoms on 8 April and died on 19 April. Even by that point, it was not known, at least publicly, that Song and her mother had been infected by SARS virus. Despite both being ill, Ms Song and her symptomatic mother had again taken a train back from Beijing to Anhui. It is inexplicable why, despite knowing that she had recently conducted SARS experiments, developed severe SARS-like symptoms and even infected her mother, Ms Song had not promptly isolated herself and sounded the alarm on a possible laboratory-acquired SARS infection. The infection of the post-doctoral researcher, Mr Yang, appeared to be a separate breach of biosafety with no laboratory incident linking him to the student. He was hospitalised on 17 April. Investigations did not identify any particular incident or accident that could explain either case of laboratory-acquired SARS infection.

The outbreak was only detected and reported by the Chinese Ministry of Health on 22 April after a nurse who had cared for Ms Song in the Beijing hospital was suspected of having SARS. This was close to a month after Ms Song had first developed symptoms from the laboratory-acquired infection and after her mother had already succumbed to the virus. The nurse had developed symptoms on 5 April, was admitted to hospital on 7 April, transferred to intensive care at a second hospital on 14 April, and had a blood test, which indicated a likely SARS infection. Contact tracing identified 171 contacts, of which five had developed fevers, including the nurse's own mother, father, aunt and another patient. The WHO was perplexed as to why the student had travelled several times on trains while symptomatic. It sent a biosafety expert team to inspect the Beijing laboratory. At the time, the WHO's Beijing spokesman said, 'We are still saying at this point that we do not see a significant public health threat because what we have seen so far is limited transmission.' The WHO's coordinator of communicable disease surveillance and response in China was less sanguine: 'We are lucky that she travelled on a train and not on an international flight. Had she landed in another country I am not sure her occupation and the fact that her mother was also sick would have been noted or rung any alarm bells.'

It later emerged that there had possibly been more laboratory-acquired infections of SARS virus in Beijing during the spring of 2004. According to the WHO, a survey of contacts identified two additional cases from the same institute with 'a history of illness compatible with SARS in February 2004'. An investigation by a panel of experts reported that the most likely source of the April infections was a batch of SARS virus that was supposed to have been inactivated (treated so as to kill the virus) before being taken into a low-safety research laboratory; however, details about the possible source of the lab leak were sparse. Each of the four laboratory-acquired SARS cases appeared separate, suggesting that SARS virus had actually leaked from the same institute four different times; yet no specific laboratory accident could be identified as leading to any one of these four leaks. After this episode, the director of the Chinese CDC and several officials resigned. The SARS laboratory was closed.

Of these multiple lab leaks of the SARS virus, only one case – the Taiwan patient – could clearly trace the laboratory-acquired infection to a specific incident in the lab. Both internal and external investigators could not figure out exactly how the West Nile sample had become contaminated by SARS virus in the Singapore lab, or how the researchers in the Beijing lab had each separately become infected with SARS virus between February and April 2004. Thus, it is possible for viruses to leak from labs without there having been an obvious, recorded accident and without anybody noticing until severe cases present at hospitals and are correctly diagnosed only several weeks later. After these episodes, the WHO strongly recommended that all research with live SARS virus be conducted at BSL-3 at a minimum. It also called for the strengthening of biosafety training and record-keeping so as to make it possible to identify the errors that led to leaks.

Linking these SARS infections to laboratories was straightforward for one simple reason: by that time, the SARS epidemic had subsided and there was no SARS virus in the human population in Singapore, Taiwan, Beijing or anywhere else. Had SARS been circulating in the general public, these leaks might have been misattributed to community spread as part of a natural epidemic. Such attribution would be even harder now in the case of Covid-19 because SARS-CoV-2 is one of the sneakiest viruses public health experts have encountered. The virus has a long incubation time of up to two weeks; it can be spread by asymptomatic or pre-symptomatic individuals; it can result in a plethora of random symptoms that resemble other afflictions ranging from allergies to the flu; and it does not produce a severe disease in most individuals, allowing for the majority of infected people to go about their daily lives or not even be aware that they have been infected.

This raises a disturbing question. If a young researcher had been working with live SARS-CoV-2 virus in 2019, or even cross-contaminated or improperly inactivated samples (similar to the cases of the researchers in Singapore, Taiwan and Beijing), and had become infected by the virus unawares, how much time might have elapsed, how many train rides or flights could this individual have taken, and how many people would have been exposed, before a severe case was diagnosed and contact-tracing performed? Unlike the 2003 SARS virus, such a laboratory-based outbreak of a stealthy virus like SARS-CoV-2 might have gone undetected for several weeks or months, under the cover of the regular flu season, before the first severe case could be recognised as a novel pathogen in a local hospital. In January 2021 the US State Department announced that it had evidence that Wuhan laboratory workers had been among the first cases of Covid-19, 'with symptoms consistent with both Covid-19 and common seasonal illness', but gave no further details. In May 2021, a previously undisclosed US intelligence report was leaked to the *Wall Street Journal*, detailing how three researchers from the WIV had been sufficiently ill in November 2019 that they went to hospital. Without further information it is hard to tell if these were just cases of flu. But three laboratory workers getting ill enough to go to hospital in the same week does sound unusual and given that most younger people generally have mild symptoms of Covid-19, there would have been more than three people infected who were not so ill – if there had been a local outbreak. According to a former State Department official, David Asher, the wife of one laboratory worker died.

Since the start of the Covid-19 pandemic, many laboratories have been handling samples of SARS-CoV-2 virus, inactivated or live, all over the world. Some of the viruses might have leaked through accidents, infecting laboratory workers. If so, nobody would know for sure that these were laboratory-acquired infections rather than transmission of the virus at a local café, on the metro, or in any other way.

We now know that early in the pandemic a senior scientist at a Beijing research institute was infected with SARS-CoV-2 while working in the laboratory. This was revealed by the non-profit investigative research organisation US Right to Know, which obtained emails exchanged between virologists at American institutions, dated 14 February 2020, via Freedom of Information (FOI) requests. The senior scientist was a former colleague of one of the virologists, who were concerned at the time to play down the possibility of a lab leak. One email read: 'I actually am very concerned for the possibility of SARS-2 infection by lab people. It is much more contagious than SARS-1 ... I actually was IBC chair at UMMS which is the only university which can do live SARS, and my lab did live SARS work. How to manage such things is very tricky. Not just PPE, but the whole design and logic.' Before the release of these emails, there had been no public reporting of the lab-acquired SARS-CoV-2 infection in Beijing.

A different kind of leak

In May 2021, the *New York Times* reported a story of leopards escaping from a safari park in Hangzhou, a Chinese city that is one of the country's most popular tourist destinations. Three of the animals got loose on 19 April 2021 when keepers who were cleaning their enclosure accidentally left the doors open. The park sent out search teams with dogs, drones and dart guns; they posted live chickens as bait and even sent people up on parachutes to try to spot the leopards, which are masters of camouflage. However, they did not notify any of the nearby residents, fearing that it would trigger panic or cause a drop in the number of visitors to the park over the Chinese Labour Day holiday.

People living near the park started to see the animals. In one case, a tea farmer spotted a leopard sitting in a field and took a photograph before continuing to work in his fields. When he saw a leopard again, he mentioned it to friends, who advised him not to report it in case it resulted in 'unnecessary hassles and interfered with work'. However, worried that the leopard could attack someone, the farmer posted his photograph on WeChat, the Chinese social media app. He received both praise and criticism, with some people saying he made too much of a fuss about leopards on the loose.

Even as the number of sightings rose and the police were called, the safari park denied to journalists that any leopards were missing. Only when the district government broke the true story did the park issue an apology on 8 May. Chinese news sites published recommendations on what to do if residents were to encounter a stray leopard: 'Whatever you do, don't panic.' And, if attacked, try punching the leopard with your fist down its throat because 'that's the only chance of saving your life'. (But don't panic, right.)

A leopard is nothing like a virus, but the saga of the escaped Hangzhou leopards illustrates how authorities and others may respond to a breach in safety protocols. Numerous people were involved in the search for these creatures and reports began to trickle in about leopard sightings, yet it required a government announcement of the escape before the safari park confessed to the incident. A leopard eyewitness was advised by friends not to speak out in case it dragged him into trouble. None of this required any conspiring. It was just human nature at play.

Breaches of containment

In 2003, in Kunming, Yunnan, laboratory rats infected students with a form of hantavirus, a potentially lethal infectious pathogen that causes haemorrhagic fever. Hantaviruses are found all over the world but human infections are most common in China. The virus is almost always caught directly from rodents. A second-year medical student studying for a master's degree at an unnamed college (College A) fell ill with fever, headache, lower back pain and conjunctivitis on 14 May 2003 and was hospitalised on 23 May. He was diagnosed as having rapidly rising levels of antibodies to hantavirus. It turned out that he had been doing an intense animal experiment and 'had not left the college for two months during which he fed the laboratory rats every day'. He had been bitten by one of the rats ten days before he developed symptoms. A scientist took sixty blood samples from rats in the laboratory and found that twenty-nine were positive for hantavirus antibodies. The infection was traced to a batch of rats that had been reared at the Center for Laboratory Animals in the city and supplied to two institutions in March that year. At both colleges and the centre, at least fifteen people tested positive for hantavirus antibodies (50 per cent and 14 per cent of the tested individuals at College A and College B respectively). All of the students including the index case recalled being bitten by the rats, but the index case was the only person to have fallen ill. The investigating team determined that the new virus isolate was a recombinant derived from human and rodent hantaviruses; in other words, this virus had never been seen before and looked like a 'natural reassortant and recombinant strain' but had leaked from a laboratory into people from three different institutes. As a result, all laboratory rats and animals housed in the same room at the Center for Laboratory Animals were eliminated; animal handlers and laboratory staff were vaccinated against hantaviruses. The team investigating recommended that the laboratory animals should be regularly tested for hantavirus antibodies, especially because the pathogen was found to cause asymptomatic infections. The interesting bit about this story is that the scientist who led the team that solved this case and pinned the source of the virus on a laboratory leak was none other than Dr Shi Zhengli of the Wuhan Institute of Virology.

In November 2019, just before the Covid-19 outbreak was detected in Wuhan, a different outbreak was unfolding in another part of China. In Lanzhou, the capital of Gansu province in the north-west of the country, a leak at a vaccine production company resulted in at least 10,528 people becoming infected by a bacterial disease. The disease, brucellosis, causes symptoms such as headaches, muscle pain, fever and fatigue in people, who typically get infected via contact with sick livestock, contaminated animal products or inhalation of the bacteria. Human-to-human transmission is extremely rare. If untreated, the disease can turn chronic and manifest as arthritis, recurrent fevers, and swelling of the heart, liver, spleen or reproductive organs. After an investigation, the Lanzhou health commission released a statement in September 2020 to clarify that the vaccine factory had used expired disinfectants in July and August 2019, resulting in live bacteria being released in its waste gas. These bacteria-containing aerosols were carried downwind to the nearby veterinary research institute, where human cases were first recorded in November 2019, culminating in at least 181 infected people by the end of December. The factory was shut down that same month, its vaccine production licences revoked in January, and the plant finally dismantled in October 2020. That month China's top legislative body passed a law for biosecurity risk prevention and response.

This is not to single out China. Biosecurity lapses happen in other countries. In 2019, research was shut down due to inadequate wastewater decontamination practices at laboratories at Fort Detrick, the United States' army biomedical and research facility (and biological weapons laboratory up until 1969, when the lab shifted to defence against such weapons). The *New York Times* reported that the US CDC cited 'national security reasons' to justify why more information was not released. In March 2021, Alison Young, a journalist who has been tracking lab leaks since 2007, published a scorching article in *USA Today* describing a history of safety lapses at elite US labs. Young described how she had begun to receive tips about lab problems at the US CDC and later obtained internal documents and emails revealing how the CDC dismissed warnings about potential structural failures that could lead to a leak of lethal viruses. At one of the BSL-3 laboratories, scientists resorted to a DIY 'enhancement', sticking duct tape to the edges of a door after a contamination risk was discovered. The names of the laboratories where such accidents happened are mostly not disclosed.

The annual report of the United States Federal Select Agent Program revealed that, in 2019 alone, there were 219 accidental releases and 13 lost samples of 'Select Agents', which include dangerous pathogens such as anthrax, Ebola, deadly strains of avian influenza, and SARS coronavirus that are deemed to pose severe threats to human health. That comes to, on average, more than four accidents involving Select Agents each week in 2019. From 177 of these reported accidents, 1,076 individuals were provided with occupational health services including medical assessments, diagnostic testing and medical treatments to prevent disease. It is important to note that these reports do not include accidents involving pathogens that are not on the US federal list of Select Agents. There is also no mandatory reporting of laboratory-acquired infections in the US. In other words, we do not know how many individuals are infected by lab pathogens within the US each year.

Elsewhere in the world, virus leaks also happen. Smallpox escaped from laboratories in the United Kingdom three separate times between 1966 and 1978, resulting in eighty cases and three deaths. Foot-and-mouth virus escaped from a laboratory at Pirbright in Britain in 2007 via a leaking pipe, which likely contaminated the wheels of a construction vehicle that later visited a farm, passing on the virus to livestock at two farms. A Russian laboratory worker caught the lethal fruit-bat-borne Marburg virus while working in a laboratory in 1990. In fact, Marburg was first discovered after it infected laboratory workers in Germany (Marburg and Frankfurt) and Yugoslavia (now Serbia) in 1967. The researchers at these three different institutes had been handling African green monkeys or their tissues imported from Uganda. In this initial outbreak, thirty-one people – the laboratory workers, healthcare workers and relatives – reportedly caught the disease and seven died. For emphasis, this was a completely novel virus and the researchers had not been aware that the virus was in their samples.

One of the worst laboratory accidents occurred in the Soviet Union in April 1979. Anthrax, a bacterium, escaped from a biological warfare laboratory in the city of Sverdlosk (now Yekaterinburg) resulting in at least sixty-four deaths. For thirteen years the Soviet regime denied that anthrax was the cause, removing the medical records of victims and blaming their deaths on contaminated meat. This cover story even survived a formal, international investigation in 1986. A team of western scientists was finally permitted to investigate properly more than a decade after the outbreak, in 1992 and 1993, after the fall of the Communist regime. Although hospital records had been confiscated by the KGB, the scientists managed to deduce that the index victims had all been downwind of the military facility. Dr Kanatjan Alibekov, former first deputy chief of Biopreparat, the civilian part of the Soviet biological weapons programme, who had defected to the United States in 1992, told the *Frontline* television programme that, 'In the Soviet Union, thousands of people were involved in developing an anthrax biological weapon' and 'the Ekaterinburg facility was responsible for continuous manufacturing [of] anthrax biological weapons. The amount of this weapon produced was hundreds of tons.' According to Dr Alibekov, it turned out that a worker had removed a filter from an exhaust system and not replaced it. He left a note for the next shift, which failed to notice the note and turned on the machine. If the wind had blown the anthrax spores towards the centre of the city, the death count could have been in the hundreds of thousands.

Less than a decade later, on 18 September 2001, following the 9/11 attacks, anthrax spores were sent through the mail to two senators and a handful of news organisations in the United States, infecting twenty-two people, five of whom died. The FBI began to investigate but were not trained to work with dangerous pathogens, and so the anthrax spores were sent to Fort Detrick for identification. The researchers there proved to be well informed about the capabilities of anthrax and how the attack might have been carried out. The letters purported to be from Islamists, but, by early 2002, federal agents had realised that the culprit was an individual with access to a rare strain of anthrax used in research labs – and was possibly one of the scientists they were working with at Fort Detrick. The investigators were initially distracted by another suspect, but finally seized samples from Bruce Ivins, one of the anthrax specialists at Fort Detrick, who was himself helping with the investigation. They found a possible match to the spores that had been used in the attacks. In 2008, Ivins was interviewed by prosecutors, disclosing mental health issues. By July of that year, he had taken his life. The FBI and Department of Justice eventually concluded, in 2010, that Ivins had probably been the culprit, although doubts remained. Independent scientists recruited by the National Academies of Sciences reviewed the evidence, concluding that it was consistent with Ivins being guilty, but that the science alone did not prove that he was. Although not a laboratory accident, the anthrax letters episode is a reminder that laboratory pathogens can be extremely difficult to trace back to a specific laboratory, person or incident even with access to the scientists and samples at the heart of the issue.

Laboratory accidents, just like natural spillovers when people catch viruses from wild animals, are rare events, but the sheer number of pathogen laboratories around the world means that lots do happen – even if they are not detected or reported. In an article in 2015, Alison Young revealed that from 2006 to 2013, laboratories in the United States had notified federal regulators regarding approximately 1,500 incidents of exposure or release, thefts or losses of dangerous pathogens; of these, 800 cases involved employees receiving medical treatment or evaluation, and 15 people were infected by laboratory pathogens. The title of Young's 2021 *USA Today* article was: 'Could an Accident Have Caused COVID-19? Why the Wuhan Lab-Leak Theory Shouldn't Be Dismissed'.

A pandemic that began in a laboratory

There is at least one example of a global epidemic that began in a laboratory. An Influenza A (H1N1) strain of flu swept the world in 1977, beginning in northern China, though it was soon named the Russian flu after the Soviet Union was the first to report it to the WHO. Curiously, the influenza virus had been isolated in three different regions of China, up to a thousand kilometres apart, around the same time that year. It was no big deal, being a mild illness – less than five people died per 100,000 infected – but it had one irregular feature: it mainly affected those under the age of twenty-six. It turned out that this 'new' strain of flu was essentially identical, genetically, to an H1N1 flu that had been common in the 1950s but had since disappeared. This meant that adults who had been exposed to the virus in the 1940s and 1950s would have developed and retained some level of immunity; not so for those in their early twenties and younger. Genome sequencing was not possible in the 1970s, of course, but flu viruses could be characterised based on whether they could bind to specific antibodies. It was also possible to perform a primitive form of genetic fingerprinting in which the RNA of the virus was separated on a gel and dazzling leopard-print patterns of RNA spots were compared between different virus strains. Using these methods in 1978, scientists at Mount Sinai School of Medicine in New York determined that the 1977 virus was an old friend, all but identical to two H1N1 viruses that had been isolated in 1950, but different from other H1N1 viruses isolated in 1947, 1950 and 1956. Seeing as how H1N1 viruses from the 1940s and 1950s had diversified extensively, and flu viruses passing through birds and horses also rapidly accumulated mutations, it was bizarre that the recent Russian strain proved so similar to a couple of strains that had been isolated in 1950. The Mount Sinai scientists commented, 'It seems unlikely that a 1950 virus survived by normal sequential transmission in the human population … it seems much more likely that the genetic information in the Russian viruses has been preserved over the last 25–27 years by some unusual mechanism.'

Ingenious explanations, including that it had lurked in a frozen body somewhere in the polar regions, seemed far-fetched. Such an origin story would have removed any blame from China or Russia where the earliest cases had emerged (it is relevant to note that some scientists in 2021 have aired the hypothesis that the SARS-CoV-2 virus had first reached Wuhan via frozen foods transported over long distances, also absolving the city of being the starting point of the pandemic). The possibility of a laboratory accident seeding the 1977 outbreak came to the fore. Scientists from the Chinese Academy of Medical Sciences published a paper in the *Bulletin of the World Health Organization* in 1978, dismissing the idea, using a similar argument to one that would be deployed for Covid-19: 'Laboratory contamination can be excluded because the laboratories concerned either had never kept H1N1 virus or had not worked with it for a long time.'

Nearly four decades later, in 2015, Dr Michelle Rozo and Dr Gigi Kwik Gronvall at the UPMC Center for Health Security in Baltimore, Maryland, reanalysed genetic sequences from the 1977 isolates versus isolates from 1947 to 1957. They found a 98.4 per cent similarity between the 1977 strain and several samples from around the world in 1948–51. After discussing various possible explanations, including those of a Soviet biological weapon targeted at soldiers in their twenties, a vaccine trial that went wrong and a laboratory accident, the researchers concluded that vaccine trials were the most likely source for several reasons. In 1977, the world had been reacting to an alarming H1N1 swine flu outbreak among 230 soldiers at Fort Dix, New Jersey, the year before – thirteen of them had developed severe respiratory illness and one other patient had died. This had caused scientists to revisit their freezers and start new, live influenza virus vaccine development projects to prepare for the worst-case scenario of hundreds of millions of people needing H1N1 vaccines. Furthermore, between 1962 and 1973, tens of thousands of children were being vaccinated in such clinical trials in both the USSR and China. Could a lab somewhere have thawed out an H1N1 strain from the 1950s to develop new vaccines in preparation for a possible pandemic?

In 2004, one of the Mount Sinai scientists, Dr Peter Palese, revealed that a Chinese scientist who had published in the WHO *Bulletin*, Dr C. M. Chu, had written to him privately to say that, despite the lack of hard evidence, 'the introduction of this 1977 H1N1 virus is now thought to be the result of vaccine trials in the Far East involving the challenge of several thousand military recruits with live H1N1 virus'. Dr Chu was a former director of the Chinese Academy of Medical Sciences and a distinguished virologist. The vaccine technology at the time was not perfect and live attenuated viruses used in vaccines sometimes recovered their ability to cause disease in humans. (It is important to note that many of the Covid-19 vaccines today are mRNA or adenovirus-based vaccines, and do not rely on live attenuated SARS-CoV-2 viruses.) The simultaneous appearance of the virus in three cities in China fitted a vaccine trial explanation. Several Russian flu viruses isolated in 1977 were strangely temperature sensitive, a possible sign that the virus had been attenuated in the laboratory.

Dr Rozo and Dr Gronvall concluded that 'the unnatural origin, mildness of presentation of the virus, widespread dissemination of cases in a short amount of time, temperature sensitivity of the samples, contemporary observations, and existence of live-virus vaccine trials which were occurring at that time' supported the hypothesis that a vaccine trial had gone awry. It does not count, therefore, as a laboratory accident, they argued. Dr Martin Furmanski of the Center for Arms Control and Non-Proliferation in Washington, DC, disagreed: 'The 1977 H1N1 virus caused a global epidemic, and as Rozo and Gronvall themselves concluded, it originated in a microbiology laboratory and its release was unintentional.'

An orchestrated statement

Against this background, it was never unreasonable to ask whether Covid-19 could have begun with a laboratory accident, perhaps even one that was not detected. The head of the Wellcome Trust, Sir Jeremy Farrar, in his July 2021 book *Spike: The Virus vs the People, the Inside Story*, gives an insider's account of how senior virologists thought about this possibility in late January 2020. It turned out to be very different from what they said in public. Dr 'Eddie' Holmes in Sydney 'was about 80 per cent sure this thing had come out of a lab'; Dr Kristian Andersen in San Diego 'was about 60 to 70 per cent convinced in the same direction'; Dr Robert Garry of Tulane University and Dr Andrew Rambaut in Edinburgh were 'not far behind'. And Dr Farrar himself 'was going to have to be persuaded that things were not as sinister as they seemed'. Dr Andersen emailed Dr Anthony Fauci on 31 January, saying of the virus's genome that 'some of the features (potentially) look engineered'. On 1 February Dr Farrar arranged a conference call for these and other scientists with the British government's chief scientist, Sir Patrick Vallance, and his US equivalent, Dr Fauci. What was said on that call remains largely a mystery: the participants have said little and emails exchanged by them afterwards and released under the freedom of information act were heavily redacted. Dr Farrar says only that it was argued that 'the ingredients were probably out there in the wild'; even so, after the call, Dr Farrar still expressed that he thought a lab release as likely as a natural origin: 'On a spectrum if 0 is nature and 100 is release – I am honestly at 50!' According to his book, Dr Andersen said, 'I was battling with the idea that, having raised the alarm, I might end up being the person who proved this new virus came from a lab.'

Three days later, on 4 February, Dr Andersen appeared to have had a complete change of mind. He wrote to Dr Daszak and a different group of scientists: 'The main crackpot theories going around at the moment relate to this virus being somehow engineered with intent and that is demonstrably not the case' and 'the data conclusively show that neither [engineering for basic research or nefarious reasons] was done'. What was the new information that had obliterated all of Dr Andersen's suspicions of an engineered virus in those three days? And why was he telling other scientists that they could confidently rule out a lab engineered virus?

You see, on 3 February 2020, an official letter seeking help in determining the origins of the novel coronavirus was published in the United States. It was from the Office of Science and Technology Policy (OSTP), an executive office of the president: 'OSTP requests NASEM [the National Academies of Sciences, Engineering and Medicine] convene a meeting of experts, particularly world class geneticists, coronavirus experts, and evolutionary biologists, to assess what data, information, and samples are needed to address the unknowns, in order to understand the evolutionary origins of 2019-nCoV and more effectively respond to both the outbreak and any resulting misinformation.'

The word 'misinformation' almost certainly referred to the suggestion that the virus had been deliberately engineered perhaps as some kind of weapon. To set the scene, a preprint titled 'Uncanny Similarity of Unique Inserts in the 2019-nCoV Spike Protein to HIV-1 gp120 and Gag' authored by researchers from Delhi, India, had been posted online three days before, on 31 January. Unsurprisingly, this manuscript spurred intense speculation that the novel SARS-CoV-2 coronavirus had been engineered with parts of the HIV virus (a hypothesis that has since been thoroughly debunked by numerous scientists, including one of the authors of this book, Alina). On 2 February, Dr Shi Zhengli from the WIV posted a furious response to these allegations on WeChat: 'I, Shi Zhengli, use my life guarantee, and it has nothing to do with the laboratory. I advise those who believe in and spread the rumours of bad media, and those who believe in the so-called academic analysis of the unreliable Indian scholars, shut your stinky mouth.'

NASEM responded to the OSTP's letter on 6 February, after consulting the following leading experts: Drs Kristian G. Andersen (Scripps Research Institute), Ralph Baric (University of North Carolina School of Public Health), Trevor Bedford (Fred Hutchinson Cancer Institute), Aravinda Chakravarti (New York University School of Medicine), Peter Daszak

(EcoHealth Alliance), Gigi K. Gronvall (Johns Hopkins Bloomberg School of Public Health), Tom Inglesby (Johns Hopkins Center for Health Security) and Stanley Perlman (University of Iowa). On the issue of the origin, their verdict was agnostic: 'The experts informed us that additional genomic sequence data from geographically – and temporally – diverse viral samples are needed to determine the origin and evolution of the virus. Samples collected as early as possible in the outbreak in Wuhan and samples from wildlife would be particularly valuable.' In other words, it was too early to tell how this virus had originated. The letter was sufficiently boring that it was soon drowned out by the endless news of novel Covid-19 cases emerging all over the world in early 2020.

However, two separate questions now became confused: whether the virus was engineered or whether it might be a natural virus that had leaked while being studied in a laboratory. Email exchanges among the convened experts were later obtained via FOI requests by US Right to Know. The group published these emails in December 2020, revealing that the experts had raised 'unanswered questions about lab origin, even as some sought to tamp down on "fringe" theories about the possibility the virus came from a lab'. One of the experts, Dr Kristian Andersen, was pointed in his remarks. 'The main crackpot theories going around at the moment relate to this virus being somehow engineered with intent and that is demonstrably not the case,' he wrote. 'If one of the main purposes of this document is to counter those fringe theories, I think it's very important that we do so strongly and in plain language.' Yet an early draft of the NASEM letter had included a revealing footnote: 'Possibly add brief explanation that this does not preclude an unintentional release from a laboratory studying the evolution of related coronaviruses.'

The footnote was to the statement: 'The initial views of the experts is that the available genomic data are consistent with natural evolution and that there is currently no evidence that the virus was engineered to spread more quickly among humans. [ask experts to add specifics re binding sites?]' On this point, the virologist Dr Trevor Bedford replied, 'If you start weighing evidence there's a lot to consider for both scenarios', adding, 'I would say "no evidence of genetic engineering" full stop.' Dr Bedford concluded that email with 'I'm not sure what the exact capacity of this group going forward will be, but I might suggest moving to more secure forms of communication.' When one of the authors of this book, Alina, tweeted about the FOI'ed emails, Dr Bedford confirmed that 'As of Feb 3, I felt that we could not say definitely whether emergence into human population occurred via zoonosis or lab escape (this is the 'both scenarios' reference).' However, at that time, his priority was to counter the 'engineered' virus hypothesis and 'work towards a more nuanced consideration of zoonosis vs lab escape'. On 20 February 2020, Dr Bedford created a Twitter thread to explain that most of the evidence pointed to a natural zoonosis, and the location in Wuhan was the only factor, in his opinion, that suggested a lab escape. On 1 December, Dr Bedford reaffirmed his stance: 'Short answer is that I don't (and never did) consider lab accident hypothesis a "conspiracy theory" and believe it should be addressed scientifically.'

Another expert on the NASEM-convened committee, Dr Peter Daszak of the EcoHealth Alliance, appeared to have a less nuanced position, pre-emptively rejecting lab escape and engineered virus speculations. Theories that the novel SARS virus had emerged due to lab activities threatened a horrific possibility – that the EcoHealth Alliance's type of well-intentioned pathogen research, rather than preparing the world for a new pandemic, might have started one instead. From a separate

batch of FOI'ed emails by US Right to Know, we now know
that, on 6 February, Dr Daszak sent an email to six other scientists asking them to co-sign a joint statement protesting at the
'rumours, misinformation and conspiracy theories' swirling
about the origin of the virus. In the statement drafted in the
early hours of 6 February, Dr Daszak wrote: 'We stand together
to strongly condemn conspiracy theories suggesting that 2019-
nCoV does not have a natural origin.' Remember that the
genome of the virus had only been made public on 11 January
so it is not clear how he could have known for certain by the
first week of February how the virus originated. The final letter
published on 19 February in the *Lancet*, and co-signed by a
total of twenty-seven prominent scientists, stated that scientists
'overwhelmingly conclude that this coronavirus originated in
wildlife', even though none of the nine references cited for this
statement provided unambiguous evidence for such a bold
assertion.

The publication gave no hint of Dr Daszak's role in organis-
ing the letter: the authors were shown in alphabetical order and
the email address for readers to contact was a generic
COVID19statement@gmail.com. Therefore, the emails
unearthed months later by US Right to Know came as a
surprise, accompanied by an article titled, 'EcoHealth Alliance
Orchestrated Key Scientists' Statement on "Natural Origin" of
SARS-CoV-2'. This raised eyebrows. After all, Dr Daszak is the
president of the EcoHealth Alliance, which, as a collaborator
and provider of funding to the WIV, had a considerable interest
in the matter of whether the virus had emerged from a lab. Yet
the *Lancet* letter included the following statement: 'We declare
no competing interests.' In addition, Dr Daszak had explained
to the co-authors of the statement on 9 February: 'Please note
that this statement will not have EcoHealth Alliance logo on it
and will not be identifiable as coming from any one organiza-

tion or person.' Four other co-signatories who had ties to the EcoHealth Alliance did not disclose these affiliations in the *Lancet* statement. Dr Daszak, who had first drafted the letter and recruited signatories holding high positions in the world of research, was the only signatory to disclose an affiliation with the EcoHealth Alliance.

Another surprise was that two of the original recipients of the 6 February email, Dr Ralph Baric and Dr Linfa Wang did not sign. Both were collaborators of Dr Shi in the work on SARS-like viruses in China. The reason for their absence was once again revealed by more emails obtained by US Right to Know. Dr Daszak emailed Dr Baric on 6 February saying that he had spoken to Dr Wang and agreed that 'you, me and him should not sign this statement, so it has some distance from us and therefore doesn't work in a counterproductive way ... We'll then put it out in a way that doesn't link it back to our collaboration so we maximize an independent voice.' Dr Baric replied, 'I also think this is a good decision. Otherwise it looks self-serving and we lose impact.' However, Dr Daszak in the end did decide to sign instead of removing his name from a letter he had written.

In June 2021, the *Lancet* published an addendum to clarify the competing interests of the twenty-seven authors of the letter. Dr Daszak's disclosure statement alone was more than four hundred words long, detailing the EcoHealth Alliance's work and collaborations in China. Dr Daszak also stated that his participation in the China-WHO global study on the origins of Covid-19 was being undertaken in a 'private capacity, not as an EcoHealth Alliance staff member'. The next month, a follow-up letter was submitted to the *Lancet* by twenty-four of the twenty-seven original authors with a disclosure statement more than eight hundred words long – compared with the five-word denial of competing interests in the first statement. This second letter claimed that an 'intent of our original Correspondence was to express our working view that SARS-CoV-2 most likely originated in nature'.

The first BSL-4 lab in China

The city of Wuhan is home to a number of institutions that contain laboratories which study coronaviruses in addition to the WIV. Near the city centre and the market there is the provincial Center for Disease Control, focused on practical surveillance but also (see Chapter 6) playing a role in bat sampling. Across the Yangtze river in the Wuchang district, Wuhan University has its own laboratories and also houses the Institute of Model Animals, which provided more than a million animals to various units at Wuhan University, as well as to more than a thousand laboratories across China. It trained tens of thousands of technicians to perform experiments on animals. Back in 2003, the BSL-3 laboratory in this institute was described in a media report as 'one of the main battlefields against SARS', where a vaccine had been developed and tested on monkeys. Nearby is Huazhong Agricultural University, which has collaborated on experiments manipulating pig coronaviruses (see Chapter 9). A short distance away is the Wuhan Institute of Biological Products, a state-owned company manufacturing diagnostics, medical devices and reagents, which boasts that it is the only enterprise to own a comprehensive high-level biosafety research laboratory and production facility in the world. The Hubei Wildlife Rescue Centre, which has handled more than two hundred thousand wild animals since its opening in 2000, has partnered with the WIV on zoonotic virus research since 2013. It is unclear which wild animal species may have been sampled and shared with WIV researchers, and what diverse animal viruses they may have found and experimented with over the years.

The Wuhan Institute of Virology was also situated nearby in Wuchang until, in recent years, it expanded on to a large new campus in the Jiangxia District, further south on the same side of the river. Both WIV campuses are connected by a regular shuttle bus. The Jiangxia campus contains the first BSL-4 laboratory (the highest biosafety level) in China, which was completed in 2015 and became operational in 2018. Experiments with SARS and MERS coronaviruses that can infect people generally do not need to be conducted at BSL-4 but require BSL-3. Before 2020, however, experiments with bat coronaviruses were sometimes conducted in BSL-2 labs. Here, the precautions are much less onerous: gloves, coats and masks should be worn but there are no negative-pressure safety cabinets or specially designed rooms. As Dr Shi clarified to *Science*

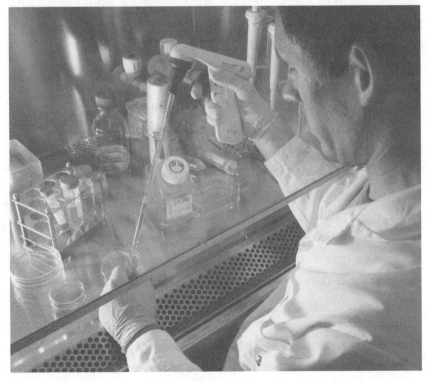

A scientist working in a biotechnology laboratory, biosafety level 2.

magazine in July 2020, 'The coronavirus research in our laboratory is conducted in BSL-2 or BSL-3 laboratories ... After the
COVID-19 outbreak, our country has stipulated that the cultivation and the animal infection experiments of SARS-CoV-2
should be carried out in BSL-3 laboratory or above.' Although
all of the SARS research at the WIV, including the animal infection experiments, had been conducted at lower biosafety levels,
stories about the WIV BSL-4 laboratory took centre stage in
early 2020. Yet even if there were no safety concerns at BSL-2
and BSL-3 laboratories at the WIV, working with a contagious
virus at these biosafety levels clearly carries a risk of escape. As
Dr Ralph Baric told the *MIT Technology Review*, 'If you study
hundreds of different bat viruses at BSL-2, your luck may eventually run out.'

How the WIV's new building with its BSL-4 laboratory came
to be built is not a happy story. China long had ambitions to
host a BSL-4 laboratory, not least because Taiwan already had
two by the time the WIV's BSL-4 was close to being opened.
Many European countries also had operational BSL-4 laboratories. North America currently has fourteen: twelve in the
United States and two in Canada. Around the time of the 2003
SARS epidemic the Chinese Academy of Sciences approached
the government of France with a request for assistance in building a high-security laboratory. Dr Chen Zhu, who was a friend
of the then president of China and had trained in haematology
(blood disorders) in France, travelled to Europe to make the
request of Jean-Pierre Raffarin, the prime minister. The French
president, Jacques Chirac, agreed to export his country's BSL-4
technology to China. On 9 October 2004 in Beijing, President
Chirac concluded an agreement that France would provide the
blueprints for a BSL-4 laboratory and several mobile BSL-3
laboratories, with French technicians overseeing the project. In
exchange, China would use a French construction firm and fifty

French scientists would work in the institute and supervise the training of their Chinese counterparts. All research conducted at these sites would be shared between the two countries. Wuhan was chosen as the site for the laboratory partly because of the city's historical and contemporary links to France. It has more French companies located there than any other city in China.

The agreement soon caused concern within the French defence and intelligence community, and even more in the United States. The fear was that such a laboratory might be used by the Chinese military. Some of the mobile BSL-3 laboratories that France had shared with China had already gone missing or been taken over by the People's Liberation Army. A Paris correspondent of the *American Spectator*, Joseph Harriss, wrote that French scientists who visited a virology lab in Harbin were alarmed by how researchers handled animals. Soon the French government found that the Chinese government was reneging on various promises. It used a Chinese construction firm (suspected to be under PLA control) rather than a French one as agreed, with little to none of the promised opportunity for French technicians to oversee the building work. The French firm that was due to certify the building refused, worried that because of the lack of information it could not be sure the construction was as secure as promised. Eventually Alain Merieux, the French co-chairman of the project, resigned, saying, 'It's a very Chinese operation. It's entirely theirs, even though it was developed with technical assistance from France.'

France's qualms had resulted in a lengthy delay, with construction only beginning in 2010. The structure was completed in 2015 at a cost of $44 million, and the four-storey building inaugurated in 2017. When the WIV's BSL-4 laboratory was close to being operational, Dr George Gao, the soon-to-be head of the CDC in Beijing, said: 'It will offer more opportunities for Chinese researchers, and our contribution on the BSL-4-level pathogens will benefit the world.' The lab began operating in January 2018. But instead of the fifty French researchers that China had initially agreed to let work in the WIV to ensure proper biosecurity training and practices, just one French scientist got to set foot on the site. As Harriss, who chronicled the saga, put it, 'Thanks to French naïveté – they actually believed Chinese promises – China got its new dual-use laboratory and the ability to do whatever it likes with it, and France got zilch.'

Dual-use means military as well as civilian projects being conducted at the facility. In January 2021, the US State Department under the Trump administration released a statement that claimed there were secret links between the WIV and military research: 'Despite the WIV presenting itself as a civilian institution, the WIV has collaborated on publications and secret projects with China's military.' In June 2021, the *Washington Post* reviewed public records and internal guidelines of the WIV and verified the existence of classified projects, although the topics remain unknown, and protocols where some research reports and dissertations could be sealed for up to two decades. The WIV had conducted regular confidentiality training for its staff, which covered best practices for managing the researchers working on classified projects. In May 2019, WIV researchers signed pledges to protect confidential information. The *Washington Post* further shared with us a WIV document from November 2019 instructing doctoral students engaged in classified research topics not to publish their papers publicly, but to submit an internal report to a review team that included the institute's confidentiality committee alongside a contract indicating the confidentiality of the project.

A warning shot

Before opening, the WIV's management sought practical advice and help from a similar facility, the Galveston National Laboratory, run by the University of Texas Medical Branch. 'In preparation for the opening of the new China [maximum containment laboratory], we engaged in short- and long-term personnel exchanges focused on biosafety training, building operations and maintenance, and collaborative scientific investigations in biocontainment,' read a joint memo from the two institutions. 'We succeeded in transferring proven best practices to the new Wuhan facility.'

Following publication of a key paper in 2017 by Dr Ben Hu and colleagues, in which it was reported that a total of three SARS-related viruses discovered in Yunnan had been isolated and grown in the lab by the WIV, the US embassy in Beijing and the US consulate in Wuhan decided to send officials to visit the new laboratory. The WIV warmly received the US officials. Its director general, Dr Wang Yanyi, emphasised that the institute had established collaborations with well-known US organisations such as the National Science Foundation, the EcoHealth Alliance and the University of Texas Medical Branch.

Unfortunately, the biosafety situation was less than ideal. The US officials sent cables back to Washington expressing serious concern about the operation of the new facility. 'The cable was a warning shot. They were begging people to pay attention to what was going on,' as one official explained. Some of these were leaked to Josh Rogin of the *Washington Post* in April 2020. One of the cables, drafted by two officials from the US embassy's environment, science and health sections and sent on 19 January 2018, read as follows: 'During interactions with scientists at the WIV laboratory, they noted the new lab has a serious shortage of appropriately trained technicians and investigators needed to safely operate this high-containment laboratory.' One of the authors of the cables told Rogin that they 'were trying to warn that that lab was a serious danger', adding, 'I have to admit, I thought it would be maybe a SARS-like outbreak again. If I knew it would turn out to be the greatest pandemic in human history, I would have made a bigger stink about it.' US assistance was denied, however, from which it is reasonable to assume that the WIV may have continued to operate with what the US officials would regard as insufficient training and expertise.

The cables also mentioned that Chinese officials had expressed a strong interest in co-leading the Global Virome Project; indeed, China had already begun to fund projects in this scheme to build a database of pathogens in China. The WIV was named as an institute contributing to this proof-of-concept database, in collaboration with the EcoHealth Alliance. However, the US officials noted that 'other countries have confidence in China's ability to build such a database but are sceptical whether China could remain transparent as a "gatekeeper" for this information'. The cables also confirmed that 'WIV scientists are allowed to study the SARS-like coronaviruses isolated from bats while they are precluded from studying human-disease causing SARS coronavirus in their new BSL-4 lab until permission for such work is granted'.

If there were issues with its first BSL-4 laboratory, the Chinese government did not let them dampen its ambition. It pressed on with its vision to build a national network of BSL-3 and BSL-4 labs. In 2016, the Chinese National Development and Reform Commission and the Ministry of Science and Technology had announced a plan for the construction of high-level BSLs across China by 2025. In September 2019, Dr Yuan

Zhiming, chair of the institutional biosafety committee at the WIV, published an article entitled 'Current Status and Future Challenges of High-Level Biosafety Laboratories in China' to review this national plan and challenges that could arise. Dr Yuan wrote that laboratory biosafety could be at risk due to compromised implementation of biosafety regulations and insufficient funds for routine processes at BSL-3 laboratories. He added that, 'Currently, most laboratories lack specialized biosafety managers and engineers. In such facilities, some of the skilled staff is composed by [sic] part-time researchers.' Dr Yuan recommended a prompt revision of existing biosafety and biosecurity regulations and standards.

Meanwhile across the river from the WIV, the Wuhan Center for Disease Control and Prevention in June 2019 conducted public bidding for a contract for the disposal of hazardous medical waste generated in its laboratories. This waste 'has not been effectively treated from 1994 to 2019', the announcement conceded, while revealing that the total amount was close to 2 tons. The Wuhan CDC itself moved into a new building by December 2019 – only a five-minute walk from the Huanan seafood market where a cluster of early Covid-19 cases would be detected that same month.

Shortly after the emergence of SARS-CoV-2, in February 2020, the Chinese Ministry of Science and Technology issued new biosafety rules for laboratories handling pathogens. These targeted chronic loopholes relating to the inactivation and disposal of biological waste from laboratories. For example, there were issues with scientists selling the animals used in laboratory experiments for profit. One top scientist had made about $1.5 million by illegally selling laboratory animals and animal products. A redrafting of the regulations covering laboratories was issued in August 2021 including the line that 'the bodies and tissues of animals used in experiments should be disposed of safely. Animals used in experiments cannot go back on the market.'

The missing pathogen database

The WIV's China-wide expeditions had gathered around twenty thousand specimens by 2019. These were catalogued in a comprehensive 61.5 MB online database, which included nasal and faecal swabs, as well as blood samples from bats and mice. From these were extracted many different virus genetic sequences. A summary paper by the WIV and the EcoHealth Alliance published in 2020 listed 630 novel coronaviruses from samples collected from 2010 to 2015. It remains unclear exactly how many SARS-like coronaviruses this includes, and even less clear how many new viruses were later found between 2015 and 2019. However, in an interview in early December 2019 – before the world knew that the novel coronavirus had already begun to spread in Wuhan – Dr Daszak stated: 'We have now found, after six or seven years of doing this, over a hundred new SARS-related coronaviruses. Some of them get into human cells in the lab, some of them can cause SARS disease in humanized mice models and are untreatable with therapeutic [antibodies], and you can't vaccinate against them with the vaccine.' After Covid-19 emerged, it should have been an urgent global priority for investigators to take a look at this inventory of more than a hundred new SARS-related viruses. Yet an up-to-date catalogue of viruses sampled by the WIV has not been shared publicly. Scientists cannot confirm whether it includes viruses that could shed light on the origins of SARS-CoV-2.

With funding since 2013 from the Chinese Ministry of Science and Technology, the WIV's database had been accessible online at batvirus.whiov.ac.cn. In 2019, with Dr Shi Zhengli as lead author, the database was described as focused on wild-life virus pathogens in a Chinese journal about scientific big data. Users could search for the host animal species, the type of sample, the distribution of the species and the viruses carried by different host species. The entries included information such as the sample name, type, source species, collection date and GPS location, plus the type, name and sequence details of any virus detected from that sample. The database also told users whether and which viruses had been detected in or isolated from any of these 22,257 specimens. The site had a password-protected section for unpublished viral sequences.

Although US government funding had contributed to the SARS-like virus research, USAID discovered when it reviewed the work in 2017 that it had no ownership of the data: 'USAID does not own data collected under PREDICT 2 ... This dramatically constrains the control over those data for future programming [*sic*], research and dissemination.' The sampling focused on five viral families: paramyxoviruses, filoviruses, influenza viruses, flaviviruses and coronaviruses. Moreover, the WIV had received funding from the European Virus Archive Global (EVAg) project, which had a mission to 'collect, amplify, characterize, standardize and authenticate viruses to develop and maintain the largest active globally accessible virus archive'.

The discovery of how the database went offline in September 2019 was made by a British open-source intelligence analyst in early May 2020. Based in Hong Kong until September 2019, Charles Small runs a business that advises journalists, business travellers and NGO workers on security risks in conflict zones and crime hotspots around the world, often at short notice. To do this he goes behind the spin to find out what is really happening on the ground in such places and his reputation rests on getting things right – if he is wrong, someone's life could be in danger. Mr Small has been coding since he was a teenager and his degree in Arabic and politics from the School of Oriental and African Studies in London and Damascus was followed by study at the London School of Economics and in Taiwan. He started investigating the Chinese wildlife trade when the pandemic began and then switched to looking into the work of the scientists at the WIV. He came upon a fact sheet describing the database but found the database itself was offline. He searched online using the name of the database and found a database monitoring website. Surprisingly, these access records were public and did not require a login to view. The records showed that the database had gone offline on 12 September 2019 between 2 a.m. and 3 a.m. local time. It stayed that way except for a few short spells from December to February, during which there was no record of external access from outside the WIV. Mr Small archived hundreds of pages of these access records, which showed a spike of significant activity in June 2019.

We know that sometime between September and December 2019, version 2 of the WIV pathogen database fact sheet was replaced with version 3, which was replaced with version 4 on 30 December. By this point the very name of the database had changed from 'Wildlife-borne Viral Pathogen Database' to 'Bat and Rodent-borne Viral Pathogen Database'. The words 'wild animal' were replaced with 'bat and rodent' in at least ten places. Comparing archived copies of versions 2 and 4, the list of keywords changes: in version 2, this read: 野生动物样本；病毒病原数据；新发传染病；跨种感染, which translates as 'wild animal samples; viral pathogen data; emerging infectious diseases; cross-species infection'. In version 4, it read: 蝙蝠；鼠；病毒；数据库, which translates as 'bat; mouse; virus; database'. It is not known whether the database itself was altered at this time, or just the fact sheet, and whether the database contained information on viruses from wild animals other than bats and rodents that may have shed light on viruses moving through the wildlife trade.

When queried, Dr Shi told the BBC in December 2020 that the database had been taken offline 'for security reasons'. But there are ways of securing data and keeping back-ups so that if a nefarious person were, say, to plant a fake sequence in your library appearing to implicate your laboratory in synthesising a virus, you could prove it was a fake. Furthermore, Dr Shi added that: 'All our research results are published in English journals in the form of papers … Virus sequences are saved in the [US-run] GenBank database too. It's completely transparent. We have nothing to hide.' If all the WIV's virus sequences were indeed saved in GenBank, why was the full genome of the RaTG13 virus (a 96 per cent match to SARS-CoV-2) only made available after SARS-CoV-2 had emerged? Recall that after intense criticism by numerous sleuths and scientists for obscuring the sample history of RaTG13, Dr Shi finally published an addendum to her Covid-19 *Nature* paper in November 2020. This addendum confirmed the findings of the sleuths – that RaTG13 had indeed been collected from the Mojiang mine where miners had sickened with a severe pneumonia likely brought on by a virus infection, and that RaTG13 had had its genome sequenced in 2018. This genome was certainly not placed on GenBank until after Covid-19 had broken out in Wuhan. Even more shockingly, the world only heard about eight more SARS-like viruses collected between 2012 and 2015 from the Mojiang mine after Dr Shi's addendum was published.

When the Drastic team of internet sleuths enquired with the WIV as to why the database was offline, it received two different answers: that it was for security reasons, and that it was in the process of being updated. Several of the Drastic sleuths tried to ask Dr Daszak if he had a copy of what was in the database. One of them, Billy Bostickson, even phrased a question for other scientists to ask (since he was blocked by Dr Daszak): 'Hello Peter, have you got a copy of the deleted 61.5 MB SQL WIV viral pathogen database administered by Zhengli Shi and its password protected section containing 100 unpublished bat betacoronavirus sequences? If not can you ask her for a copy?' By a stroke of luck, in March 2021, Dr Daszak was a participant in a public webinar and was asked on the spot whether he had requested to see this database. His response was: 'We did not ask to see the data and as you know a lot of this work is work that's been conducted with the EcoHealth Alliance and I'm also part of those data and you know we do basically know what's in those databases ... there is no evidence of viruses closer to SARS-CoV-2 than RaTG13 in those databases.'

Curiously, even after SARS-CoV-2 had emerged, Dr Daszak seemed unaware of when RaTG13 had been sequenced and studied in the lab. He told the *New York Times*, 'We sequenced a bit of the genome, and then it went in the freezer; because it didn't look like SARS, we thought it was at a lower risk of emerging. With the Virome project, we could have sequenced the whole genome, discovered that it binds to human cells and upgraded the risk.' He also told *Wired* that 'a lack of funding meant they couldn't further investigate the virus strain now known to be 96 per cent genetically similar'. Dr Daszak seemed not to know that this sample was linked to the severe respiratory illness in the Mojiang miners, and that the WIV had already sequenced the genome of RaTG13 in 2018.

Undoubtedly, this extensive pathogen database holds a wealth of information about the types of animals and their viruses that have been sampled across China over the years. It may or may not shed light on the origin of the pandemic. Sharing this information privately with international members of the scientific community, if not publicly, immediately after SARS-CoV-2 had emerged would have been a quick and effective way to dispel theories about the virus having escaped from the WIV. The failure to share the database, even confidentially with investigators, therefore deepened the suspicion that there was something to hide.

8.

Gain of function

'Nuclear weapons need large facilities, but
genetic engineering can be done in a small lab.
You can't regulate every lab in the world. The
danger is that either by accident or design, we
create a virus that destroys us.'

STEPHEN HAWKING

Bird flu is a disease of wild birds that can transmit into poultry.
One of the most lethal strains, known as influenza A H5N1,
has periodically devastated poultry farms in Asia. In 1997, in
Hong Kong, it began occasionally infecting human beings with
an alarming mortality rate of 60 per cent. Fortunately, the
H5N1 virus is usually caught directly from birds during the
slaughter and handling of diseased birds or their carcasses.
Sustained chains of person-to-person infection are uncommon,
occurring only through very close contact. If that were to
change, and the virus gained the ability to jump easily between
people, it could cause a frightening pandemic.

How easy would it be for the virus to change so that it could
spread efficiently between people? In May 2012, Dr Yoshihiro
Kawaoka's research group at the University of Wisconsin and

the University of Tokyo published a paper reporting experiments in which they had enabled H5N1 viruses to spread through the air between mammals. A few weeks later, another team of scientists led by Dr Ron Fouchier at Erasmus University in Rotterdam published a similar paper with similar results. The news sent shockwaves through the scientific community because the research seemed to involve deliberately making viruses more dangerous to mammals, including human beings.

These were not rogue teams of researchers working in secret. Both studies had received funding from the US National Institutes of Health (NIH). The papers were published in prestigious scientific journals: one in *Nature*, the other in *Science*. Each manuscript had been submitted for peer review almost a year earlier in August 2011. The motive behind the work was good: scientists like Dr Kawaoka and Dr Fouchier wanted to know if and how H5N1 could mutate into a more contagious form among mammals and which emerging variants or mutations should raise the alarm. Given the number of high-density chicken farms in the world, there is a concern that the virus could cause widespread outbreaks and mutate to become more transmissible between people – similar to the seasonal flu – causing a devastating pandemic. At the time, there was no flu shot for H5N1 readily available for large populations.

In December 2011, Dr Fouchier told the *New York Times*, 'There are highly respected virologists who thought until a few years ago that H5N1 could never become airborne between mammals … I wasn't convinced. To prove these guys wrong, we needed to make a virus that is transmissible.' So Dr Fouchier's team deliberately encouraged the virus to become transmissible through the air in a secure laboratory (BSL-3+), by first generating mutations in a key gene (haemagglutinin protein, or HA, the influenza equivalent of the coronavirus spike protein) and then 'passaging' the virus through animals.

Passaging is a common practice and one that, as we will describe, also played a role in various experiments with coronaviruses. The word simply means to infect a first batch of animals, then take the virus from that batch to infect a second batch of animals, repeating this process as many times as desired. The effect is to encourage the virus to evolve, by not-so-natural selection, to suit the new host species or to spread between hosts in a specific way. Both Dr Fouchier's and Dr Kawaoka's teams decided to use ferrets, which are easier to handle in the laboratory than monkeys and have an immune response to flu more similar to that of humans than mice do. In order to become more contagious in mammals, the virus would have to switch its attack from the virus entry receptor found in birds to the version of the same receptor found in mammals. Could it do so, and how easily could this happen?

Dr Fouchier's team wrote: 'We designed an experiment to force the virus to adapt to replication in the mammalian

Ferrets were used in influenza virus gain-of-function research.

respiratory tract and to select virus variants by repeated passage.' The final virus was sufficiently infectious that ferrets housed in cages near but physically separate from an infected animal started to fall ill. Dr Fouchier's findings suggested that as few as five mutations could enable airborne transmission of the virus between mammals. H5N1 thus had the potential 'to evolve directly to transmit by aerosol or respiratory droplets between mammals, without reassortment in any intermediate host, and thus pose a risk of becoming pandemic in humans'.

In contrast, Dr Kawaoka's group did not largely rely on serial passaging to evolve a more transmissible variant of H5N1. Instead, they showed that they could introduce random mutations in the H5N1 HA gene; create chimeric viruses with genes from H1N1 swine flu and H5N1 bird flu; identify the best performing variants in the 'library' of mutants in host cells (at BSL-2; the scientists reasoned that they had replaced a HA cleavage site with a non-virulent sequence to attenuate the viruses so the work could be performed at this lower biosafety level); and then confirm that one of these variants was indeed efficiently transmissible by air in ferrets (at BSL-3+). Dr Kawaoka's group also found that only three or four mutations were required to make the H5N1 HA capable of supporting efficient transmission in ferrets. In both approaches, scientists had harnessed techniques in the lab simulating and accelerating the process of evolution by artificial selection.

Before doing the experiments in negative-pressure safety cabinets in BSL-3+ laboratories, Dr Fouchier's team took every possible precaution, acquired every requisite permit and welcomed inspectors not just from the Dutch government but from the US CDC. They even offered H5N1 influenza vaccines to the researchers and prepared quarantine hospital rooms. Dr Kawaoka's lab also performed their work in BSL-3+ spaces approved by the US CDC and Department of Agriculture. Their facility had ongoing biosecurity monitoring and all personnel had undergone an assessment by the US Criminal Justice Information Services Division and completed rigorous biosafety and Select Agent training. Nothing went wrong. Rather, the reason for the shockwaves was that the experiments succeeded too well. It was too easy for scientists to create dangerous viruses that could spread through the air between animal models of human disease.

In late 2011, before publication, the media picked up on these radical pieces of work and amplified the concern of some scientists that such 'gain of function' experiments were too risky to allow. The phrase 'gain of function' typically refers to any experiment that results in a pathogen acquiring some new or improved capability, such as the ability to infect a new type of cell or a new host species. Why deliberately change a bird-transmissible virus so that it could transmit through the air among mammals? On 20 December 2011, the US National Science Advisory Board for Biosecurity, which advises the Department of Health and Human Services, recommended that both scientific teams and the editors of both journals should omit the methods and details of their work that could allow nefarious actors to develop dangerous pathogens. However, as a professor of pathology, microbiology and immunology at the University of California at Davis, Dr Nicole Baumgarth, pointed out: 'If two research labs have done this already, nobody is going to stop a third and fourth lab from doing the same. These are routine procedures done in many labs around the world.'

Scientists debating the issue crystallised into two camps, one asserting that such experiments must continue to help the world prepare for pandemics, the other arguing that the experiments might themselves trigger pandemics and should cease. Citing the rapid emergence of the 2009 swine flu, Drs Francis Collins, Anthony Fauci and Gary Nabel from the NIH posited that 'new data provide valuable insights that can inform influenza preparedness' and that 'the engineered viruses developed in the ferret experiments are maintained in high-security laboratories'. Their opinion was published in the *Washington Post* in December 2011, under the headline 'A Flu Virus Risk Worth Taking'.

This optimism was not shared by some other scientists and journalists. 'We cannot say there would be no benefits at all from studying the virus. We respect the researchers' desire to protect public health,' said a *New York Times* editorial on 7 January 2012, 'but the consequences, should the virus escape, are too devastating to risk.' The chairman of preventative medicine at Vanderbilt University School of Medicine in Nashville, Dr William Schaffner, said, 'People in that lab need to have a careful discussion on how to keep that virus in the lab secure. Viral escape is quite real.'

By 20 January 2012, still before the results had been published, Dr Fouchier and Dr Kawaoka joined other scientists around the world in declaring a voluntary and temporary sixty-day moratorium on further experiments to generate highly pathogenic viruses that are more transmissible between mammals. 'Despite the positive public health benefits these studies sought to provide, a perceived fear that the ferret-transmissible H5 HA viruses may escape from the laboratories has generated intense public debate in the media on the benefits and potential harm of this type of research,' they wrote. 'We realize that organizations and governments around the world need time to find the best solutions for opportunities and challenges that stem from the work.'

However, it was too late. This gain-of-function work was too tantalising, too novel, and had been published in high-profile scientific journals. Dr Baumgarth was right on the money. The very next year, a laboratory at the Harbin Veterinary Research

Institute in China successfully mixed a duck isolate of H5N1 with the 2009 H1N1 flu virus, creating 127 recombinant viruses to show that the avian virus could become transmissible among mammals in hypothetical agricultural scenarios. The work was once again published in *Science* in June 2013. The ease with which laboratories around the world could design and engage in their own unique gain-of-function studies to create highly transmissible viruses was becoming clear. Some scientists claimed that these were possible natural trajectories of virus recombination or evolution that could occur on farms or in the wild without laboratory intervention. If true, how could one distinguish a lab-derived recombinant strain from a natural recombinant strain of virus?

An ineffective moratorium

Two years later, a run of three mistakes involving smallpox, anthrax and the bird flu virus in US laboratories poured oil on the fire of the gain-of-function debate. Live anthrax was accidentally shipped under conditions that exposed up to seventy-five workers to the bacteria. Several vials of forgotten 1950s-era smallpox were discovered in a cold-storage room of a Food and Drug Administration lab; two were tested and found to still contain live virus despite the predictions of smallpox experts. And the US CDC accidentally sent a highly virulent H5N1 influenza strain that can sicken humans (initially mistaken as a milder H9N2 strain) to another lab. In the last case, the CDC lab had been notified of the error on 23 May, but the leadership only got wind of it on 7 July, a delay the CDC chief, Dr Thomas Frieden, called very troubling: 'I'm disappointed by what happened and frankly I'm angry about it.'

In response to these shocking incidents at a 'superb laboratory' (Dr Frieden's words), the FDA and NIH searched their labs to check for more forgotten smallpox samples, and the CDC temporarily closed the anthrax and influenza labs that had shipped those dangerous pathogens, alongside halting all pathogen shipments from its BSL-3 and BSL-4 laboratories. A CDC deputy director was appointed to review and supervise biosafety at the CDC, and those who had violated protocols or reporting rules were disciplined.

Despite these measures, the accidents reignited fears about the safety at other labs working with dangerous or even enhanced human pathogens, such as those of Dr Kawaoka and Dr Fouchier. In July 2014, Dr Marc Lipsitch had organised the Cambridge Working Group, based in Cambridge, Massachusetts, to lobby for a moratorium on gain-of-function research. They described themselves as a 'group of concerned scientists and experts in legal, ethical and other dimensions of Potential Pandemic Pathogen research' formed to 'enhance public understanding of the biosafety risks of such research'. The group pointed out that laboratory incidents had been growing in frequency, 'occurring on average over twice a week with regulated pathogens in academic and government labs across the country'. If similar incidents were to occur with newly created, highly transmissible pathogens, the risks would be substantially increased, leading to outbreaks that could be impossible to control. Dr Lipsitch argued that gain-of-function experiments 'unjustly require unconsenting populations to bear pandemic risk while promising them no realistic prospect of benefit'. Dr Steven Salzberg, an influenza researcher at Johns Hopkins University, agreed, adding that 'the benefits of gain-of-function research are minimal at best' and 'could easily and far more safely be obtained through other avenues of research'.

By October 2014, the White House Office of Science and Technology Policy had announced a funding moratorium on gain-of-function research. Some scientists were surprised to hear that the pause on new federal funding applied not just to influenza but also to two other viruses: SARS and MERS. Dr Ralph Baric of the University of North Carolina at Chapel Hill, a legend in coronavirus research, was attending a daughter's wedding that weekend and only heard about the moratorium when he returned to his laboratory on the Monday and opened his emails. It was devastating news. 'It took me 10 seconds to realize that most of them were going to be affected,' he said of his research projects. He wrote to the NIH arguing that 'this decision will significantly inhibit our capacity to respond quickly and effectively to future outbreaks of SARS-like or MERS-like coronaviruses ... Emerging coronaviruses in nature do not observe a mandated pause.'

Yet Dr Baric had himself already raised concerns about the kind of work in which he had made ground-breaking advances. In a lengthy 2006 review of reverse genetics (a method to understand the function of a new gene by altering it or inserting it into a different genetic context) experiments on viruses, entitled 'Synthetic Viral Genomics: Risks and Benefits for Science and Society', he detailed how a clever bioterrorist could create infectious pathogens undetected by any existing surveillance system: 'It is conceivable that a bioterrorist could order genome portions from various synthesis facilities distributed in different countries throughout the world and then assemble an infectious genome without ever having access to the virus. To our knowledge, no international regulatory group reviews the body of synthetic DNAs ordered globally to determine if a highly pathogenic recombinant virus genome is being constructed.'

The coronavirus experiments

The coronavirus experiments that raised concern deserve close scrutiny if we are to understand what two laboratories in particular, Dr Baric's in Chapel Hill, North Carolina, and Dr Shi's in Wuhan, Hubei, were capable of doing with SARS-like viruses. Dr Baric had joined the University of North Carolina, Chapel Hill, in the 1980s to work on coronaviruses, then somewhat low-profile creatures that were responsible for some common colds but had never apparently caused serious outbreaks of human disease. In 1995, he published a book chapter about mouse hepatitis virus (MHV), arguing, in its title, that 'coronaviruses may be potentially important emerging viruses'. He described experiments in which MHV gained the ability to infect the cells of hamsters, by using serial passaging, as described earlier. The virus could 'rapidly alter its species specificity and infect rats and primates'.

After the epidemics of SARS, then MERS, coronavirus research was no longer backwater. Baric now held professorships in epidemiology, microbiology and immunology, and was an acknowledged world expert on the genomes of coronaviruses. His lab had a reputation for continually breaking new ground in manipulating coronavirus genomes. This included inserting new sequences into them using a 'no-see'm' method, patented in 2006, that Dr Baric named after a small biting insect sometimes encountered on beaches in North Carolina. The beauty of this technique, compared with traditional genetic engineering, was that it allowed a seamless insertion with no tell-tale 'restriction site' sequence attached because 'the sites are removed during reassembly, leaving only the desired mutation in the final DNA product'. This is a vital point, given the argument often still made against laboratory-leak hypotheses for Covid-19, that restriction site 'scars' in its genome would be expected if genetic manipulation had occurred. Dr Baric's no-see'm method was used from at least 2016 by the WIV scientists in some of their experiments and has since been enhanced by other seamless methods developed since 2006. In fact, once the SARS-CoV-2 genome was made public in January 2020, it took coronavirus experts in Europe and America just one to two months to separately devise and demonstrate reverse genetic systems to create synthetic SARS-CoV-2 genomes in the lab. The synthetic genome could have been used to create an infectious virus that would have left no signs of having been constructed in a laboratory.

In 2007, Dr Shi published a paper announcing her arrival at the forefront of this field of reverse genetics in coronaviruses. She and her colleagues used a 'pseudovirus' made from HIV, the virus that causes AIDS. Their aim was to test the ability of various spike receptor-binding domains found in SARS-like bat viruses to use the ACE2 receptors from a human, a civet or the horseshoe bat *R. pearsonii*. A pseudovirus has some of the attributes of a virus but is not capable of making more copies of itself even after entering a host cell; this makes the research safe because even if the virus were to leak, it cannot transmit from host to host.

The reason that Dr Shi's group had to test the spike of a novel bat SARS-like virus in a pseudovirus was that they could not isolate the SARS-like virus itself. Often specimens from the wild will show evidence of the presence of a virus genome, but in such a degraded state that an infectious virus itself cannot be recovered. Using the pseudovirus system, they found that a bat SARS-like virus's spike was 'unable to use ACE2 proteins of different species for cell entry'. However, when they replaced just the receptor-binding domain of the bat virus spike with that of the 2003 SARS virus that had infected human beings, the hybrid protein had the ability of the 2003 SARS virus to use human ACE2 as a receptor. This indicated that this relatively short segment of the spike was both necessary and sufficient to convert a virus from bats into one that could infect human cells via the ACE2 receptor.

Remarkable though this experiment was, it was not altogether surprising. Seven years before, in 1999, researchers at Utrecht University in the Netherlands had done a similar reverse genetics experiment using cat and mouse viruses. They replaced the part of the spike that sticks out from the virus membrane in a mouse coronavirus, MHV, with the equivalent section from a cat coronavirus, called infectious peritonitis virus. The chimeric virus they thus created could infect cat cells but not mouse cells. The scientists inferred that the quality of the fit between the coronavirus spike and its entry receptor on host cells was what determined which host species or cell types a coronavirus can infect.

Scientists are sometimes collaborators, but they can also be competitors. The Baric group published an even more ground-breaking experiment in the same year, 2007, as Dr Shi's team, though not involving bats. They manufactured five coronavirus strains with their genome backbones derived from the genome of a SARS virus. The group used one of the first SARS virus variants to have been sequenced in 2003, named after Carlo Urbani, a courageous WHO epidemiologist who died after treating some of the first patients and raising the alarm on the new infectious disease. But Dr Baric's team replaced the spike with the equivalent from viruses extracted from infected palm civets, raccoon dogs, and the early, middle and late phases of the human SARS epidemic. They found that the human variants grew well in cultures of human 'airway' cell cultures, but the animal versions did not. This suggested that the animal versions are less pathogenic in humans but can evolve, while spreading among humans, to become more effective human pathogens. The reference to human airway cell cultures is a significant one. One of Dr Baric's collaborators, Dr Raymond Pickles, had by now ingeniously managed to culture the cells that line the human respiratory tract, so that they formed a two-layered floating tissue complete with tiny waving hairs called cilia, to simulate the lining of the human lung. This was an ideal set-up in which to test the virulence of viruses in a model of a human being's respiratory tract or lungs.

Another important new technique developed around this time was the creation of 'humanised' mice to accurately and rapidly study diseases such as SARS. In 2007, three separate teams of scientists succeeded in developing lines of healthy mice that expressed human ACE2 (hACE2) receptors to study the effect of SARS viruses. The reason was that SARS viruses do not infect mice easily because of an incompatibility with the mouse ACE2 receptor. The word 'express' in this context has a specific meaning: that a gene has been inserted and is actively being decoded to produce a protein. In most cases the mouse version of the ACE2 gene remains functional, but in some cases the human version replaces it. When infected with SARS virus, the hACE2 mice at the University of Iowa lost weight, became lethargic with laboured breathing and died within a week. By contrast, mice that did not express hACE2 displayed no clinical disease or mortality. The mice developed at the University of Texas Medical Branch in Galveston, Texas, exhibited 'ruffled fur, lethargy, rapid and shallow breathing, and persistent weight loss' when similarly infected. The hACE2 mice developed at the Institute of Laboratory Animal Science in Beijing suffered from severe lesions in the lungs alongside degeneration and necrosis in other organs including the brain. Almost a decade later, in 2016, Dr Baric's team also developed a line of humanised mice and injected SARS into their noses, whereupon they 'exhibited rapid weight loss and death between days 4 and 5'. These mice were shared with Dr Shi's group at the WIV. After SARS-CoV-2 had emerged, it was revealed that a team in Beijing had been developing hACE2 mice in which the human ACE2 receptor replaced the mouse version of ACE2; in other words, this mouse was biologically as close to humans with regards to ACE2 as scientists could get. The scientists in Beijing inoculated their young and old hACE2 mice with the SARS-CoV-2 virus and found that the aged mice developed pneumonia.

Armed with these new techniques, the Baric lab could anticipate significant advances against coronaviruses. A seminal 2008 paper by Dr Baric and Dr Mark Denison of Vanderbilt University, and their colleagues achieved the feat of effectively creating a synthetic virus. At the time, no bat coronavirus had yet been successfully cultured in the laboratory, making it challenging to understand how coronaviruses were jumping species. Dr Denison's team embarked upon a daring experiment. Using the sequences from the 2003 SARS virus and the four known bat SARS-like coronaviruses at the time, the scientists designed a hypothetical 'consensus' sequence, that is, a sequence that uses the most frequent letter at each position. Surprisingly, this worked. Chimeric viruses with spikes that were part SARS and part bat SARS-like virus were able to infect cells. This was a spectacular advance, and the team proudly announced 'the design, synthesis, and recovery of the largest synthetic replicating life form'. They had laid the foundations for future generations of scientists to create infectious pathogens not found in nature by weaving natural virus sequences together, even without access to samples of actual, physical virus particles. The scientists were optimistic, claiming that their work represented 'an approach for rapid recovery and testing of newly identified pathogens, and which may improve public health preparedness and intervention strategies against natural or intentional zoonotic-human epidemics'. Adopting their

methods, scientists could more quickly and precisely understand what makes a new virus able to infect human cells, how
it causes disease, and ultimately how to develop therapeutics or
vaccines against these viruses.

In 2013, Dr Baric met Dr Shi at a scientific meeting and
asked if she would share one of the new, unpublished virus
sequences closely related to the original SARS virus that her
group had discovered. He wanted to create and study a chimeric
virus using the spike of the new sarbecovirus in the backbone
of his laboratory mouse-adapted strain of the 2003 SARS virus.
Dr Shi graciously shared the sequence with Dr Baric. The result
was that during the moratorium on gain-of-function research,
in 2015, Dr Baric published a landmark paper with Dr Shi
showing that a SARS-like virus from Chinese horseshoe bats
could potentially lead to an outbreak in humans. The reason
that the two groups came to work together is clear. Baric had
the greater expertise in synthesising coronavirus backbones and
creating chimeras, but Shi had a growing library of bat coronavirus genomes collected during her collaborations with Dr
Daszak's EcoHealth Alliance. Unlike in 2007, this library now
included some viruses that were closely related to SARS – *R.
sinicus* bat viruses that had been collected in Yunnan in the
years running up to 2015. In the words of Dr Daszak, describing this collaboration to another virologist in December 2019
before news of the Covid-19 pandemic had broken, 'You can
manipulate [coronaviruses] in the lab pretty easily … you can
get the sequence [of the spike gene], you can build the protein,
and we work with Ralph Baric at UNC to do this, insert into
backbone of another virus.'

Dr Vineet Menachery, then a postdoctoral researcher in
Baric's lab, took the backbone of a SARS virus that had been
slightly adapted to grow in mice and replaced its spike protein
with that of a bat SARS-like virus sampled by Dr Shi's group.

The resulting chimeric virus could use human ACE2 to enter and infect human airway cells that had been cultured in the laboratory. When tested in mice, the virus caused severe damage in the lung, and was resistant to treatment by antibodies. It was also resistant to an early design of a vaccine against the SARS virus, in the form of an inactivated whole SARS virus – not a vaccine that was yet considered safe for use in the human population. Worryingly, the scientists found that, compared with having the spike of the 2003 SARS virus in this mouse-adapted SARS backbone, the bat spike in the same backbone showed a gain in virulence. The main conclusion they drew from the experiments was that bat viruses on their own were potentially dangerous. They did not need to be refracted through an intermediary species to start an epidemic. As the paper put it, 'our results suggest that the starting materials required for SARS-like emergent strains are currently circulating in animal reservoirs.' Dr Daszak said the study raised the profile of a wild virus 'from a candidate emerging pathogen to a clear and present danger'.

Yet if the experiment underlined the risk of natural zoonosis, it also emphasised the risk of laboratory experiments on such viruses. The paper itself warned that 'scientific review panels may deem similar studies building chimeric viruses based on circulating strains too risky to pursue, as increased pathogenicity [virulence] in mammalian models cannot be excluded'. It added: 'The potential to prepare for and mitigate future outbreaks must be weighed against the risk of creating more dangerous pathogens.' The work had been initiated before the funding moratorium and reviewed by the NIH, which funded and approved its continuation. It was nonetheless contentious. An article in *Nature* quoted virologists questioning whether the results of the experiment could justify the risks. Dr Simon Wain-Hobson, of the Pasteur Institute in Paris, said: 'If the virus escaped, nobody could predict the trajectory.' Dr Richard Ebright of Rutgers University said: 'The only impact of this work is the creation, in a lab, of a new, non-natural risk.'

The work shifts to China

Several months before the gain-of-function moratorium began, the EcoHealth Alliance won a five-year, multi-million-dollar grant from the US National Institute of Allergy and Infectious Diseases (NIAID), with Dr Daszak as the principal investigator. Its purpose was 'understanding the risk of bat coronavirus emergence'. The research would be carried out by the EcoHealth Alliance's overseas partners, including the WIV: 'Predictive models of host range (i.e. emergence potential) will be tested experimentally using reverse genetics, pseudovirus and receptor binding assays, and virus infection experiments across a range of cell cultures from different species and humanized mice.' In all, according to the EcoHealth Alliance, the Wuhan institute received about $600,000 from this grant.

One project funded by this grant resulted in a 2017 paper in the journal *PLoS Pathogens* by Dr Ben Hu, one of Dr Shi's doctoral students who had stayed on in the group as a researcher at the WIV. It described the creation of recombinant viruses that carried the spike genes of eight novel SARS-like coronaviruses sampled from bats, spliced into the genetic backbone of another SARS-like virus that the group had successfully isolated from bats and named WIV1. Two of these recombinant viruses (as did the natural WIV1 virus) proved capable of infecting and replicating in cells expressing human ACE2.

The paper summed up the achievements of the scientists' visits to the Shitou cave near Kunming, Yunnan, yielding fifteen full-length genomes of SARS-like coronaviruses found in bats, as well as a total of three viable viruses isolated from the samples: WIV1, WIV16 and Rs4874. The genomes of WIV16 and Rs4874 were 99.9 per cent identical, and both came from samples collected on 21 July 2013, though WIV16 was published in 2016 and Rs4874 in 2017. The spike proteins of the two viruses were the most closely related yet to that of the 2002–3 SARS virus, with more than 97 per cent protein-sequence identity, and could use human ACE2 as an entry receptor. This discovery suggested to the scientists that some of the diverse SARS-like viruses found in that cave could be capable of transmitting directly from bats to humans. Based on the SARS-like viruses that had been found in the cave over the course of five years, 2011 to 2015, the authors hypothesised that the direct progenitor of the 2002–3 SARS virus may have arisen from recombination among the viruses found in that region in Yunnan. The scientists pointed out that the closest village to the cave was just over a kilometre away and, in 2003, there was a civet farm in Kunming that sold civets to Guangdong for consumption. The scientists did not know whether bats had transmitted the SARS virus to civets in Yunnan that were transported to Guangdong, or whether it was bats in Guangdong that had carried the virus and passed it to civets in the same province. However, they proposed continued monitoring of the SARS-like viruses to probe for and prevent spillover in these areas.

It is not clear if Dr Hu's work, funded by the NIH, fell under the type of gain-of-function research for which new US federal funding had been paused. Dr Anthony Fauci, head of the NIAID, had helpfully defined gain-of-function research back in 2012 in a speech: 'What historically investigators have done is to actually create gain-of-function by making mutations, passage/adaptation or other newer genetic techniques such as reverse genetics and genetic re-assortment.' This did seem to describe Dr Hu's work. Yet in 2021, in response to a question from US Senator Rand Paul, Dr Fauci categorically denied that NIH had funded gain-of-function research in Wuhan. Certainly, there is room for disagreement here depending on the definition of gain of function in use. In particular, the funding pause did not apply to the study of natural viruses unless there was a 'reasonable expectation that these tests would increase transmissibility or pathogenicity'. In experiments in which only natural bat SARS-like viruses, not known to infect humans, were being recombined, it could be reasoned that the resulting viruses would not exhibit increased transmissibility or virulence in humans – after all, the scientists were only mixing and matching parts of bat viruses and not of human viruses. For instance, Dr Baric had argued that the mouse-adapted human SARS virus (from the 2003 epidemic) 'actually makes the germs less able to infect human cells' and are safer to work with in the laboratory. Bats are more distantly related to humans than mice are. The US government said that the point at which gain-of-function research must stop would be determined for individual grants via discussions between the scientists and their funding officers.

Aside from this confusion over what was considered gain-of-function research, the moratorium only targeted new funding. On funding that had already been granted, it merely called for a pause, meaning that there was no enforcement. This loophole was further enlarged by a footnote that read: 'An exception from the research pause may be obtained if the head of the funding agency determines that the research is urgently necessary to protect the public health or national security.' The NIH, responding in early 2021 to questions from Fox News, argued that the government-funded EcoHealth Alliance-WIV research did not breach the moratorium because the work 'characterized the function of newly discovered bat spike proteins and naturally occurring pathogens and did not involve the enhancement of the pathogenicity or transmissibility of the viruses studied'.

In 2017, the moratorium ended under the Trump administration, and the NIH announced that funding of gain-of-function experiments involving influenza, MERS and SARS would resume under a new framework. Each research proposal would be assessed on a case-by-case basis by the Department of Health and Human Services. 'If we are pursuing this research in an active way, we will be much better positioned to develop protection and countermeasures should something bad happen in another country,' argued Dr Carrie Wolinetz of the NIH. Dr Lipsitch of the Cambridge Working Group disagreed: 'I still do not believe a compelling argument has been made for why these studies are necessary from a public health point-of-view.'

Some scientists expressed concern that the wording of the new framework was too narrow, affecting only those research projects that were 'reasonably anticipated to create, transfer, or use enhanced PPPs [potential pandemic pathogens]'. To count as an enhanced PPP, a pathogen had to be highly transmissible between humans, highly virulent and have resulted from an enhancement of either transmissibility or virulence. Notably, the new framework said that 'Enhanced PPPs do not include naturally occurring pathogens that are circulating in or have been recovered from nature, regardless of their pandemic potential.' The scientists were experimenting with viruses collected from bats, with some, albeit unknown, potential for transmitting to humans. Even if pandemic potential was suspected – and indeed repeatedly emphasised in numerous EcoHealth Alliance-WIV publications – the experts appointed by the NIH determined that this type of research fell outside the scope of the funding pause and the 2017 framework. They stressed that the EcoHealth Alliance application had been rigorously reviewed and 'judged to be very high priority' because of the emergence of the original SARS virus. In any case, this research would have almost certainly continued with

Chinese government funding even if the NIAID grant had been paused. In today's world of expensive research, $600,000 of funding over five years is small beer. A top laboratory in China such as Dr Shi's group at the WIV, holding at any time numerous such grants and sharing funding with other top labs, would not have folded if a measly $120,000 did not make its way from the US into its coffers each year.

The vaccine hope

What was the point of all this virus hunting and genetic manipulation? By the end of 2017, the Wuhan scientists led by Dr Shi had achieved most of their original goals on the trail of SARS-like viruses. They had tracked down the likely bat origin of the 2003 SARS virus in Yunnan, although it was still a mystery how the virus had made its way to Guangdong. They had isolated three viruses from bat samples collected up to 2013. They had made chimeric viruses to test the infection abilities of diverse spikes in SARS-like viruses found in the wild. They had challenged these viruses against human cells and humanised mice. They had gathered huge amounts of data on the pandemic threats that these SARS-like viruses presented, indicating a probable spillover zone in southern China where people who handled wild animals or lived near bat caves were found to carry antibodies against SARS viruses. In addition to assessing the risk from viruses, the obvious priority now was surely to use this knowledge to prevent a pandemic occurring.

In this area in recent years a new word has begun to crop up with growing frequency: vaccine. The dream of designing a vaccine against all SARS-like viruses was very much within the ambitions of those studying SARS viruses. In this way, people living in areas prone to the spillover of SARS viruses could be vaccinated to pre-empt an outbreak of novel coronaviruses. As early as 2008 Dr Baric and Dr Denison had declared that 'to protect against future emerging zoonotic pathogens, it is crucial to develop cell culture and animal models to test vaccines and therapeutics, ideally against entire families of organisms, such as [coronaviruses]'. In 2019, one of the reasons Dr Daszak cited for the laboratory work was to help develop a broad-spectrum vaccine against new coronaviruses: 'The logical progression for vaccines is, if you're going to develop a vaccine for SARS, people are going to use pandemic SARS [that is, the virus that caused the 2003 epidemic] ... but let's try and insert some of these other related [viruses] and get a better vaccine.'

So was a vaccine against SARS viruses on the horizon? It is not mentioned in any of the key grant programmes. The National Science Foundation of China announced a new grant to support continuing work on the 'Evolution mechanism of the adaptation of bat SARS-related coronaviruses to host receptor molecules and the risk of interspecies infection'. It was to run for three years from the start of 2018 with Dr Shi as the principal investigator. That same year, Dr Hu in Dr Shi's lab received a further 250,000 yuan grant under the WIV's Youth Science Fund beginning in January 2019 to study the 'pathogenicity of two new bat SARS-related coronaviruses to transgenic mice expressed ACE2'. This probably refers to WIV1 and either one of WIV16 or Rs4874 (near identical viruses), which had been isolated and characterised in cell cultures, but not yet in laboratory animals. Incidentally, Francisco de Ribera (a tenacious data sleuth who we will meet later in the book) has been able to identify and account for all the published viruses isolated at the WIV, from WIV1 to WIV19, with one exception. He could find no published account of WIV15. When details of WIV6 were published in June 2021, Ribera commented: 'We found WIV6. A little bit strange, but unrelated with SARS-like viruses ... This leaves WIV15 as the only missing live isolate. It was always the most suspicious due to the range of dates in which it should have been isolated.' Based on the date range, Ribera speculated that WIV15 might have possibly been isolated from a Mojiang sample.

In 2018, a further large five-year grant was announced by the Chinese Academy of Sciences under a 'Pathogen host adaptation and immune intervention' special project. This project focused on discovering new viruses in animals, as well as humans, afflicted with unknown pathogens, understanding how these viruses could jump between species and cause disease, and how they were able to evolve to defeat both the immune system and treatments. Using this knowledge, the scientists aimed to develop and test novel antibody therapy and vaccines against emerging infectious diseases. They would enlist cutting-edge technologies, including human organs-on-a-chip and humanised mice, as well as in-laboratory directed evolution. There were five sub-projects in total. Dr Shi co-led the first alongside a different Dr Shi. And it seems that Dr Shi's own group was also to receive funding from this strategic priority research programme of the Chinese Academy of Sciences (grant number XDB29010101) valued at about $1.35 million, from July 2018 to June 2023.

Thus, in the year preceding the Covid-19 outbreak, research was ramping up in China, and was clearly aimed at comprehensively cataloguing natural viruses with pandemic potential, including SARS-like viruses, and devising vaccines against them. With the impressive track record of Dr Shi's group, it seemed inevitable that vaccines and therapeutics against SARS viruses would soon be developed, giving scientists the upper hand against emerging coronaviruses.

In America, in October 2018, Dr Baric's group published a paper suggesting how to design a live vaccine that would work against SARS viruses. Briefly, vaccines work by raising immunity against real infections and to do this they are either inert, dead versions of (usually parts of) a virus or they are 'live' but attenuated viruses: still capable of causing an infection, but a very mild one. Live vaccines have been used successfully against measles, mumps, rubella, yellow fever and chickenpox. The power of live vaccines is that, although the virus is weakened, they cause an infection similar to the natural virus and can stimulate strong immunity against it that can last a lifetime. The problem with live vaccines is that sometimes they revert to the untamed form, usually via recombination with a natural, closely related virus that happens to infect the vaccinated person or animal. For instance, sporadic cases of vaccine-derived poliovirus have occurred when the weakened poliovirus used in the oral vaccine continues to circulate in a particular population for a long period of time, allowing it, through mutation, to regain its ability to cause severe disease. For this reason, the US only uses the inactivated polio vaccine and not the live attenuated virus vaccine.

Live attenuated African swine fever vaccines sometimes mutate back into a pathogenic form. In early 2021, reports began to surface that some of the recent African swine fever outbreaks in China had been caused by unlicensed vaccines being administered to large numbers of pigs. However, due to the issue's political sensitivity, 'reporting of the recent African swine fever outbreaks was extensively covered up', according to Channel News Asia. These attenuated viruses are difficult to detect and can cause long-term and widespread infections across pig farms, with the potential to devastate the pork industry in China – particularly because some farms feed their pigs with kitchen waste that may still carry live viruses.

Dr Baric's team was trying to get around this limitation of live vaccines by making sure that their live attenuated SARS virus would not recombine with other SARS-like viruses and revert to virulence. They were going to do it by rewriting a small section of special text in the genome of the virus. This text, known as a transcription regulatory sequence (TRS), triggers the expression of genes. The team showed that within each of the nine regulatory sequences of the SARS genome there was a six-letter text that was the same – ACGAAC – and if they replaced this in each case with a slightly longer and markedly different text – UGGUCGC – they had effectively changed the password on the whole network. This attenuated the virus, weakening its virulence by messing up the efficiency of its gene transcription. Encouragingly, Dr Baric and his colleagues found that inoculating aged mice with this TRS mutant could protect against a lethal SARS virus challenge. This change in TRSs also meant that an attenuated version of the virus would not revert to virulence if it found itself in the same cell as a virulent version of the virus since its ability to start swapping sequences and repairing faulty parts had been compromised. Thus, they had made a weakened and recombination-resistant virus as 'a candidate strategy for a broadly applicable, rapidly implementable CoV vaccine platform'. There is a note of hope, even triumph, in the final words of the paper's discussion: 'This attenuation strategy ... could bring live-attenuated CoV vaccines within the reach of realization in the face of the ever-growing threat of new human and animal CoV-based epidemics.' Sadly, this technology was not developed in time to contribute to ending the Covid-19 pandemic. It would be other kinds of novel vaccines, messenger-RNA (mRNA) vaccines and recombinant vector vaccines, rather than live attenuated vaccines, that would come to the rescue of humanity during 2021.

Alongside raising hopes for a vaccine, the research in both Dr Baric's and Dr Shi's laboratories held a darker promise, as a demonstration of how to synthesise novel viruses. Once other labs saw that it was possible to stitch together sequences from different SARS-like viruses from bats and bring the final chimera to life, they soon began to copy these techniques or invent their own unique approaches. The public sees only the research that scientists decide to publish, not the projects that fail, that remain unfinished, or that they choose to keep secret. This is particularly troubling in recent years when it has become possible for labs to install their own gene-synthesis machines that print out any sequence. Enforcing worldwide surveillance of the gene synthesis performed by each of these machines is impossible.

Yet it is important to note that it remains impossible even today to magically conjure up a virus based on no sequences. Access to virus samples is not needed, but scientists invariably have to use the sequences of natural virus genomes as a starting point for their genetic designs. For this reason, it can be difficult to distinguish a natural from a synthetic virus without knowing what natural viruses could have been used as a template and what laboratory techniques had been deployed. In the absence of both pieces of knowledge, it is just about impossible to know from the sequence alone whether a given genome is a product of natural recombination or of laboratory engineering.

A whistleblower emerges

The idea that these techniques could have been used to make bioweapons, and that Covid-19 might have been caused by such a weapon, was especially inflamed when, in April 2020, a young Chinese scientist-turned-whistleblower fled to the United States in fear of being 'disappeared' in China, and began to give interviews about the early events of the pandemic. Dr Limeng Yan had previously worked on influenza vaccines and is a co-first author on a *Nature* paper about how SARS-CoV-2

causes disease in hamsters. She blew the whistle on two of the
senior authors of the paper, one of whom was her supervisor,
Dr Leo Poon, claiming that they had known about the human-
to-human transmissibility of SARS-CoV-2 but failed to relay
this crucial information to the WHO and the world in a timely
manner. According to stories on Fox News (July 2020) and in
the *New York Times* (November 2020), the saga unfolded as
follows. The Hong Kong laboratory in which Dr Yan worked
was a WHO reference laboratory, one required to inform the
WHO of developments concerning influenza viruses and
pandemics. In December 2019, Dr Poon asked Dr Yan to look
into the SARS-like cases that were being reported in Wuhan. Dr
Yan, who had trained as a doctor in mainland China, tapped
into her network in various medical facilities across the country. On 31 December, a friend at the Chinese CDC told her that
the virus was transmitting from person to person, weeks before
China or the WHO confirmed human-to-human transmission,
and she reported this to Dr Poon. However, by early January,
her contacts were going silent. The doctors said: 'We can't talk
about it, but we need to wear masks.' Dr Yan again reported
this to Dr Poon on 16 January, but was told 'to keep silent, and
be careful'. By 19 January, Dr Yan had leaked this information
to Lu De, a YouTuber and China critic, who told his hundred
thousand followers that the novel coronavirus was transmissible between people. The very next day, the Chinese National
Health Commission (and the WHO in turn) confirmed human-
to-human transmission of the virus. In his video, Lu De added
that the virus had been deliberately engineered and released by
the Chinese government, pointing to research on bat SARS-like
viruses in Zhoushan, in Zhejiang province, published in
September 2018 by scientists at the Third Military Medical
University, Chongqing, and the Research Institute for Medicine
of Nanjing Command.

Dr Yan's escape story revealed a problem that one of us (Alina) pointed out on Twitter: 'If there is one thing that this entire saga has made clear – it is that whistleblowers (as it pertains to SARS-CoV-2) have no obvious safe route of sharing their information.' When she fled to America, Dr Yan had to rely on the contacts of Lu De, an anti-Chinese Communist Party billionaire named Miles Guo and his friend Steve Bannon, ex-advisor to Donald Trump. (Steve Bannon was arrested on Guo's yacht in August 2020 for allegedly defrauding donors in a crowdfunding campaign to build the US border wall with Mexico; he pleaded not guilty.) According to the *New York Times*, Dr Yan was packaged by Guo and Bannon as 'a whistle-blower they could sell to the American public', and Guo recounted on his show telling Dr Yan: 'Don't link yourself to Bannon, don't link yourself to [Miles] Guo Wengui ... Once you mention us, those American extreme leftists will attack and say you have a political agenda.' Dr Yan gave a series of inter-views to right-wing media in the United States, deviating far from her original whistleblower report about human-to-human transmission of the virus to talk about topics such as hydroxy-chloroquine treatment, the controversial anti-malarial drug championed by Donald Trump as a treatment for Covid-19. As Alina pointed out on Twitter, the Yan episode showed how easily a whistleblower can be tainted by their dependency on their host or saviour, someone who they now have to rely on for security for the rest of their life.

In early September, a meeting was arranged between Dr Yan and Dr Daniel Lucey, an infectious diseases expert from Georgetown University. In a blog post at the end of June 2020, Dr Lucey had himself published questions about whether the virus might have emerged from a laboratory. According to the *New York Times*, Dr Lucey 'said that Dr. Yan appeared to genu-inely believe that the virus had been weaponized but struggled to explain why'. Undeterred, Dr Yan published a lengthy preprint that month, setting out her claims that SARS-CoV-2 had been created in a laboratory. The day after publication, on 15 September, she went on *Tucker Carlson Tonight*, one of the most popular Fox News shows. Asked if she believed that the Chinese government had released the virus on purpose, Dr Yan replied: 'Yes, of course, it's intentionally.' Inevitably, Dr Yan's report and interview were harshly criticised and debunked by scientists.

Alina, as well as other scientists, soon found numerous problems with the scientific claims in the Yan preprint. Most crucially, the Zhoushan viruses found by the Third Military Medical University, Chongqing, on which the entire report hinged were more than three thousand mutations away from SARS-CoV-2 – therefore they were highly unlikely to have been the template virus genome for SARS-CoV-2. This destroyed the report's credibility, and had diverted attention away from RaTG13, the Mojiang miners and the missing WIV virus database. Within a month, Dr Yan had released another preprint, this time titled 'SARS-CoV-2 Is an Unrestricted Bioweapon: A Truth Revealed through Uncovering a Large-Scale, Organized Scientific Fraud'. This one did not mince words: 'Records indicate that the unleashing of this weaponized pathogen should have been intentional rather than accidental. We therefore define SARS-CoV-2 as an Unrestricted Bioweapon and the current pandemic a result of Unrestricted Biowarfare.'

Despite the drama surrounding these preprints, the possibility cannot be dismissed that laboratories in America, China and elsewhere have been developing cutting-edge genetic-manipulation technologies, partly to be ready to fend off their use by terrorists, but also to use them against their enemies if necessary. 'The technology to synthetically reconstruct genomes is fairly straightforward and will be used, if not by the United States, then by other nations throughout the world,' Ralph Baric had warned in 2006 – and technology has advanced considerably since then. In 2021, the US State Department published a statement that included this claim: 'The United States has determined that the WIV has collaborated on publications and secret projects with China's military. The WIV has engaged in classified research, including laboratory animal experiments, on behalf of the Chinese military since at least 2017.'

However, at this stage, in the absence of harder evidence, we think allegations that SARS-CoV-2 is a bioweapon or a vaccine trial that went wrong are a distraction. If the virus came from a laboratory, it is much more likely that it was a leak from experiments designed to understand viruses that pose potential pandemic threats. The SARS-CoV-2 virus does not appear to have the features of an attenuated vaccine. If anything, it seems to have features that promote increased transmissibility, virulence and ability to evade the immune system, and it would have made no sense to release a bioweapon in the very city in which a world-renowned lab was performing such similar research, if the plan was to do so without raising suspicions.

9.

The furin cleavage site

'Often what people don't say or leave out, tells
the real story.'

SHANNON ALDER

In all the 29,903 letters of RNA that comprise the genome of a
typical SARS-CoV-2 virus, there is one little segment, just
twelve letters long, spelling out the recipe for four amino acids,
that has since become a miniature but fiercely contested battle-
ground in the war to understand the origin of the pandemic.
There are three reasons that this short sequence matters so
much: it appears to play a big role in making the virus more
infectious; it seems to be unique to this virus among all the
sarbecoviruses; and it is the sort of sequence in exactly the location that scientists have been deliberately inserting into other
coronaviruses.

It is called a 'furin cleavage site'. Though 'furin' might sound
like a drink taken by a Viking before a raid, it is in fact short
for the boring and unilluminating phrase 'FES Upstream Region
(FUR) Protein'. Furin is a vital protein doing steady work in
human cells every minute of every day, going around cleaving
proteins in two to change their shapes, the better to enable

them to do their myriad different jobs. Furin is lured to do this in the right place in each protein by special sequences of amino acids. The furin cleavage site in SARS-CoV-2's genome reads CGG-CGG-GCA-CGT, which is the recipe for the amino acids arginine-arginine-alanine-arginine, RRAR. It lies in a key spot in the spike gene of the virus, immediately upstream of the point where the furin will cleave the spike in two.

When the genome sequence of SARS-CoV-2 was first released to the world by Dr Edward Holmes and Dr Zhang Yongzhen on 11 January 2020, scientists quickly spotted this novel feature. A team of scientists in France and Canada, including Dr Bruno Canard and Dr Etienne Decroly in Marseille, were the first into print outside China. In a paper submitted for publication on 3 February 2020, they wrote that the furin cleavage site may provide a gain of function to SARS-CoV-2 compared with other SARS-like viruses, increasing the efficiency with which the virus spreads among people. Two weeks earlier, on 21 January, a team of scientists from four universities across cities spanning China from north to south – Tianjin, Jinan, Nanjing, Kunming – published a manuscript in the Chinese *Journal of Bioinformatics*, titled 'A Furin Cleavage Site Was Discovered in the [Spike] Protein of the 2019 Novel Coronavirus'. Their paper claimed to be the first to report the 'very important mutation'; they too thought it was likely to enhance the infectiousness of the virus.

What is so special about a furin cleavage site? The spike of a coronavirus determines which host species the virus can infect and how many different types of cells in the body the virus can invade. In order for the spike protein to help the virus get inside a cell, it has to be cleaved by cutting at two sites. This sounds like an accident, but it is not. The spike protein is cut, but it does not fall apart. Spikes cluster in groups of three, like a bouquet, and after cleavage, each spike opens out into a new

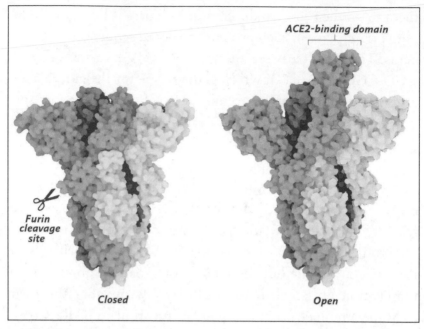

Three spike proteins, each shown in a different shade of grey, fitted together on the surface of a SARS-CoV-2 virus. The closed form (left) can shift to the open form (right) once a cleavage is made by furin at the point indicated. This exposes the ACE2 binding domains, as shown by one of the spike proteins (top right).

configuration like an origami puzzle. One half of each spike, called S1, clutches the entry receptor (ACE2 in the case of SARS-CoV-2), while the other half, S2, entices the cell membrane to fuse with the virus membrane. Cleavage at the boundary between S1 and S2 causes a change in the shape of S2. Cleavage at a second site within the S2 domain then forces the fusion of virus to cell so that the virus genome slips into the cell. (Pause to admire the fact that we live in an age when science can tell us such intricate and beautiful facts about invisibly small entities!)

These cutting events take place on entry into a cell but they also occur as virus particles are being made inside a cell that has already been hijacked, priming the new viruses to invade other cells. Human cells manufacture the little protein scissors called proteases whose day job is to prepare lots of proteins for work by cleaving them. Furin is one such protease, found in the brain, lung, digestive tract, kidney, pancreas and reproductive organs. Each protease recognises a specific sequence in its target proteins at which to make the cut. Some viruses have acquired the ability to exploit this tool, using furin, or other proteases, for their own purposes – to reshape their own proteins. Several other coronaviruses that infect humans attract furin, including MERS, HKU1 and OC43. These are also betacoronaviruses, but they are not in the sarbecovirus subgenus. Other well-known viruses such as avian influenza, HIV and measles also rely on a similar furin-based mechanism.

In May 2020, Drs Markus Hoffmann, Hannah Kleine-Weber and Stefan Pöhlmann in Göttingen, Germany, listed the spike sequences of fifty-five SARS-like viruses, including SARS-CoV-2, five strains of human SARS, two civet strains, one raccoon dog strain, one pangolin version of SARS-CoV-2 and forty-five different bat viruses. There is no sign of a furin cleavage site at the S1/S2 junction in any of them except one: human SARS-CoV-2 stands out like a sore thumb. In this virus's spike, four extra amino acids – PRRA – interrupt the otherwise similar sequence in the S1/S2 junction. Since the next letter is R (standing for arginine), this means there is a sequence reading RRAR in just the right place to attract the attention of furin scissors. In this group, even the bat virus RaTG13 and the pangolin virus that have the spikes most similar to SARS-CoV-2 have no furin cleavage site.

Facts like these have led some scientists to think that the insertion of a sequence with such import into the gene at just the right place is a strong sign that SARS-CoV-2 was deliberately altered in a laboratory. Others furiously disagree, arguing that the virus probably acquired this feature naturally. In April

2021, scientists from the NIH found long inserts had appeared naturally in the genomes of some SARS-CoV-2 variants during the pandemic. They wrote that the furin cleavage site resembled these inserts, although they could not identify a precise mechanism by which it originated in the virus. Both sides of this debate can mount a decent case.

One thing is for certain: furin cleavage is a big reason that SARS-CoV-2 has such pandemic potential. Several groups of scientists have found that when they remove the furin cleavage site from the spike of SARS-CoV-2, the virus replicates much less efficiently in the respiratory tract and causes less severe disease in both hamsters and humanised mice. In SARS-CoV-2 viruses sampled from human beings, mutations of the Rs that define the PRRAR motif have been exceedingly rare: only about one in ten thousand has mutations at one of the first two Rs in the motif, and essentially none have mutations at the last R, which is where the cleavage occurs. This implies that it is a key feature of the virus.

The discovery of furin cleavage sites

The history of this niche field of research goes back thirty years. In 1992, a team at the Institute of Virology in Marburg, Germany, discovered that a type of furin cleaves the haemagglutinin protein on the surface of an avian influenza virus, enabling the virus to get inside cells. A particular sequence of four amino acids at the cleavage site was necessary to attract the attention of furin: R-X-K/R-R, where X can be anything and the third amino acid can be either lysine (K) or arginine (R). Just a few months later, a different group of scientists led by the same senior authors demonstrated that such a sequence motif in HIV also serves as a furin cleavage site that renders the virus infectious.

Since then, spotting furin cleavage motifs in virus proteins has become a bit of a hobby among virologists, a potentially useful way of gauging how dangerous a virus might be. They always begin and end in arginine (R). The cut happens just after the last R. In SARS-CoV-2, the motif is RRAR; in mouse hepatitis virus, it is RRAHR; in bovine coronavirus, RRSRR; in OC43, a common cold coronavirus, RRSR.

Yet other coronaviruses, including SARS, manage to be infectious without furin cleavage at the S1/S2 boundary. The truth is that scientists still do not fully understand what is going on. This is why several labs around the world have been deliberately inserting furin cleavage site sequences into the spike genes of different coronaviruses to see how this changes the virus's ability to infect different types of cells. We will briefly describe five experiments carried out over fifteen years in which scientists manipulated furin cleavage sites in the spike genes of coronaviruses. Our purpose is to show how widespread, indeed fashionable, it has been in virology laboratories to insert or alter these sites in coronaviruses in recent years.

Dr Jack Nunberg, a biologist whose career has included spells at a pharmaceutical firm and two biotech firms before he returned to academia at the University of Montana, carried out the pioneering experiment in 2006. He wanted to see what the spike of the 2003 SARS virus could do if it was given a furin cleavage site at the S1/S2 boundary. It should be stressed that he did not do the experiment with whole viruses, only the spike protein molecules that cannot make more copies of themselves. It was not a dangerous gain-of-function experiment. Dr Nunberg, working with his colleagues, took a SARS spike and made a crucial change to it so that it had a new RRSRR motif at the S1/S2 boundary, markedly enhancing the tendency of an infected host cell to fuse with another potential host cell containing the ACE2 receptor. However, pseudoviruses fitted with these spikes were no better at infecting human kidney cells carrying the ACE2 receptor.

In 2009, a team at Cornell University introduced a furin cleavage site into the SARS spike at the S1/S2 junction, but this time with a second site added where the S2 domain is usually cleaved. They observed a dramatic increase in cell fusion. Again, the experiment did not use whole, infectious viruses so there was no risk of a laboratory escape of an enhanced SARS virus.

In May 2015, a scientific collaboration between the Huazhong Agricultural University in Wuhan and Utrecht University in the Netherlands published a study showing that the creation of a furin cleavage site in the spike of Porcine Epidemic Diarrhea Coronavirus conferred increased ability to enter cells and trigger cell fusion. As the name implies, this virus gives pigs diarrhoea and has recently caused devastating epidemics on farms in Asia and America. This experiment actually used live coronavirus, albeit not one that infects human beings. The sequence was not inserted at the S1/S2 junction but at the secondary S2 cleavage site in the spike gene. The only funding acknowledged was a grant from the Natural Science Foundation of China and two of the experimenters came from Huazhong Agricultural University. By changing just one amino acid in the spike, the scientists rendered the viruses capable of triggering fusion in cells. The authors concluded that introducing furin cleavage sites into a coronavirus in the lab could help a virus to infect more cell types both in a Petri dish and in an animal.

Also in 2015, Drs Shi Zhengli and Ralph Baric were co-authors on a paper describing a similar experiment to those of Dr Nunberg and others. The virus in this case was MERS. Comparing the MERS genome to the genome of HKU4, a MERS-like coronavirus discovered in bats in Hong Kong, Dr Fang Li's University of Minnesota team hypothesised that two tiny differences in the spike gene were responsible for the fact that MERS could cross the species barrier into human beings while HKU4 could not. One of the two differences was an S1/S2 furin cleavage site that is present in the MERS spike but not that of HKU4. Using a pseudovirus, they made just two small changes in the sequence of the MERS spike gene, which rendered MERS largely incapable of entering human cells because it could no longer be cleaved by a protease. They also engineered the HKU4 spike so that it now had the two cleavage sites, including a novel S1/S2 furin cleavage site. The result was pseudoviruses more capable of infecting different types of human cells. Once again, the risk of this work was very low. The pseudoviruses had no coronavirus genome inside so they could not generate more live coronaviruses even after infecting cells. Interestingly, the scientists also found that bat cells had a different protease system from human cells – perhaps explaining why some of these cleavage sites differ in bat viruses and human viruses, although some bat coronaviruses do possess furin cleavage sites in their spikes. The conclusion of the study was that 'the two functional human protease motifs in MERS-CoV spike played a critical role in the bat-to-human transmission of MERS-CoV'. It was possible therefore that just two mutations would be necessary to turn HKU4, a relatively harmless bat virus found in China, into a pathogen that could take its first steps into human beings.

We need to recount one more furin cleavage site experiment to bring the story up to date. In October 2019, the Key Laboratory of Animal Epidemiology of the Ministry of Agriculture in Beijing published the results of an experiment on infectious bronchitis virus. This disease of chickens was the earliest coronavirus disease to be identified, back in the 1930s in America, decades before the name 'coronavirus' was coined. The experiment showed that putting a furin cleavage site into the S2 domain resulted in a more lethal virus and one that could damage the blood-brain barrier and infect brain cells to cause encephalitis. The scientists feared that such a sequence could evolve naturally in infectious bronchitis virus and result in terrible losses in the poultry industry.

It is safe to say that by 2019 the practice of artificially introducing or removing furin cleavage sites in the spike genes of coronaviruses, or their equivalents in other viruses, had become a routine experiment in virology.

CGG CGG

On 6 May 2021, in a lengthy online essay, the veteran science journalist Nicholas Wade, former deputy editor of *Nature* and science writer on the *New York Times*, quoted the eminent virologist Dr David Baltimore as saying, 'When I first saw the furin cleavage site in the [SARS-CoV-2] viral sequence, with its arginine codons, I said to my wife it was the smoking gun for the origin of the virus … These features make a powerful challenge to the idea of a natural origin for SARS2.' This sent a tremor through the scientific community, which until that point had largely decried any notion of a laboratory origin of the virus as deeply unlikely or even a conspiracy theory.

Dr Baltimore won the Nobel Prize in Physiology or Medicine in 1975 for his discovery that RNA tumour viruses, specifically retroviruses, can insert their genomes into the DNA genomes of host cells, using a 'reverse transcriptase' enzyme that makes a DNA copy of the RNA virus genome. His biography includes the presidency of both the California Institute of Technology and Rockefeller University. He won the US National Medal of Science. So, although other less well-known scientists had suggested that the furin cleavage site in SARS-CoV-2 was possibly a product of genetic engineering, Dr Baltimore's statement gave this conjecture much more weight. Later Dr Baltimore would partly retract his comment, telling the *Los Angeles Times* he 'should have softened the phrase "smoking gun" because I don't believe that it proves the origin of the furin cleavage site but it does sound that way. I believe that the question of whether the sequence was put in naturally or by molecular manipulation is very hard to determine but I wouldn't rule out either origin.' He also told *Caltech Weekly* that 'the fact that evolution might have been able to generate SARS-CoV-2 doesn't mean that that's how it came about. I think we very much need to find out what was happening in the Wuhan Institute of Virology.'

'The SARS-CoV-2 furin cleavage site is yet again in the news – this time because of a quote by Nobel laureate David Baltimore,' tweeted Dr Kristian Andersen from the Scripps Research Institute in La Jolla, California, on 9 May 2021. Dr Andersen had been the lead author on a highly influential article, 'The Proximal Origin of SARS-CoV-2' published in *Nature Medicine* in March 2020, which had concluded that: 'Our analyses clearly show that SARS-CoV-2 is not a laboratory construct or a purposefully manipulated virus.' At the time and even till today, the 'Proximal Origin' paper has been cited by many other scientists to rule out all laboratory-based hypotheses. Indeed Dr Andersen had been quoted in the Scripps press release as saying that two features of the virus 'rule out laboratory manipulation as a potential origin for SARS-CoV-2'. It therefore came as a surprise in early June 2021 when FOI'-ed emails were published showing Dr Andersen telling Dr Anthony Fauci, the head of NIAID, on 31 January 2020 that 'the unusual features of the virus make up a really small part of the genome (<0.1%) so one has to look really closely at all the sequences to see that some of the features (potentially) look engineered' and that his discussions with other experts found the genome to be 'inconsistent with expectations from evolutionary theory'. Dr Andersen explained later that it was a normal part of the scientific process to begin with doubts and then assuage them. As more related virus genomes emerged, their team judged that SARS-CoV-2 had evolved naturally. We will return to this developing story on the early 2020 conversation between Dr Andersen and other prominent scientists later in the book.

The basis on which Dr Andersen and his colleagues publicly rejected a laboratory origin was that any genetic engineer would have, first, designed a different-looking spike receptor-binding domain, based on existing data, and, second, used a known genetic backbone. Mr Wade thought both arguments were weak yet 'his conclusion, grounded in nothing but two inconclusive speculations, convinced the world's press that SARS2 could not have escaped from a lab'. In response, Dr Andersen pointed out Mr Wade's 'troubled history of misrepresenting (and/or misunderstanding) the very basics of evolutionary biology', a reference to Wade's contentious 2014 book *A Troubling Inheritance: Genes, Race and Human History*, which was criticised by numerous scientists who co-signed a letter arguing that 'there is no support from the field of population genetics for Wade's conjectures'. Wade was also co-author of *Betrayers of the Truth: Fraud and Deceit in the Halls of Science*, a 1982 book about scandals in science and their cover-ups. One of the whistleblowers, whom the book described as exposing a scandal in immunology involving his supervisor, was Dr Steven Quay, who also authored a lengthy essay in 2021 arguing that SARS-CoV-2 originated in a laboratory. This is a reminder that until the middle of 2021, the story of the search for the origin of Covid-19 predominantly relied on outsiders and scientists or journalists who were willing to risk their reputation and challenge the scientific consensus.

What are these arginine codons that made Dr Baltimore think that the furin cleavage site is a smoking gun for genetic manipulation? In the universal genetic code for all organisms – the dictionary that translates DNA or RNA language into protein language – there are six different words encoding the amino acid 'arginine' (R): AGA, AGG, CGG, CGC, CGA and CGT (the last being CGU in RNA). These three-letter words are called codons. Within the genome of SARS-CoV-2, the commonest codon for arginine is AGA and the rarest is CGG. Yet the furin cleavage site has two of the latter in tandem: CGG-CGG. In contrast to viruses, human and animal cells use CGG rather more frequently. In 2004, a mainly Boston-based team of scientists with collaborators in China altered the spike gene of the 2003 SARS virus to use codons preferred by human cells – they call this process 'codon optimisation' – and found that the codon-optimised spike was much more abundantly produced by the human host cell. So it is known that optimising the codons in a SARS-like virus may improve its ability to infect specific host cells. What was the furin cleavage site in SARS-CoV-2 doing with not just one but two codons that are rare in coronaviruses but preferred in human or animal cells? Was the tandem CGG-CGG a sign that the furin cleavage site had been genetically engineered into SARS-CoV-2 in a similar manner to the other experiments where they had been introduced into other viruses? One of the earliest to raise this question was the Russian-Canadian biotech entrepreneur Yuri Deigin, who, like Dr Baltimore, saw this rare appearance of a CGG doublet as a clue that SARS-CoV-2 may have had its furin cleavage site artificially inserted in the lab.

Yuri Deigin is a founding member of Drastic, under its original name of 'Daszak's fan club' – because they had all been blocked on Twitter by Dr Daszak. (The name was changed when Billy Bostickson joined.) Mr Deigin got his bachelor's

degree in computer science and mathematics from the
University of Toronto, followed by an MBA from Columbia
Business School in 2010. After overseeing the R&D and clinical
trials at a biotech start-up for six years, Mr Deigin founded
Youthereum Genetics in 2017 to explore the use of gene therapy to reprogramme cells to fight the ageing process. The field
of ageing science, with its many snake-oil theories, gave Mr
Deigin an acute ability to detect when scientists are making
weak arguments, he said. On a Russian Facebook debate club,
in March 2020 he began to examine the arguments in Dr
Andersen's 'Proximal Origin' paper. Initially, Mr Deigin was
inclined to support a natural origin and assumed that the WIV
was in Wuhan because it was where bat coronaviruses were
found, but the more he looked into the issue, the more his
doubts grew and the more he realised that the lab origin
hypothesis was not a conspiracy theory. In particular, he
concluded that the points put forward by Dr Kristian Andersen
reminded him of the story of the emperor who had no clothes.
To make sure you really understand something, give a talk or
write about it, Mr Deigin thought, so he wrote what was to
become an influential essay first in Russian and then in English
on the *Medium* website. This essay was one of the catalysts for
the formation of Drastic and signalled the start of a continuous
guerrilla campaign on social media about furin cleavage sites
and other aspects of the virus genome. Dr Rossana Segreto then
approached Mr Deigin to collaborate on a scientific paper
analysing the Andersen et al. 'Proximal Origin' article and
setting out the argument for the laboratory-leak hypothesis,
which was ultimately published in November 2020 under the
title 'The Genetic Structure of SARS-CoV-2 Does Not Rule Out
a Laboratory Origin'. Later Mr Deigin continued to debate
with people on Facebook and Twitter, saying that 'Nothing
moves science forward like the right scientific opponent.' He

has recently collaborated on a project to search all sarbecoviruses for CGG-CGG doublets and found none, except for the one in the furin cleavage site of SARS-CoV-2.

Against this, the frequency at which CGG appears on its own in SARS-CoV-2 overall is comparable to that of other closely related coronaviruses. CGG is used at a 3 per cent frequency in SARS-CoV-2 compared with 5 per cent in the 2003 SARS virus. In Dr Andersen's words, 'Nothing unusual here'. And the probability of a CGG doublet appearing depends on what sequence it had mutated from – was a single letter mutation required or more? Dr Andersen highlighted a feline coronavirus where a furin cleavage site shared the same 'PRRAR' motif and the first two Rs were encoded by CGG CGA, just one RNA letter different. Dr Andersen's conclusion was firm. On 9 May 2021, he tweeted: 'Baltimore's quote *is* shocking – however, not because it's true, but because it's wrong. There's nothing mysterious about the FCS or the codons – anybody who'd care to take a close look at the data would realize this. That, admittedly, requires a little more than his "first" look.'

Our view is more agnostic. Yes, it is true that virologists altering the sequences of viruses to make them more compatible with human cells are more likely to use CGG codons for arginine than nature does, which makes the CGG-CGG doublet in the furin cleavage site of SARS-CoV-2 suspicious at least. But the argument is suggestive, rather than conclusive, and nature is clearly capable of using these codons. We think – as Alina said on Twitter in May 2020 – 'there is zero evidence that confirms that the SARS-CoV-2 S1/S2 PRRA(R) FCS arose naturally or artificially, but neither scenario can be ruled out.' Dr Nunberg likewise was quoted in June 2020 as saying, 'There is no way to know whether humans or nature inserted the site.'

Natural insertions

In their March 2020 'Proximal Origin' paper, Dr Andersen's group had confidently predicted that a bat virus with a similar furin cleavage site to SARS-CoV-2's would soon turn up: 'Given the level of genetic variation in the spike, it is likely that SARS-CoV-2-like viruses with partial or full polybasic cleavage sites will be discovered in other species.' Shortly afterwards, a bat virus was found and given a lot of publicity for at least seeming to prove that it did appear to possibly have a natural insertion at the S1/S2 boundary of its spike, although not one that acted as a furin cleavage site. It was a sarbecovirus, in which some parts of its genome were very closely related to their counterparts in SARS-CoV-2.

Its name is RmYN02. Rm stands for *Rhinolophus malayanus*, the Malayan horseshoe bat; YN stands for Yunnan. The story behind RmYN02 is that between May and October 2019, scientists from the Key Laboratory of Etiology and Epidemiology of Emerging Infectious Diseases in Shandong and the Center for Integrative Conservation at the Xishuangbanna Tropical Botanical Garden caught 227 bats of twenty species in Mengla County in the far south of Yunnan, very near the border with Laos. This is yet another team of scientists that was catching bats in southern China in search of viruses. This far south in China, as well as two widespread species of horseshoe bat, *R. sinicus* and *pearsonii*, they found horseshoe bat species that are more typical of Indochina and not found further north, including *malayanus*. The bats were sampled for viruses and the samples sequenced and analysed. Different tissue and faecal samples were merged into pools for sequencing, with each pool containing as many as eleven samples. Although the approach was cost effective, this meant that when the team found signs of a SARS-like virus, which they dubbed RmYN01 and

RmYN02, they had to then figure out which of the eleven samples in the pool contained these two viruses. Only one of the samples, number 123, collected on 25 June 2019, tested positive for both sequences. From this sample, the scientists were only able to verify parts of the RmYN02 genomic sequence but not that the sequenced parts were necessarily from the same virus. RmYN02 looks very similar to SARS-CoV-2 over most of its genome but its spike gene is very different – only a 72 per cent genetic match. Could the spike gene or part of it have come from some other virus lurking in the same sample?

It is in this distantly related spike of RmYN02 that another sequence, 'PAA', is found, which the Shandong scientists called a natural insertion similar to that of 'PRRA' in SARS-CoV-2. For this reason, they called the similarity 'strongly suggestive of a natural zoonotic origin of SARS-CoV-2'. However, upon a closer look, this S1/S2 segment of RmYN02's genome is shorter even compared with most of the close relatives to SARS-CoV-2, let alone SARS-CoV-2 with its four amino acid insertion. As Mr Deigin and Dr Segreto argued in a December 2020 preprint (later published in *BioEssays* in May 2021), 'to support the claimed PAA insertion not only a 9-nucleotide insertion, but also a 15-nucleotide deletion must have occurred ... the claimed PAA insertion is more likely to be the result of mutations'. In other words, RmYN02's spike is so different from that of SARS-CoV-2 and other closely related viruses that it is difficult to be certain that the 'PAA' even constitutes a natural insertion in RmYN02.

Note how ambiguous genetic code can be at this level. Text written in three-letter words with no gaps between words gets garbled if a single letter is removed, and you have to guess how to read it. We are like secret agents trying to decipher coded messages over a faint radio from a spy working at a missile

```
                    SARS-CoV-2  A S Y Q T Q T N S P R R A R S V A S Q S   I I A Y T M S L G A E N S V A Y
             Yunnan Bat RaTG13  A S Y Q T Q T N S       R S V A S Q S     I I A Y T M S L G A E N S V A Y
         Cambodia Bat RShSTT200 A S Y Q T Q T N S       R S V T S Q S     I I A Y T M S L G A E N S V A Y
      Guangdong Pangolin MP789  A S Y Q T Q T N S       R S V S S Q A     I I A Y T M S L G A E N S V A Y
               Zhejiang Bat ZC45 A S Y H T A S I L      R S T S Q K A     I I A Y T M S L G A E N S I A Y
                 Japan Bat Rc-o319 A T Y H T P S M L    R S A N N N K R   I V A Y V M S L G A E N S V A Y
              Yunnan Bat RmYN02  A S Y N S P A A        R   V G T N S     I I A Y A M S I G A E S S I A Y
          Thailand Bat RacCS203  A S Y N S P V A        R   V G T N S     I I A Y E M S I G A E S S I A Y
```

```
                              Y   Q   T   Q   T   N   S   P   R   R   A   R   S   V   A   S   Q   S
               SARS-CoV-2  tat cag act cag act aat tct cct cgg cgg gca cgt agt gta gct agt caa tcc
        Yunnan Bat RaTG13  tat cag act caa act aat tca             cgt agt gta gcc agt caa tct
   Cambodia Bat RShSTT200  tac cag act caa act aat tca             cgt agt gta acc agt caa tcc
Guangdong Pangolin MP789   tat cag act caa act aat tca             cgt agt gtt tca agt caa gct
        Zhejiang Bat ZC45  tac cat acg gct tct ata tta             cgc agt aca agc cag aaa gct
          Japan Bat Rc-o319 tat cac acg cca tct atg cta            cgt agc gca aac aac aat aag
        Yunnan Bat RmYN02  tac aac tca cct gca gcg                 cgt gta ggt act aat tcc att
     Thailand Bat RacCS203 tat aac tca cct gta gca                 cgt gta ggt act aat tct att
```

An alignment of the S1/S2 region of the spike gene of SARS-CoV-2 as compared to other closely related sarbecoviruses listed in descending order of approximate similarity: protein amino acid sequence (top) and RNA nucleotide sequence (bottom). The PRRA (cct-cgg-cgg-gca) insertion is shown in bold. The furin cleavage site is underlined. Letters that deviate from the SARS-CoV-2 sequence are in the lightest shade of grey.

base. The spy has sent a message reading 'Saw the cat and red fox' but we only received 'Saw tec ata nre dfo x'. If we are clever, we spot that an 'h' and a 'd' are missing. But we might come up with a different interpretation 'Saw ten cat and red fox'. 'Cat' might be code for 'missile', say, and we may be badly misled as to how many missiles the spy has seen on the base. Thus we cannot know for sure if RmYN02 has deletions, insertions or both. As for the significance of the text itself, true, 'PAA' looks a bit like 'PRRA' but it does not have any of the Rs that are critical to the creation of a furin cleavage site.

As the pandemic continued, scientists from different countries went back to their freezers and dug out samples from horseshoe bats to test them for coronaviruses. In December 2020, scientists from the University of Tokyo published the genome of a SARS-CoV-2-like virus, Rco319, from a horseshoe bat of the species

Rhinolophus cornutus, captured in 2013 in a cave in Iwate Prefecture in Japan. Its genome was 81 per cent similar to that of SARS-CoV-2. It had no S1/S2 furin cleavage site insertion.

In early 2021, French and Cambodian scientists found two nearly identical viruses, RshTT200 and RshTT182, in two *Rhinolophus shameli* horseshoe bats. The samples had been collected in a cave in Stung Treng province in Cambodia in 2010, and stored at the Institut Pasteur du Cambodge until being tested after the pandemic began. These were 93 per cent similar to SARS-CoV-2, so not as close as RaTG13, although in some short sections they were closer. Their spike proteins had receptor-binding domains that were slightly less similar to SARS-CoV-2 than RaTG13's. And they also had no S1/S2 furin cleavage site insertion.

The next new virus to be reported was in Thailand. In June 2020, a team of scientists visited a colony of three hundred horseshoe bats that were living in a large irrigation pipe in a wildlife sanctuary in Chachoengsao province in eastern Thailand. They captured a hundred of the bats and identified them as belonging to the species *Rhinolophus acuminatus*, one not reported to be found in China. Viruses were detected in thirteen of the bats' rectal swabs, yielding a single sequence that resembled RaTG13. Full genome sequencing revealed that this virus, which the scientists named RacCS203, was most similar to RmYN02 at the S1/S2 boundary; it had the letters PVA where RmYN02 had PAA. Still no furin cleavage site insertion. Furthermore, the spike of RacCS203 could not use human ACE2 as an entry receptor to infect cells.

These studies showed that SARS-CoV-2-like viruses are widespread in horseshoe bats in Asia. By the time of writing, seven bat species so far have been found to harbour them: *R. affinis*, *malayanus* and *pusillus* in Yunnan, *sinicus* in Yunnan and Zhejiang, *cornutus* in Japan, *shameli* in Cambodia and *acumi-*

natus in Thailand. No doubt more will be found and in more locations. The same studies also hinted at the mosaic nature of their genomes, showing evidence of recombination. But the fact that none of them was as genetically close to SARS-CoV-2 as RaTG13, which had been found inside a Mojiang mineshaft by the WIV and brought to Wuhan in 2013, hardly helped the cause of those arguing for a natural origin of the virus. Rather, each discovery of a slightly less similar virus than the one from Mojiang strengthened the case for taking the laboratory-leak hypothesis seriously. As for the furin cleavage site, if dozens or even hundreds more bat SARS-like viruses are collected and still none has a furin cleavage site insertion, the perception of SARS-CoV-2 having originated from a laboratory will go up.

It is worth remembering that the furin cleavage site debate is all about whether the virus was manipulated once in a laboratory; it may or may not clarify whether a natural, non-engineered virus was in a laboratory or had infected researchers during fieldwork. Sure, if the furin cleavage site proves to have been inserted artificially, it confirms that the virus was in a laboratory and was altered. But if, on the other hand, the furin cleavage site proves to be natural, it still says nothing about where the virus came from. A natural bat virus with a natural furin cleavage site could still have leaked from a lab. Indeed, its possession of such an aid to infectivity might be precisely what made it challenging to contain in a BSL-2 or BSL-3 laboratory. The same logic applies to a natural infection: a virus that possessed an advantageous furin cleavage site rather than one less well-endowed would be more likely to spark a natural outbreak once it had spilled over into humans. Almost by definition, a virus that starts a pandemic must have been especially infectious. So finding a natural origin of the furin cleavage site will not clear up the question of whether the virus first jumped into people in the wild or in the course of research activities.

The dog that didn't bark in the night

The first published paper to analyse and discuss the newly discovered genome sequence of SARS-CoV-2 in great detail was the one authored by Drs Shi Zhengli, Peng Zhou, Ben Hu and twenty-six other colleagues in Wuhan. Several of the authors were experienced virologists who specialised in coronaviruses. This was posted as a preprint on 23 January 2020 and published in *Nature* on 3 February. In their discussion of the novel coronavirus genome, they zeroed in on the spike, noting that 'the major differences in the sequence of the [spike] gene of 2019-nCoV are the three short insertions in the N-terminal domain as well as changes in four out of five of the key residues in the receptor-binding motif compared with the sequence of SARS-CoV'. They paid careful attention to other features and insertions in the spike gene sequence but did not even once mention the feature that stands out like a sore thumb: the never-seen-in-a-sarbecovirus-before furin cleavage site insertion. We were not the only scientists to be surprised by this omission. Even without experiments to prove it, this unique feature would have been predicted to affect the infectiousness of the virus and its ability to hijack different host species and cell types. Indeed, a figure in the paper's 'extended data' shows the amino-acid sequence of the S1 section of the spike alongside those of six other viruses, but it stops at position 675, just short of the furin cleavage site (positions 681–685).

It is one of the strangest omissions in a scientific paper. The sarbecovirus specialists were clearly paying extremely close attention to this part of the genome, but the most remarkable feature of all escaped their attention. Not one of them appears to have said, when reading a draft, are you sure you are not leaving out the most interesting bit? It is as if you discover a unicorn and you compare it with other horses, describing in detail the hair and the hooves, but you don't mention the horn. This was also the paper that first mentioned RaTG13 by its new name and did not connect it to the 4991 SARS-like virus sequence published in 2016 or to the mysterious pneumonia cases in 2012 that had spurred Chinese research teams to scour the Mojiang mine for viruses. If this were the plot of a novel, the reader would think something was up.

A week after the preprint had been posted, another paper was published online, on 31 January 2020, in the journal *Emerging Microbes & Infections*. This time it had just three authors, of which only one was from the WIV: Dr Shi. In this paper, the authors noted that the genome of the novel corona-virus was 89 per cent similar to the two Zhoushan bat viruses (ZC45 and ZXC21) but did not mention RaTG13. The authors included a diagram showing the location of the S1/S2 junction in the spike. Yet once again they made no mention of the novel furin cleavage site insertion that is absent from ZC45 and ZXC21, and all other sarbecoviruses, yet present in SARS-CoV-2, or 2019-nCoV as it was then known. These three authors – Drs Shibo Jiang, Lanying Du and Shi Zhengli – had been part of the 2015 MERS study described earlier in this chapter, in which an S1/S2 furin cleavage site had been intro-duced and found to enhance the capabilities of a MERS-like coronavirus spike. Despite saying then that such cleavage sites 'played critical roles in the bat-to-human transmission of MERS-CoV, either directly or through intermediate hosts', in 2020 they did not even mention the uniqueness of the furin cleavage site insertion in SARS-CoV-2. The only thing they said was: 'By aligning 2019-nCoV S protein sequence with those of SARS-CoV and several bat-SL-CoVs, we predicted that the cleavage site for generating S1 and S2 subunits is located at R694/S695.' For context, the 2003 SARS virus also has a cleav-age site at this location. What is novel about SARS-CoV-2 is the insertion directly upstream of the R forming a PRRA(R) furin cleavage site. Even having inspected this very site in the spike, the three scientists said not a word about how the furin cleavage site insertion was missing from the 2003 SARS and bat SARS-like viruses but present in SARS-CoV-2 – a fact that, in our view, should have rung alarm bells about the virulence and infectivity of the new virus. These two papers by Dr Shi were submitted to the journals on the day that the Chinese authorities conceded to the world that human-to-human transmission was happening.

Thus it was left to another Chinese team and a team of French and Canadian scientists, as mentioned at the beginning of this chapter, to publicly point out the obvious in January and February 2020 respectively.

How could the Wuhan scientists have missed this out-of-place feature, which was not only obvious but, as both other papers' comments made clear, ominous? Could these experienced coronavirus researchers really have missed such a critical discovery in their careful characterisation of the virus's genome? Even after recently publishing work on introducing an S1/S2 furin cleavage site in the spike of a MERS-like virus? Or did they see the furin cleavage site but decide against drawing attention to it? These are questions we would love to put to Dr Shi and her colleagues if they would respond to emails.

Some scientists have pointed out that two other early papers by Chinese virologists describing SARS-CoV-2's genome had also missed the furin cleavage site insertion. However, these two other papers had not been looking at the S1/S2 region. One was the paper that had been submitted to *Nature* on 7 January 2020 by Dr Zhang Yongzhen's group, describing the first SARS-CoV-2 genome to be made public. Remember that his team had only obtained the sequence two days before submitting their manuscript to the journal, and Dr Edward Holmes who was an author on the paper only had the genome for about an hour before posting it online on 11 January. Their paper had zoned right in on the spike receptor-binding domain, which continues to be the region of greatest interest in the SARS-CoV-2 genome – many of the variants of concern are defined by mutations in this area. The second paper appeared to perform a cursory analysis of the spike, pointing out that the novel coronavirus spike had 'only a few minor insertions or deletions' and was 'longer' compared to closely related virus spikes.

What we do know is that, for more than a decade, scientists had been doing experiments deliberately designed to insert or delete furin cleavage sites with a view to seeing whether they made viruses more or less able to invade various types of host cells under different conditions. We do not know if the WIV or another laboratory in Wuhan was engaging in similar work using bat coronaviruses, but the omission of the furin cleavage site in two keystone Covid-19 papers of the WIV is curious to say the least.

In our view, it is this fact, more than anything else, that lends this funny little genetic insertion its walk-on part in the Covid-19 play. Given that Dr Shi's group had made chimeric SARS viruses and Dr Shi and co-authors had recently collaborated on a project studying parallel sites in MERS-like viruses, their silence on the unique furin cleavage site with critical implications when they published the first sarbecovirus genome is the dog that did not bark in the night-time.

10.

The other eight

'Even if all parts of a problem seem to fit
together like the pieces of a jigsaw puzzle, one
has to remember that the probable need not
necessarily be the truth and the truth not
always probable.'

SIGMUND FREUD

In January 2020, when the WIV scientists published their first
manuscript about SARS-CoV-2, revealing that the most closely
related virus, RaTG13, had been sequenced in their laboratory,
many readers of the paper began to ponder the possibility that
the pandemic might have begun with a lab leak. Such suspicions deepened with the revelation, in the addendum to the
paper published the following November, that Dr Shi's group
had been studying at least eight other viruses very closely
related to SARS-CoV-2 – and that these eight had also been
collected from the same mineshaft in Mojiang where workers
had sickened with a mysterious respiratory disease in 2012. It
was almost a year since SARS-CoV-2 had been detected, and in
that time the Wuhan scientists had not once mentioned these
viruses. Even in the addendum they did not describe them,

share their sequences or say what they had done with them. As we were writing this book, in the middle of 2021, they had still published very little about these viral cousins of the cause of the pandemic.

The addendum, as readers may recall (see Chapter 1), had cleared up quite a lot. It at last confirmed the internet sleuths' findings about the illness of the Mojiang miners, the WIV scientists writing 'we suspected that the patients had been infected by an unknown virus'. It confirmed that thirteen blood samples from four of the patients had been sent to the WIV, but whether they tested positive for antibodies to SARS was unclear. The addendum said no; the 2016 doctoral thesis from the Chinese CDC director's laboratory said yes, and the 2013 medical thesis said the WIV had found virus antibodies in the patient samples, concluding a probable infection of the miners by a SARS-like virus from bats. The addendum also corroborated the internet sleuths' finding that the virus labelled 4991 had been renamed RaTG13, to 'reflect the bat species, the location, and the sampling year'. It confirmed that 4991 had been described in a 2016 publication and deposited in the GenBank online database in 2016, although neither were cited in the original *Nature* paper – making it challenging for outsiders to know if the new RaTG13 sequence and the old 4991 sequence had come from the same sample. And finally, once again validating the findings of the internet sleuths, the WIV scientists confirmed that RaTG13 had actually been sequenced in 2018 and not after SARS-CoV-2 had emerged in Wuhan as their paper had implied. Even the admission of the existence of eight other viruses was not entirely surprising: some of the internet sleuths had already made informed guesses as to their existence and identities. So the addendum had not added much that was truly new – it had confirmed the predictions of internet sleuths and concerned scientists.

Genomic jigsaw puzzles

Thanks to hints in genomic databases, some clues to the identity of the eight coronaviruses had emerged well before the addendum appeared. A digression is necessary here into how genetic sequences are assembled. Think of a bat faecal sample as a huge box in which the pieces of more than one jigsaw puzzle are mixed. These days, to extract the genome sequences of viruses from such samples, a scientist would employ a process called next-generation sequencing (NGS). This generates hundreds of thousands of mixed puzzle pieces, called 'reads'. These reads have to be carefully pieced together to see which puzzles (genomes) can be more or less completed. Unless you have a great deal of a particular virus in your bat faecal sample, there will often be gaping holes and regions of uncertainty that need checking. The problem is even worse, you can imagine, if the box contains several puzzles that are largely similar to each other, equivalent to several different viruses in the same sample: which pieces belong to which puzzle?

One way to fill these holes is to generate more puzzle pieces from your sample using another approach that specifically seeks out sequences that are close to the spot in the genome where the gap is. This is called amplicon sequencing. (Amplicon sequencing can also be performed if scientists want to look at particular genes, and is not necessarily evidence of full genome sequencing and assembly.) To explain what this means, consider that the way genetic sequences are read is by first rapidly copying lengthy fragments of DNA in a chain-reaction fashion: each new copy serving as a template for more copies, and so on. This generates many copies of the same fragment, effectively amplifying its content so that it can be reliably read by the sequencing machine to generate a letter-by-letter sequence of nucleotides, the basic building blocks of DNA and RNA. The amplified sequence of each fragment is therefore known as an amplicon. The amplicons should overlap with other known pieces of the genome, allowing the reads to be assembled to give a complete end-to-end sequence of the whole genome.

Many scientists are fastidious and like to check whether they can assemble a published genome by starting with the mess of puzzle pieces – using both the NGS and amplicon data. In other words, can they reproduce the final genome from scratch without knowing what the puzzle should look like? Such careful researchers are not satisfied by being simply told that someone else has already correctly assembled the genome so there is no need for anyone else to try their hand at it. When the RaTG13 genome was published, some scientists who were trying to reproduce the assembly discovered that only the NGS reads from the sample had been published, but not the amplicon reads. Alina confirmed that at least one US scientist had reached out to the WIV, asking for the amplicon reads in April 2020, to verify the genome sequence of this closest relative to SARS-CoV-2.

On 19 May 2020, the amplicon sequences were deposited into the GenBank database without any fanfare. Internet sleuths soon discovered these new reads and got down to analysing them, sharing their insights on Twitter. The first thing that popped out was the dates attached to the reads. Some had been obtained in June 2017, others in September and October 2018. This was odd at the time (before the addendum) because the wording of the *Nature* paper had implied, and been taken by many to mean, that the sequencing of the RaTG13 genome had occurred after the outbreak of Covid-19. The key sentences read: 'We then found that a short region of RNA-dependent RNA polymerase (RdRp) from a bat coronavirus (BatCoV RaTG13) – which was previously detected in *Rhinolophus affinis* from Yunnan province – showed high sequence identity to 2019-nCoV. We carried out full-length sequencing on this RNA sample.' This impression was confirmed as late as July 2020, when Dr Peter Daszak – who, remember, had been a long-time collaborator and funder of the virus-hunting work at the WIV – told the *Sunday Times* that the WIV scientists 'went back to that sample in 2020, in early January or maybe even at the end of last year, I don't know. They tried to get full genome sequencing, which is important to find out the whole diversity of the viral genome … I think they tried to culture it but they were unable to, so that sample, I think, has gone.' On 5 July 2020, Alina pointed out Dr Daszak's mistake in the *Sunday Times* on Twitter: 'I think Daszak was misinformed because the amplicon sequencing data on NCBI clearly shows that the WIV accessed the sample repeatedly in 2017 and 2018 … It wasn't just sitting forgotten in a freezer for 6 years.'

Alerted by the tweets among the internet sleuths – namely Francisco de Ribera and Babarlelephant, who we will introduce shortly – Alina now also analysed the amplicon sequences for RaTG13. She noticed that the bulk of these reads, coincidentally or not, matched the gaps in the genome (the holes in the puzzle) that had been assembled based on the NGS reads. The short 4991 fragment had been sequenced and published in 2016. Was it possible that this sample had been chosen for next-generation sequencing in the following years, and that the amplicon sequencing was done to patch the holes in the genome?

Later that same month, Dr Shi revealed the sample history of RaTG13 in an interview with *Science* magazine: 'As the sample was used many times for the purpose of viral nucleic acid extraction, there was no more sample after we finished genome sequencing, and we did not do virus isolation and other studies on it.' And, contrary to the impression that readers had received from her *Nature* paper, 'in 2018, as the NGS sequencing technology and capability in our lab was improved, we did further sequencing of the virus using our remaining samples, and obtained the full-length genome sequence of RaTG13 except the 15 nucleotides at the 5' end.' (A convention in genetics is that sequences are usually written in one direction, from the so-called 5-prime end to the 3-prime end, designated 5' and 3', the terms referring to different atoms in the pentagonal structure of sugar molecules.)

Why, if Dr Shi had a full genome for RaTG13 all along, did she not mention it in the paper, but instead first focus on the short fragment? As for whether the sample was 'gone', in Dr Daszak's words, one very small section of the genome, fifteen letters long at the front (5') end, was not uploaded until 13 October 2020. How did they obtain this new sequence data if the sample had been used up?

Ribera's big sudoku

One thing about the May upload of RaTG13 amplicon data made the hairs stand up on the back of the neck of one careful observer. The amplicons sequenced in 2017 and 2018 had various labels. Many of the reads from 2017 and 2018 are labelled 'RaTG13'. However, some of the earliest reads from June 2017 were labelled '7896'. What did this mean?

Here began a trail of clues leading to further revelations. It started with a tweet on 2 July 2020 from a Spaniard called Francisco de Ribera, tweeting under the name @franciscodeasis and who is a key member of the Drastic team. Who is Ribera? A forty-year old resident of the Chamberí district of the Spanish capital, Madrid, Ribera lost his job as a technology consultant on 24 March 2020 at the start of the pandemic. Like a lot of people he found he had time on his hands and decided to put it to good use. Drawing on his savings, he resumed work on his

PhD in economics at Comillas Pontifical University, to add to his bachelor's degree in industrial engineering and his master's degree in business. He also began to dabble in modelling the data about the spread of the virus during the first wave. Handling data came naturally to Ribera who has worked for banks and investment managers and proved a fiend when it came to Microsoft Excel. On 13 April, he read a CNN news article that said China was restricting academic publications on the origins of SARS-CoV-2. Around the same time, he read Matt's *Wall Street Journal* essay on 'the bats behind the pandemic' and was struck by the sentence: 'Unless other evidence emerges, it thus looks like a horrible coincidence that China's Institute of Virology, a high-security laboratory where human cells were being experimentally infected with bat viruses, happens to be in Wuhan, the origin of today's pandemic.' The possibility that the virus might have leaked from that laboratory intrigued Ribera and he began to follow the trail of breadcrumbs online. He knew that an old trick used by company auditors is to pay attention to serial numbers on invoices: a missing number may indicate a missing document. So he started to painstakingly assemble a huge spreadsheet, which he called his 'big sudoku', that lists all that is known about every virus sample referred to by the WIV scientists in papers, seminars and genetic databases. By July he had an unrivalled knowledge of the identification data on bat coronaviruses.

Ribera's crucial 2 July tweet is not immediately understandable to those not following the details: 'One of the inconsistencies of the GenBank upload of Latinne et al. (2020) is precisely an ID collision for sample 7896: – *Rhinolophus* bat coronavirus HKU2 isolate 7896 … Bat betacoronavirus isolate 7896 … Could it be somehow related?' Let us explain what he meant. Ignoring the reference to HKU2, which relates to a different issue, Latinne et al. is a paper that appeared as a preprint on 31 May 2020, with Drs Linfa Wang, Peter Daszak and Shi Zhengli as senior authors and Alice Latinne of the EcoHealth Alliance as the lead author. Entitled 'Origin and Cross-Species Transmission of Bat Coronaviruses in China', the Latinne et al. paper was a comprehensive review of the 630 novel bat coronavirus sequences that the WIV-EcoHealth Alliance project had collected from 2010 to 2015 across numerous Chinese provinces, including ninety-seven new SARS-like viruses from thirty-one species of bat. The paper had been submitted on 6 October 2019, before the pandemic began. It was released as a preprint on 31 May 2020 and finally published on 25 August 2020. The data files behind it were processed by GenBank on 7 November 2019, with an embargo until 1 June 2020. Because one of the genes (RdRp) of these viruses had been partially sequenced in each case, the supplementary material included a vast trove of these sequences.

Just as Ribera was getting his head round the data, another Twitter user called Babarlelephant uploaded a tree diagram that he had created from the Latinne et al. data. 'Babar' has retained his anonymity. He is French and has a background in maths and computer science; at the start of the pandemic he knew little biology. He spent several months reading biological papers till he knew how to interpret and create the tree diagrams of gene sequences that are a key tool of genomic analysis. In the tree that Babar created, Ribera spotted a surprise. Nestled in the middle of the SARS-like viruses in the branches of the tree right next to the pangolin viruses was a small group of eight very similar viruses. One of them was numbered 7896, the very number that had appeared among the amplicons of our old friend RaTG13.

'There is a mysterious "7896" label in the reads of RaTG13 that may match with that virus,' said Ribera on Twitter, implying that the virus named 7896 might be somehow connected to

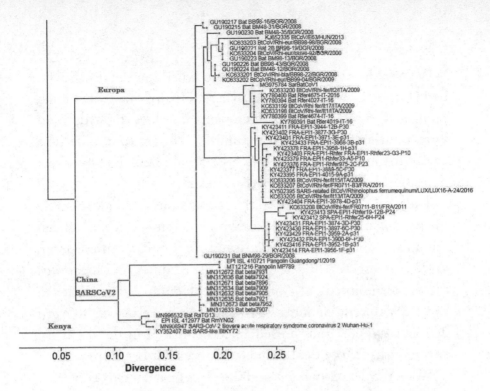

Twitter user @babarlelephant's tree from 7 July 2020.

RaTG13. Finding a run of eight similar numbers – 7896, 7905, 7907, 7909, 7921, 7924, 7931 and 7952 – attached to viruses so close to SARS-CoV-2 in the family tree surely merited further inquiry. On 7 July, he mused that 7896 was 'suspiciously close to RaTG13 in the sarbecovirus tree. Could it be that the 7896 was used somehow as an aid for sequencing RaTG13?' The next day, 8 July, Ribera asked Dr Daszak, one of the senior authors of Latinne et al., to explain: 'Hi Peter, do you know why some reads of RaTG13 have a "7896" label in their filenames? There is a "7896" sample in a clade quite close to 4991 that was recently uploaded (only RdRp) by Latinne et al. (2020). Why there is no study on that new lineage within SL CoVs?' (A clade is a group of creatures and their common biological ancestor.) Dr Daszak's response was to block Ribera on Twitter.

This only made Ribera more intrigued. He began to follow the trail. In his big sudoku of all the samples collected by the

WIV over the years, he had begun to work out details of who had collected each sample and where. He did this by combing published papers and other documents for clues to the dates of virus hunting trips to different sites. Each trip to a mine or a cave corresponded to a group of serial numbers. But the exact dates of the visits were not always clear.

Then, on 8 July, up popped the anonymous and indefatigable internet sleuth called the Seeker with a helpful clue. The official database of the Chinese National Genomics Data Center, known as BigD, included more details of dates and locations, the Seeker pointed out. Sure enough, this told Ribera that the sample with serial number 7896 had been collected on 30 April 2015 in Yunnan and the sequence deposited in the database two months before the Latinne et al. preprint had appeared.

Using the details in BigD, and the clues given in two published papers that provided the dates and locations of some visits, Ribera was able to identify which samples came from which location in Yunnan on which date. On 2 August, he found there was a 'cluster of consecutive samples from 7895 to 7966 (72 samples, 47 positives, 3 co-infections) suggesting a same trip'. It was unlikely to be from the Shitou and Yanzi caves in Jinning, where the bats are mostly *R. sinicus* and where a different set of serial numbers had just been used for a visit. The profile of bat species looked more like those from the mineshaft in Mojiang. From this and other clues, Ribera gradually became all but certain that the eight new viruses came from the copper mine in Mojiang County, where they had been collected two years later than the RaTG13 sample, in the spring of 2015.

Three months after Ribera figured this out and tweeted publicly about it, Dr Shi's group published the addendum to their *Nature* paper in which they confirmed almost all the previous discoveries of the Drastic sleuths and added that eight other viruses had been collected from the same location as

RaTG13. Eight, note. Given what Ribera had already worked out in his big sudoku, that a mysterious group of eight viruses had been collected from bats in the Mojiang mine in 2015, which were very closely related to RaTG13 and to SARS-CoV-2, we could conclude with something approaching certainty that the 7896 group of viruses were the eight referred to in the addendum. We do not know why the addendum finally brought up the eight viruses. It is possible that as a result of Ribera's tweets, the existence of the eight SARS-CoV-2-like viruses from the Mojiang mine could no longer be concealed. The addendum mentioned that in total 1,322 samples had been collected from the Mojiang site between 2012 and 2015, of which 284 were alphacoronaviruses and nine betacoronaviruses. Going through his own spreadsheets, Ribera realised he could pin down 1,320 of them, which gave him confidence that his method was working.

On 3 December 2020, Dr Shi spoke at a webinar organised by the French national academy of medicine and the Academie Veterinaire de France. She showed a slide with eight viruses clearly displayed in a family tree close to SARS-CoV-2, with their serial numbers exactly matching the eight that Ribera had pinpointed from Latinne et al.'s data. This was the first time that the existence of the 7896 group of viruses had been acknowledged publicly, as opposed to buried in a database or mentioned but not named in an addendum. The slide also appeared to suggest that all eight were taken from *Rhinolophus affinis*, the same species of bat as carried RaTG13: they all now had the prefix Ra in front of their serial numbers. This was yet another revelation smuggled out in the most low-key way imaginable. Despite appearing on the screen, the eight were not mentioned in the talk.

Intriguingly, Dr Shi's slide showed that the eight are not identical. Her chart showed 7909 and 7952 to be at least near identical but different from the other six. This differed from the information that had been given in the partial sequences of the eight uploaded with the Latinne et al. paper, and used to make Babarlelephant's tree, which showed 7952 and 7931 to be the ones that stood out from the other six. It was not clear which parts of the genomic sequences were being used to create Dr Shi's tree, but it was probably not the partial RdRp sequences made available by Latinne et al. A bioinformatics consultant, Moreno Colaiacovo, who spotted the slide, pleaded without success on social media for Dr Shi to publish the genome sequences of these eight viruses.

Two weeks later, on 18 December, Dr Shi gave another online lecture at a conference of the European Scientific Working Group on Influenza. She showed the same slide as before but this time she showed another slide that gave away a crucial clue. In a table, two of the eight viruses, 7909 and 7924, were listed as 'Mojiang bat CoVs'. This confirmed what Ribera had worked out: that the 7896 cluster of viruses had been collected in the Mojiang mine. On 24 February 2021, Dr Shi gave another online talk, at a conference about viruses called 'Dangerous Liaisons' organised by Hong Kong University and the Pasteur Research Pole. She showed this very slide again – but this time without the two viruses or the word 'Mojiang'. On 23 March, Dr Shi gave yet another talk in which one of her slides showed a family tree that included the eight viruses. All the viruses on the tree had their location labelled – either the country or province – except for the eight. Now, however, seven of the eight had a different prefix: Rst, meaning *Rhinolophus stheno*. The exception, which was still named Ra for *R. affinis*, was 7909. This was curious because as late as 24 February Dr Shi's slides had used the prefix 'Ra' for all eight. In answer to a question submitted during the seminar, we heard Dr Shi say that she had sequenced their full-length genomes and would publish these soon.

The partial sequences of the eight viruses in Latinne et al. had been embargoed on the GenBank database until June 2020, but they had been uploaded more than seven months earlier on 7 November 2019. Ribera had worked this out from the identification numbers attached to the samples, but it was eventually confirmed to him by GenBank in an email. At the time, the significance of these samples would have seemed small. They were just another set of bat viruses among hundreds and were not especially closely related to the 2003 SARS virus. That month people in Wuhan fell ill with a disease that would in due course make these eight viruses much more relevant. Indeed, in early 2020, these eight viruses would have been the most closely related to SARS-CoV-2, excepting RaTG13. Yet the WIV and EcoHealth Alliance scientists never drew attention to the eight viruses, even as they revised the manuscript of Latinne et al. through the early months of the pandemic. Indeed, looking at a key figure in the Latinne et al. paper that summarised their data, the 7896 cluster juts out right next to SARS-CoV-2, distinct from all other bat viruses the group had uncovered. Despite these eight novel SARS-like viruses being visibly pegged as very closely related to the pandemic virus, the scientists had neglected to label them in the figure or even discuss their provenance in the paper.

Had Babar not taken the trouble to build a family tree of the sequences, the relationship of the eight viruses to SARS-CoV-2 and the Mojiang mine might never have been noticed, as needles in a huge haystack of data. Indeed, it is hard to escape the conclusion that if the WIV-EcoHealth Alliance team had not uploaded these sequences in November 2019, before the pandemic, they might never have published the eight at all. 'They were forced to disclose the 7896 clade and to concede going to the mine until 2015,' said Ribera.

Finally, in May 2021, Dr Shi and colleagues did at last publish in a preprint the full genome of one of the eight, number

7909, simultaneously renaming it RaTG15, adding that the genomes of all eight viruses were near identical. This genome was now reported to be less similar to SARS-CoV-2 than both RaTG13 and RmYN02. The date of collection of the eight viruses was clarified in this preprint to have been 29 May 2015.

Pause here to reflect on the situation. A Spanish business consultant working in his spare time painstakingly worked out, no thanks to Dr Shi and Dr Daszak, that they found eight viruses five years ago that are very closely related to the virus causing the pandemic and brought them more than a thousand kilometres to Wuhan from the Mojiang mine. Yet not only did the scientists fail to announce this potentially crucial information at the start of the outbreak, but when they did slip out the existence of the eight in a tranche of data in June 2020, they failed to draw attention to it. Indeed, by leaving off the words 'SARS-like' in the database descriptions they obscured the viruses' relevance, whether intentionally or not. And then when, in November 2020, they did belatedly and obscurely refer to the eight extra viruses in an addendum, they gave no details and disclosed nothing about how closely related they are to SARS-CoV-2, let alone published their genome sequences. This had to be left to ordinary people to figure out for themselves. Belatedly, eighteen months into the pandemic, they published a genome sequence for the eight viruses.

We repeat for emphasis, because even as we write this, the lack of urgency and transparency is so extraordinary that we can barely believe it: the laboratory had been studying eight viruses (in addition to RaTG13) that are very closely related to a then emerging killer virus and yet did not tell the world important details about the viruses and their connection to the Mojiang miners that could shed light on the origin and characteristics of SARS-CoV-2. Imagine how differently the world would have reacted if, in January 2020, the WIV had announced that the closest relatives of the novel coronavirus came from a place where half of the infected people had died from a mysterious respiratory disease – and that nine closest relatives to SARS-CoV-2 had been under study at the Wuhan Institute of Virology long before the Covid-19 outbreak.

The WIV theses

What put the final pieces of the 7896 jigsaw in place was the discovery of another batch of doctoral and master's theses. A source had shared these theses with the Seeker and Francisco de Ribera, who then worked tirelessly to translate and decode their contents. This time, each of the four theses had been supervised by Dr Shi from the WIV. Two were of particular interest and described the Mojiang mine and the 7896 group.

A Master of Engineering thesis submitted by Ning Wang in May 2014 recounted the immediate follow-up of bat viruses found in the Mojiang mine. It described the cases of the Mojiang miners, albeit the year was incorrect: 'In 2011, Yunnan, Mojiang, three miners died from pneumonia ... the 6 miners were probably infected with pathogens carried by bats.' The thesis confirmed that the scientists investigated the bat-borne viruses in the mine, collecting seventy-seven and ninety-three bat swab samples in August and September of 2012 respectively. Besides obtaining partial RdRp sequences, they also managed to partially sequence the N (nucleocapsid) and S (spike) genes of some of the samples. They found that a whopping 64 per cent of their samples (109 out of 170) were positive for coronavirus. These sequences spoke of a variety of coronaviruses from the Mojiang mine. One was named 4991 and they successfully amplified its N gene and even translated the nucleocapsid protein from it. This had never been revealed by the WIV, not in the *Nature* paper, nor in the addendum.

The other relevant chapter of this 2014 thesis described the development of a method for detecting coronavirus from human samples based on the nucleocapsid protein of each virus – these included, as references, a bat SARS-like virus with 98 per cent nucleocapsid protein similarity to the 2003 SARS virus, the MERS virus and common cold coronaviruses. 'The research method in this paper can also serve as a reference for the study of serological detection methods for novel coronaviruses that are potentially infectious to humans.' Although the 2014 thesis did not produce a fully reliable test method, it revealed that dozens of blood samples had been taken from patients with fever in different provinces, including Yunnan.

Among the individuals thanked by Ning Wang in the thesis was Dr Linfa Wang, for the guidance he had provided during the course of Ning Wang's master's studies. Did Dr Wang know about the Mojiang miners? He told *National Geographic* in June 2020: 'When we have prevented small outbreaks, people don't care. It doesn't get media attention. In Wuhan, if three people died and it was controlled, would we know it? No. This is happening all the time, it's just in remote villages where people die. You bury them and end of the story, right?' Was this an insider reference to the three Mojiang miners who had died in a remote county? In September 2021, it was revealed that Dr Wang had helped with the analysis of the miner cases.

Another thesis, from June 2019, by Yu Ping for a Master of Natural Science in Biochemistry and Molecular Biology was supervised by Dr Cui Jie and Dr Shi Zhengli. This thesis described SARS-like viruses sampled from bats across twenty regions in China between 2011 and 2016 – 2,815 samples from Yunnan (92 SARS-positive); 612 samples from Guangxi (11 SARS-positive); 1,979 samples from Guangdong (33 SARS-positive); and 1,383 samples from Hubei, where Wuhan city is located (3 SARS-positive). They found a total of 170

SARS-like viruses, predominantly from Yunnan province. A key takeaway from the thesis was that SARS-like viruses 'tend to cluster together more by geographic location than by host species'. The only SARS-like viruses they had found to be capable of using the ACE2 receptor were located in Yunnan province, 'where the ancestor of the S gene of human SARS coronaviruses is presumed to have originated'. In other words, one was unlikely to find a SARS-like virus that could use ACE2 as far north as Wuhan. Dr Shi said as much in her July 2020 interview with *Science* magazine, when she emphasised that many years of surveillance in Hubei province had turned up no coronaviruses that were closely related to SARS-CoV-2. A scientist from the Wuhan CDC had also sampled nearly ten thousand bats from Hubei before the pandemic and had not turned up any SARS-CoV-2-like relatives or indeed any SARS-like virus that could use ACE2 as an entry receptor.

Strikingly, the thesis revealed that four SARS-like viruses of key interest – all from southern Chinese provinces – had been selected for whole genome sequencing and the sequences carefully compared against other SARS-like viruses in the literature. Ra4991_Yunnan was among the four. In a table comparing different SARS-like viruses found in China, it was one of two outliers, considerably dissimilar from the 2003 SARS virus. The other was named Rs8561_Guangdong. What is this other novel SARS-like virus that the WIV had found in Guangdong and deemed a top priority for full genome sequencing? Will we ever see its genome sequence?

The 2019 thesis also shone much-needed light on the work that had been done on the 7896 group. It revealed that different parts of the genome had already been sequenced: parts of the spike of 7896, 7905, 7909, 7924 and 7931; as well as the gene known as ORF8 of 7896, 7909 and 7952. In particular, the full RdRp gene (2,757 RNA letters) had been sequenced for all these strains, not just the short 440-letter partial sequences deposited on GenBank for the Latinne et al. paper. It also revealed that primers (a short nucleotide fragment used for sequencing) named after 7896 had been used to sequence the spike genes of other SARS-like viruses, although the RaTG13 amplicon reads labelled with the mysterious '7896' had been from outside the spike gene. This confirmed that the 7896 sequence had been used to help obtain the sequences of other closely related SARS-like viruses. Remember that Francisco de Ribera had asked, as long ago as July 2020, 'Could it be that the 7896 was used somehow as an aid for sequencing RaTG13?'

The 2019 thesis also showed that over the years, the WIV scientists had identified four different lineages of SARS-like viruses. Although the thesis was written before the emergence of SARS-CoV-2, we now know that SARS-CoV-2 belongs to lineage 4; the 2003 SARS virus is in lineage 1. In a family tree showing the four different lineages of SARS-like viruses, only nine viruses were in lineage 4: Ra4991 and the eight from the 7896 group. A map of China showed the geographical location of this special lineage. Unlike the others, lineage 4 was found in only one province: Yunnan. Remember that these had all come from a single mine in which people had sickened with severe respiratory disease that was suspected to be caused by a virus from bats.

Another curiosity now emerged. The author of the thesis, Yu Ping, had also published her work in the journal *Infection, Genetics and Evolution* shortly before she defended her thesis. The paper had been submitted to the journal in November 2018 and published in February 2019. A near replica of the family tree and map of China from the thesis was published in this manuscript. Except there was no lineage 4. The 2019 journal paper showed only three lineages of SARS-like viruses, even though all nine of the viruses in the fourth lineage had been sampled between 2012 and 2015. There was no discussion of a mine from which top Chinese labs were collecting bat viruses in a quest to understand how human beings had contracted a mysterious pneumonia. There was no discussion of an entirely novel group of SARS-like viruses.

Was the mandate of these virus surveillance programmes not to find pathogens with pandemic potential and alert the world in a timely manner? Why did they obscure the discovery of eight novel sarbecoviruses in the February 2019 paper, in the Latinne et al. paper submitted in November 2019, and even post-Covid in the January 2020 WIV paper?

That the emergence of SARS-CoV-2 did not catalyse the release of information regarding the Mojiang miners and the 7896 group is inexplicable. The broken-down lorries, surveillance cameras and excuses about wild elephants used to prevent journalists from visiting the mine only add to the impression that a part of SARS-CoV-2's origin story is being covered up. What is new and even more extraordinary is the hint – from the omission of lineage 4 from the early 2019 publication – that the withholding of information perhaps even began before the pandemic.

11.

Popsicle Origins and the World Health Organization

'That which can be asserted without evidence, can be dismissed without evidence.'

CHRISTOPHER HITCHENS

On 30 January 2020, at a time when the Chinese government was silencing whistleblowers and ordering samples destroyed, the World Health Organization's director general Dr Tedros Adhanom Ghebreyesus said China's speed in detecting the virus and sharing information was 'very impressive, and beyond words', adding 'so is China's commitment to transparency and to supporting other countries'. Two weeks earlier, on 14 January, the WHO had even issued a tweet in line with China's messaging insisting that human-to-human transmission was not happening: 'Preliminary investigations conducted by the Chinese authorities have found no clear evidence of human-to-human transmission of the novel #coronavirus (2019-nCoV) identified in #Wuhan, #China.' Reinforcing the point the same day, 14 January, the acting head of the WHO's emerging diseases unit, Dr Maria Van Kerkhove, announced that although there were clusters of cases in families, 'it is very clear right now that we have no sustained human-to-human transmission'. Thus, at this stage the official WHO position was that the main way a person could catch this disease was if they came into close contact with an infected animal.

A WHO team visited Wuhan on 20 January 2020 to understand the response to the outbreak. They concluded that human-to-human transmission of the virus was likely, as the Chinese authorities confirmed that day, and obtained the design information of the diagnostic kit that the Chinese government was using to test for infection. 'This follows China's rapid identification of the virus and sharing of the genetic sequence,' read the mission summary of the WHO field visit. 'The delegation commended the commitment and capacity demonstrated by national, provincial, and Wuhan authorities and by hundreds of local health care workers and public health specialists.'

Behind the scenes, however, things were less friendly. WHO officials were complaining that information was not forthcoming. Dr Mike Ryan, executive director of the WHO Health Emergencies Programme, was secretly recorded as comparing the situation to the 2003 SARS outbreak: 'This is exactly the same scenario, endlessly trying to get updates from China about what was going on in Guangdong.' He added: 'General "there has been no evidence of human-to-human transmission" is not good enough. This would not happen in Congo and did not happen in Congo and other places.' The WHO did not confirm that these remarks had been made.

In February, Dr Lawrence Gostin of Georgetown University complained: 'We were deceived.' He elaborated to the *Washington Post*: 'Myself and other public health experts, based on what the World Health Organization and China were saying, reassured the public that this was not serious, that we could bring this under control. We were giving a false sense of assurance.' To the sharp surprise of many, the WHO decided, on the advice of an emergency committee, on 23 January not to declare an international emergency. This had followed intense lobbying by the Chinese government. China's ambassador made it clear that an emergency declaration would be regarded as a vote of no confidence in the country's handling of the outbreak. The decision was reversed on 30 January, with only the Chinese delegate voting against. A delegation from the WHO was sent to Beijing in mid-February. Three international experts went to Wuhan alongside three Chinese experts. They stayed for a day and visited two of the hospitals but did not visit the Huanan seafood market. 'It was an absolute whitewash,' said Dr Gostin to the *New York Times*. 'But the answer was, that was the best they could negotiate with Xi Jinping.'

It was not until 11 March that the WHO declared a pandemic. Even after that, Dr Gauden Galea, the WHO representative in China, lamented publicly that his repeated requests that the WHO be allowed to observe the Chinese government's investigations were being refused. 'We know that some national investigation is happening but at this stage we have not been invited to join,' he said. We attempted to contact Dr Galea in January 2021 but were told by the WHO that he was 'on leave and not reachable'.

Terms of reference

During the spring and summer of 2020, the WHO negotiated with the Chinese government to be allowed to engage in a joint study. Dr Peter Ben Embarek, a Danish food scientist and WHO official, visited China in July for talks with various relevant agencies, but, according to later leaked documents, was dismayed to find that 'little had been done in terms of epidemiological investigations around Wuhan since January 2020'.

Finally, in November 2020, the WHO released a 'Terms of References for the China Part' detailing its current understanding of the origin of the pandemic and setting out how it intended to follow up. The document was dated 31 July, so it had been in preparation for three months when released. It frankly conceded that the origin was still a mystery: 'Little is currently known about how, where and when the virus started circulation in Wuhan. Preliminary studies have not generated credible leads to narrow the area of research, and studies will therefore focus on developing comprehensive study plans to help generate hypotheses on how the outbreak may have started in Wuhan.' The document also conceded that there was still no evidence to support a natural origin hypothesis: 'There is no evidence to demonstrate the possible route of transmission from a bat reservoir to human through one or several intermediary animal species.' And it admitted that there was no evidence either for an origin within the food markets: 'There is no evidence that contaminated food items may have contributed to transmission.' It did not mention the possibility of a laboratory leak or steps that would be taken to gather information about the research that had been carried out on SARS-like viruses in Wuhan.

The final composition of the international team was to be agreed by both China and the WHO, with a list of candidates submitted to the Chinese authorities for approval. We would later learn that half of the team comprised scientists in China, and that all team documents had to be signed off by each member. One of the Chinese members was Dr Feng Zijian, the deputy director general of the Chinese CDC and the man who co-signed a gag order in February 2020 that was leaked to the Associated Press. Another was Dr Tong Yigang, a collaborator of Dr Daszak and the WIV in virus hunting, who was also a co-author on one of the pangolin papers. Dr Feng co-led the epidemiology subgroup, and Dr Tong and Dr Daszak led the animal and environment subgroup.

Dr David Relman of Stanford University, who raised the possibility early on that the virus might have leaked from a laboratory experiment, told us at the time: 'Based on the scant information that has been shared publicly about the WHO investigation, it doesn't appear that WHO has adequately represented the range of views and perspectives of key stake-holders or incorporated all needed forms of expertise. And it does not appear that the investigation will receive adequate, independent oversight and review. Needless to say, it doesn't make sense to allow any one nation to dictate the terms of the investigation or dominate membership of the committee.'

Of particular concern was the presence on the WHO-convened team of Dr Peter Daszak, given his close collaboration with the WIV over several years and his having organised a statement in the *Lancet* in February 2020 to 'condemn conspiracy theories suggesting that Covid-19 doesn't have a natural origin'. That same month he wrote an opinion article in the *New York Times* claiming that the outbreak vindi-cated his prediction that a new pandemic 'would likely result from a virus originating in animals and would emerge some-where on the planet where economic development drives people and wildlife together'. In June 2020, Dr Daszak showed he had not changed his mind, writing an opinion piece for the *Guardian* headlined, 'Ignore the Conspiracy Theories: Scientists Know Covid-19 Wasn't Created in a Lab'. Neither the *Guardian* nor the *New York Times* noted his links with the WIV when they ran these articles, though the *Guardian* later added a note to 'make clear the writer's past work with researchers at the Wuhan Institute of Virology'. As a reminder of those links, in an email sent in early January 2020 to a contact at the NIH, later revealed under FOI requests, Dr Daszak had said: 'I spent New Year's Eve talking with our Chinese contacts ... I've got more information but it's all off the record.'

Dr Daszak's inclusion in the WHO team brought strong criticism. 'The independence of the WHO investigation may be seriously compromised by the process used to choose investigators,' said Miles Pomper, one of the authors of a substantive guide to investigating outbreak origins, published in 2020 by the James Martin Center for Nonproliferation Studies. 'In particular, the choice of Dr Daszak, who has a personal stake in ensuring current Chinese practices continue and who is a long-time collaborator of a scientist at the center of the investigation, is likely to taint its results.' Later the journalist Nathan Robinson asked: 'Why would the WHO retain an investigator who seems to have made up his mind about the pandemic's origins before investigating them, and whose entire personal reputation rests on reaching one conclusion over another?'

The first phase of investigation, focused on the epidemiology of early cases, was to rely heavily on Chinese work to 'build on existing information and augment, rather than duplicate, ongoing or existing efforts'. Chinese scientists would conduct all the primary research, including analysing sewage and blood donations, going back through hospital records and interviewing early patients. They would also map and trace the cases of people who visited the market and categorise the food on sale there. As the months passed, however, no information emerged about such investigations and by January 2021, when the WHO team was due to travel to China, it was still wholly in the dark about what had been learned.

By then the WHO team was due to be in Wuhan but its arrival had been delayed by the Chinese government, resulting in what Bloomberg called 'a rare rebuke' from the WHO on 5 January 2021. After some of the team had already set out for China, visas were unaccountably not issued so they had to return home. Dr Tedros, the WHO director general, said: 'I am very disappointed in this news. I have once again made it clear that the mission is a priority for WHO and the international team. I have been assured that China is speeding up the internal procedure. We are eager to get the mission under way as soon as possible.' The Chinese foreign ministry spokeswoman responded with 'There might be some misunderstanding in this' and 'There's no need to read too much into it.'

The press conference

Eventually, on 14 January, some but not all of the international scientists arrived in Wuhan, led by Dr Ben Embarek. They spent the first two weeks in quarantine in a hotel. Only then, for the first time, did the Chinese half of the team share the results of analysis they had done over the previous months. Only on 28 January did the team embark on twelve days of tightly controlled visits to relevant places in the city, going to hospitals, a museum with an exhibition celebrating China's heroic efforts to defeat the pandemic, a food wholesaler, the seafood market, its cold storage, an animal health centre, the Center for Disease Control and the WIV. The team did not visit the Wuhan Central Hospital where the two most prominent doctor whistleblowers had been working: Dr Ai Fen and Dr Li Wenliang. They did visit the Jinyintan hospital where the severe cases were sent but were told that there had been no measures to sample from families of early cases or to store samples for longer than a week. When the Jinyintan staff were asked about the origins, they referred to early cases in other countries, imported food products and the cold chain. The team visited the closed Huanan seafood market and found no evidence of live mammals having been sold, though they did raise the fact that the virologist Dr Edward Holmes had taken a photograph seven years before showing a caged mammal at the market. Witnesses produced for the team said they had not seen any live mammals being sold, but there was no opportunity to question ordinary market workers or passers-by to resolve this contradiction. Even as the visit continued, the Chinese government played it down. It was 'part of a global study, not an investigation,' said a foreign ministry spokesman, Zhao Lijian, pointedly on 1 February.

On 9 February, the WHO-China Joint Mission on Coronavirus Disease held a press conference in Wuhan to announce its findings. It took close to three hours but answered only five questions. The team announced that the hypothesis that the virus arrived in Wuhan on frozen animal products or seafood was plausible and should be followed up, while the hypothesis that it leaked from a laboratory was so 'extremely unlikely' that it would not be pursued in the second phase of the study. In August, an interview of Dr Ben Embarek for Danish television would reveal that he had negotiated with his Chinese counterpart for the inclusion of the laboratory-leak theory in the China-WHO report; this inclusion was made on the condition that the report would not recommend any specific studies to follow up on the laboratory-leak hypothesis. Instead, more work had to be done to understand 'the possible role of the cold chain frozen products in the introduction of the virus over a distance,' said Dr Ben Embarek, adding that the virus 'could have taken a very long and convoluted path involving also movements across borders, travels, etc. before arriving in the Huanan Market'.

In saying this, the WHO team appeared to brush aside the exoneration of the market announced the previous May by the Chinese CDC director. Dr Ben Embarek told *Science* magazine: 'Some traders at the Huanan market were trading in farmed wild animals – badgers, bamboo rats, rabbits, crocodiles and many others. Several of these animals are known to be susceptible to SARS viruses. Some of them come from farms in

中国—世界卫生组织
新型冠状病毒溯源研究联合专家组
新闻发布会
WHO- convened Global Study of Origins of SARS-CoV-2: China Part
WHO-China Joint Study Press Conference

WHO-China origins of Covid-19 joint study news conference in
Wuhan, 9 February 2021.

provinces where coronaviruses have been isolated from bats: Guangdong, Guangxi, Yunnan. Potentially, some of these animals were infected at those farms and then brought the virus into the market.' Interviewed on CNN, one of the members of the team, Dr Daszak, said: 'There was a really striking piece of evidence that was mentioned today in the press conference that in those products were included wildlife meat and carcasses from animals that we know are susceptible to coronaviruses and also that the supply chains come from places in China where we know the SARS-coronavirus-2-related viruses are.' In his opinion, this was 'a direct link' from the potential bat origins of the virus to the Wuhan market. Yet Dr Daszak conceded that all the frozen carcasses that were tested had proved negative for coronaviruses. This left some viewers and his CNN interviewer baffled as to quite what was so striking or what specific evidence supported the link with frozen products.

It is worth thinking through this frozen-food theory, which Alina named the 'Popsicle Origins' hypothesis as a pun on the widely discussed 'Proximal Origin' paper. The WHO-convened experts and the Chinese government were asking us to believe that somewhere in southern China or Southeast Asia a ferret-badger or some other susceptible animal came into contact with a horseshoe bat, contracted a disease, was killed, butchered, frozen and shipped to a market many hundreds of miles away in Wuhan. Miraculously, the virus not only survived on or in the frozen meat for an extended period of time but proved capable of efficiently infecting people on arrival in Wuhan. Yet it caused no illness in people at the source, the local community, or other destinations supplied with these frozen products, though given what we know of viruses it would have been much more infectious at the start of its journey, before being frozen, than at the end. Despite never having infected a human being before, this rare virus, once in Wuhan, proved so well suited to infecting human beings as to be able to readily transmit from a popsicled animal to its first victim and then between people. Then, to cap it all, having achieved this feat once, the virus mysteriously vanished, leaving no trace in all other samples of meat in the same market or elsewhere in China. One international expert on the China-WHO team, Dr Fabian Leendertz from Germany, later told the *New York Times* that he thought the frozen-food hypothesis was a 'very unlikely scenario' but the team had agreed to include it 'to respect, a bit, the findings' of their Chinese counterparts.

As for the possibility of a laboratory leak, the total amount of time the WHO-convened team had spent at the WIV on 3 February was only two to three hours. The team was also walked through the BSL-4 laboratory at the WIV even though all the SARS research had been conducted at BSL-2 and -3. And they only visited the new Jiangxia campus, whereas at least until some time in 2019 most of the relevant laboratory work had been done at the old Wuchang campus, as shown by the address given in the relevant publications (see map on page xi).

Whether and when various work streams shifted to the new location has not been divulged, so it is not clear that the WHO investigators saw a laboratory that had been used by Dr Shi's team at all. Dr Ben Embarek said that during their visit to the WIV they had 'a very long, frank, open discussion with the management and the staff'. He also told reporters, apparently without irony, that WIV officials were 'the best ones to dismiss the claims and provide answers to all the questions'. He later added in an interview with *Science* magazine, 'We discussed: what did you do over the past year to dismiss this claim? What did you yourself develop in terms of argumentations? Did you do audits yourself? Did you look at your records? Did you test your staff?' Also, said Dr Ben Embarek, they had looked at the BSL-4 laboratory and 'it was very unlikely that anything could escape from such a place'. Which was revealing, because most of the work on SARS-like viruses happened in BSL-2 and BSL-3 laboratories, as someone seriously looking into the lab-leak hypothesis should have known. This approach of determining whether a laboratory leak had resulted in Covid-19 was so transparently naive that the host of *60 Minutes*, Lesley Stahl, later called it out in an interview with Dr Daszak: 'But you're just taking their word for it.' To which, Dr Daszak replied, 'Well, what else can we do?'

Access to raw data

The WHO-convened team encountered a roadblock when they enquired about raw data, particularly those describing the first Covid-19 cases in Wuhan. 'We did not collect any samples ourselves, we didn't carry out any laboratory studies there, we just analysed what we were being shown,' said one of the international experts on the team, Dr Vladimir Dedkov, an epidemiologist from the St Petersburg Pasteur Institute in

Russia. The international half of the team only had access to what were largely summary reports from their Chinese counterparts, although the leader of the Chinese half of the team said that the international members could view the data and materials. 'They showed us a couple of examples, but that's not the same as doing all of them, which is standard epidemiological investigation,' said another member of the team, Dr Dominic Dwyer, a microbiologist from Australia. It is notable that the report's map of confirmed early cases in Wuhan, plotted by their home addresses, closely resembles the density of the elderly population in the city. However, the international team members were not shown raw data, making it difficult to tell if there were clusters in homes or retirement communities to explain the explosiveness of the outbreak in December 2019 in places with many elderly residents.

This lack of access meant that the analysis could not be independently reproduced by other experts. Later in 2021, the *Washington Post* reported several important errors in the China-WHO joint report relating to the early patients and clusters. Independent experts told the newspaper that it was imperative to find out how such mistakes of critical importance had been made, warning that these fed into public distrust of the China-WHO study. In response to questions about conflicting details in their report, the WHO clarified that the first family cluster of infections in Wuhan had no exposure to the Huanan seafood market. In addition, the map in the joint report depicted the earliest Covid-19 case as living (home address) on the north-west side of the Yangtze river where the Huanan seafood market was located. However, this conflicted with the Wuhan government's account that the first patient who sickened on 8 December 2019 actually lived on the south-east side of the river and in the Wuchang district, where one of the WIV's two campuses is based. (It is important for us to note here that, due to the lack of access to patient data, the dates of symptom onset for early cases are often difficult to verify.) The China-WHO report had detailed that this patient, an accountant by profession, had only visited one market: an RT-mart, which belongs to a chain of hypermarkets comparable to a Costco or Walmart, in other words, not a place where live wild animals would be sold. Furthermore, the RT-mart that the Wuchang accountant had visited was in the Jiangxia district, where the second WIV campus was located and even further away from the Huanan seafood market.

The WHO described these 'unintended errors' or 'editing errors' in their report as not being relevant to the origin of the virus because 'the current first known patient is most probably not the first case'. However, by the time these errors had been identified, a group of prominent scientists had already attempted to extract the approximate home addresses of early patients, using Adobe Illustrator, from the low resolution map in the China-WHO report. Despite not being able to find an animal source of the virus at even a single market, the same scientists speculated that there could have been not one but multiple spillover events in Wuhan. Their multi-market hypothesis was based on the China-WHO report's vague description of some of the early cases' exposure to other Wuhan markets – which we now know includes supermarkets.

The China-WHO report also revealed the criteria by which early cases had been identified, which likely suffered from ascertainment bias. Specifically, exposure to the Huanan market was one of the key factors in determining whether a patient was a suspected case of Covid-19. Cases were marked as suspect if they were linked to 'related markets in Wuhan' and presented at least two clinical manifestations: fever, pulmonary imaging characteristic of pneumonia, reduction in white blood cell count or lymphocyte count, or no significant improvement in response to three days of antibiotic treatment. Eventually, these criteria were expanded to evaluate cases with links to other fever or respiratory disease patients or Covid-19 clusters. This means that the initial approach of identifying Covid-19 cases may have been biased towards those with links to the Huanan market and its cluster of cases, as well as the elderly who are concentrated in the area north-west of the Yangtze and are more likely to exhibit multiple clinical manifestations when infected by SARS-CoV-2.

As for early cases of illness in 2019 that might have later proved to be Covid-19, these were first selected by doctors in Wuhan, then evaluated by six Chinese experts, and the summary was presented to the international experts. The triaging of these suspected cases was stringent, requiring at least two non-mild symptoms such as fever and pneumonia – diagnostics were still in development and not widely used at the time – which may have led to the exclusion of real Covid-19 cases. The early criteria also focused on exposure to markets, but not laboratories. Out of a total of 76,253 episodes of illness retrospectively identified, only ninety-two were considered clinically compatible with Covid-19. Even so, further review deemed all these cases not to be SARS-CoV-2 infections and the patients were only followed up in January 2021. Unsurprisingly, more than a year post-illness, none of them were found to still carry SARS-CoV-2 antibodies.

Tantalisingly, Dr Marion Koopmans, a Dutch virologist on the team, told the *Wall Street Journal* that even the data they saw could possibly indicate infections as far back as September 2019. Dr Ben Embarek estimated that there had been most likely thousands of cases in Wuhan in December 2019. But, showing a remarkable lack of urgency, the Chinese scientists had only tested stored patient samples in January 2021, right before the WHO team arrived. The possibility of analysing Wuhan blood banks, with blood donated and stored since early

2019, for evidence of early undetected cases of infection was under negotiation, but the Chinese side cited issues with obtaining appropriate permissions to test the blood. In addition, no samples from adults exhibiting influenza-like illness in the last three weeks of December 2019 were available for testing. Samples from the first Covid-19 patients had also been discarded early in the outbreak for safety reasons. Dr Rasmus Nielsen of the University of Berkeley expressed surprise that this work of testing blood samples and contact tracing the first patients had not been done to find the origin of the virus. Without access to these samples and data, it was not possible for the team to determine when the virus had first started spreading in Wuhan, or who the earliest Covid-19 cases were. Another member of the team, Dr Thea Fischer, a Danish epidemiologist, told the *Wall Street Journal*: 'I am a scientist and I trust data. I trust documented evidence based on data, I don't just trust what anyone tells me.'

The China-WHO study did at least glean some new information about wildlife sampling. The leader of the Chinese half of the team detailed that sampling of bats in Hubei province had 'failed to identify evidence of SARS-CoV-2-related viruses and sampling of wildlife in different places in China has so far failed to identify the presence of SARS-CoV-2'. Tens of thousands of animal samples from different species, wild or farmed, and from at least thirty-one provinces in China collected in 2019 and 2020, had all tested negative for SARS-CoV-2. Within hours, Dr Daszak was retweeting Chinese state-owned media: 'Aside from China, global experts are looking at southeast Asia, including Cambodia for potential origins of [the] coronavirus.' That his tweet might cause undue alarm in Cambodia did not seem to occur to him.

A unique and original dataset

To many scientists' surprise, in June 2021, researchers published an original study of the wild animals sold in Wuhan's markets before the pandemic (as we described in Chapter 5). This was information directly relevant to the search for the intermediate host of SARS-CoV-2, conscientiously collected between May 2017 and November 2019 by Chinese scientists. The Huanan seafood market was among the sites they had surveyed on a monthly basis. Contrary to what the international experts had been told, there were shops at Huanan that sold live mammals that could have been possible hosts of SARS-CoV-2 although no bats or pangolins had been observed at the Wuhan markets. Some of the traded animals, including Chinese bamboo rats, raccoon dogs and hog-badgers, were housed in stacked cages and poor conditions. There was little observed enforcement of the requirement for the animals to be quarantined, disease-free and have origin certificates. It appeared that, despite the absence of SARS-CoV-2-positive animal samples, the Huanan seafood market had indeed carried potential animal hosts.

Yet the world was only hearing about this vital information a year and a half post-outbreak. In response to this long-awaited information on the live animals sold in Wuhan, one of the China-WHO team members, Dr Koopmans from the Netherlands, remarked on Twitter that these data 'would potentially have made a difference during our visit'. Adding that had they been provided with this information, 'I think we would have dwelled less on frozen meat.'

We noticed that the authors of this unique study were international, spanning China, the UK and Canada. We reached out to understand why the study had been released so late and were taken aback to hear that the manuscript had initially been submitted to a scientific journal in the spring of 2020 but was rejected several months later after the editor found it hard to believe that there were not pangolins and bats sold in Wuhan markets. An author told us that, despite having been subjected to two rounds of review, the paper was ultimately deemed by the editors to be of insufficient general interest despite its implications for the origin of Covid-19. After submission to *Scientific Reports* on 21 October 2020, it took another seven months before it was accepted for publication. In our view, this seemed an extraordinary case of journal editors not perceiving the significance and urgency of reviewing and releasing the study's findings to help the search for the origin of the worst pandemic in a century.

The authors elaborated to us that the seasons were the main factor determining the amount of wildlife sold at the markets. In the winter, including November 2019, it was the quiet season of the wildlife trade in the markets so not many animals, including those susceptible to SARS-CoV-2, were on sale. Contrary to speculations that a shortage in pork might have fuelled the wildlife trade in China in 2019, the authors told us that there had been no noticeable increase in the sale of wild animals in Wuhan. It also remains to be seen how many of these species could be capable of transmitting SARS-CoV-2 to humans. Based on the numbers provided by their study, only a handful of Covid-susceptible animal hosts were sold by Wuhan stalls on average each day. For instance, on average, only an estimated eleven civets were sold by all seventeen wet market stalls combined each month, and stall keepers did not store more animals in the market than they expected to sell. This description of the wildlife trade in Wuhan is hardly comparable to the roaring civet trade in southern China. Remember that ten thousand civets were killed in Guangdong restaurants, farms and markets following the connection of the first SARS virus to civets. Furthermore, the Wuhan wildlife trade paper was staunch in its claim that no pangolin or bat had been found among the animals for sale: 'Our comprehensive survey data corroborates that pangolins are unlikely implicated as spill-over hosts in the COVID-19 outbreak.' If this information had been available to the world in February 2020, would so many people have been so quick to embrace the hypothesis of the pangolin as intermediate host or the idea that the novel virus had come from illegally sold wild animals at the Huanan market in the middle of the winter?

While this study languished behind the curtain of peer review, videos of Chinese people consuming bat soup circulated in the media; the pangolin papers were published; the media went full-pangolin; and the China-WHO team assembled, made their plans, visited Wuhan and delivered their final report.

Incredulity in the west

Reaction in the west to the WHO's dismissal of the laboratory leak hypothesis and endorsement of the frozen-food hypothesis was highly sceptical. Not all went as far as Dr Bruno Canard, a virologist at France's National Centre for Scientific Research, who said: 'The WHO investigation is a masquerade. There are so many conflicts of interest and obfuscation ... The WHO is committing credibility suicide.' Or *The Australian* newspaper, which wrote: 'It would be hard to view the report by World Health Organization investigators into the origins of COVID-19 as anything other than an outrageous whitewash that has found exactly what Beijing has been trying to get the world to believe about the start of the pandemic. The report's barefaced conclusion that it is "extremely unlikely" the virus emerged from the Wuhan Institute of Virology, just because Chinese scientists say so, beggars belief.' Even moderate voices were critical. Jamie Metzl, a senior fellow at the Atlantic Council and a WHO advisor, called the press conference 'a low point' for the organisation. And the US national security advisor, Jake Sullivan, said: 'We have deep concerns about the way in which the early findings of the COVID-19 investigation were communicated and questions about the process used to reach them.'

In response to such widespread criticism, the WHO director general, Dr Tedros, changed course: 'All hypotheses remain open and require further study.' Dr Ben Embarek also rowed back: 'Let me be clear on this: the fact that we assessed this hypothesis as extremely unlikely doesn't mean it's ruled out.' He explained that 'if you want to explore such a hypothesis further, you need a different mechanism. You need to do a formal audit, and that's far beyond what our team is mandated to do or has the tools and capabilities to do', adding: 'I don't think the press conference was a PR win for China.' Dr Daszak went another way. He tweeted, '@JoeBiden has to look tough on China. Please don't rely too much on US intel: increasingly disengaged under Trump & frankly wrong on many aspects. Happy to help WH w/ their quest to verify, but don't forget it's "TRUST" then "VERIFY"!' He criticised the US government's response to the WHO announcement: 'I'm disappointed that a statement came out that might undermine the veracity of this work even before the report is released.' He then proceeded to make numerous statements himself before the report was released, and even participated in a public Chatham House webinar with two other members of the China-WHO team on 10 March 2021 titled 'Inside the WHO-China Mission'.

The Wuhan press conference and the interviews of the international experts on the China-WHO team only generated more questions. CNN reported that some scientists expressed disbelief, finding it implausible that Chinese scientists had not yet performed rudimentary investigative work to find the origins of the virus. This would have included more extensive contact tracing of the first known Covid-19 patient in Wuhan and checking the supply chains of traders at the Huanan seafood market. Dr Daszak told CNN that the Chinese scientists had not yet visited the farms in southern China that supplied the Huanan seafood market: 'No one has been there to test the animals ... the farms are now closed.' A *Wall Street Journal* report corroborated this story, interviewing farmers in southern China who had bred or trapped wild animals, who said that their animals had been bought and killed by the Chinese government and their farms shut down in response to a government campaign. It remains unclear whether animal samples had been collected from these traders and farmers in 2019 or early 2020, and whether tests were run to determine if the virus might have been present in the supply chain to Wuhan markets. However, Dr Maureen Miller, an infectious disease epidemiologist at Columbia University, commented to CNN: 'It's implausible that this research has not been done. It's not realistic, given they have world-class scientists there, and the technology invested in over the last twenty years. They are sophisticated, they understand transmission pathways, and have been working on them for years.'

Earlier, in December 2020, Dr Daniel Lucey of Georgetown University in Washington, had told the BBC that he was sure that China had already searched for the virus in human samples from hospitals and in animal populations because Chinese scientists had the capability, resources and motivation to identify the origins of the virus. 'I think it's quite possible it's already been found,' he said, 'but then the question arises, why hasn't it been disclosed?' If they thought the Huanan seafood market was the source of a contaminated animal or frozen product, why had they not tracked these back to suppliers located in the likely spillover zone of SARS-like viruses? After the Wuhan press conference, Dr Lucey told the *South China Morning Post* that it was 'frankly implausible' that these tests had not been done one year after the outbreak: 'My question is why would it not have been done? It was known to be necessary and it's in China's scientific interest, it's in their public health interest and it's in their national security interest.'

One concern, expressed by Alina in a tweet after the press conference, was that the handling of the investigation showed the world that a country could get a free pass if a virus escaped from one of its laboratories so long as it took care not to give outsiders access to records and data, but simply asserted that all was well. The lack of international pressure, indeed the degree to which westerners and the WHO were willing to overlook a lack of transparency, will have given governments where accidents might happen in future – or even those contemplating making bioweapons – a clear steer on how easy it is to deflect awkward questions.

Ready to deploy

Under mounting pressure from the media, the China-WHO team scrapped its plan to publish an interim report. This was announced on 4 March 2021 after an open letter calling for a 'Full and Unrestricted International Forensic Investigation into the Origins of COVID-19', signed by two dozen scientists and experts, had been circulated among journalists and shared with the WHO days before publication in the *Wall Street Journal*. The authors had been meeting once a month, thanks mainly to the efforts of scientists based in France. This 'Paris group' included some of the most passionate investigators of the origins of Covid-19: Francisco de Ribera, Dr Rossana Segreto and Dr Monali Rahalkar, who we have described in earlier chapters as some of the key sleuths who uncovered the true story behind RaTG13; Dr Richard Ebright of Rutgers

University who had been sounding the alarm on a possible lab origin of SARS-CoV-2 since January 2020; Dr Bruno Canard and Dr Etienne Decroly, the French virologists who had been the first outside of China to publish the observation of the unique furin cleavage site in SARS-CoV-2; Dr Filippa Lentzos, a science and international security expert from King's College London, who had been consistently publishing articles about the possibility of a lab leak since the early days of the pandemic; Dr Nikolai Petrovsky, an Australian professor of medicine and one of the first scientists to point out that SARS-CoV-2 was surprisingly well adapted to utilising human ACE2 as a receptor; and one of the authors of this book, Alina.

After the press conference in Wuhan, Jamie Metzl, a WHO advisor, had galvanised the group to write the letter. It pointed out that the China-WHO team did not have the mandate, the independence or the access to do a thorough investigation into all plausible origin hypotheses. It also expressed concern that the media had given the impression that there had been a thorough investigation and that the results represented the official position of the WHO. The *Wall Street Journal*, which published the letter, quoted China's foreign ministry criticising the letter as '"old wine in new bottles" that assumed guilt and lacked scientific credibility'. The *Global Times*, a Chinese state organ, promptly quoted China's foreign ministry spokesperson Wang Wenbin: 'The scientists involved in the so-called open letter know very well whether the letter was advice on coronavirus origins based on scientific and professional attitude or presumption of guilt and politicisation.'

On 30 March 2021, the full report of the China-WHO joint study was released. It stressed that 'the final report describes the methods and results as presented by the Chinese team's researchers'. This time, Dr Tedros said: 'The team also visited several laboratories in Wuhan and considered the possibility that the virus entered the human population as a result of a laboratory incident. However, I do not believe that this assessment was extensive enough. Further data and studies will be needed to reach more robust conclusions.' He added that 'although the team has concluded that a laboratory leak is the least likely hypothesis, this requires further investigation, potentially with additional missions involving specialist experts, which I am ready to deploy'.

Governments from around the world weighed in on the report. Those of Australia, Canada, Czechia, Denmark, Estonia, Israel, Japan, Latvia, Lithuania, Norway, the Republic of Korea, Slovenia, the United Kingdom and the United States issued a joint statement expressing 'shared concerns' over the study and arguing for 'a swift, effective, transparent, science-based, and independent process' for future outbreaks. The European Union published a separate statement demanding 'further and timely access to all relevant locations and to all relevant human, animal and environmental data available'.

Needless to say, this did not please the Chinese government. An unidentified Chinese scientist on the China-WHO study team was reported by Chinese state media as accusing Dr Tedros of being 'extremely irresponsible' for pursuing the lab-leak theory and warned that the WHO would be responsible if these comments jeopardised future work on tracking the origins of Covid-19.

The *Science* letter

Dr Tedros's remarks led Alina to contact Dr Jesse Bloom of the Fred Hutchinson Cancer Research Center in Washington State and Dr David Relman of Stanford University – both highly respected experts in their respective fields of virology and microbiology – to see if they agreed that a letter by top experts calling for a proper investigation of a possible lab leak was now warranted. They had previously expressed and heard similar concerns about the lack of a credible investigation into lab-origin hypotheses from other scientist colleagues. Within a week, the three had drafted a short statement in support of a rigorous and credible investigation into the origins of Covid-19, early co-authors had been recruited, and the letter sent to *Science* magazine. It was published on 14 May with eighteen authors hailing from the Howard Hughes Medical Institute, Fred Hutchinson Cancer Research Center, the Broad Institute of MIT and Harvard, Caltech and the universities of Harvard, Yale, Stanford, Berkeley, Arizona, Chicago, North Carolina, Toronto, Basel and Cambridge. The letter stated: 'We must take hypotheses about both natural and laboratory spillovers seriously until we have sufficient data.' It acknowledged that 'at the beginning of the pandemic, it was Chinese doctors, scientists, journalists, and citizens who shared with the world crucial information about the spread of the virus – often at great personal cost'. It urged that the world now 'show the same determination in promoting a dispassionate science-based discourse on this difficult but important issue'.

The editor-in-chief at *Science* wrote an accompanying blog post saying: 'My opinion is that the zoonotic origin of COVID-19 is far more likely, but good science requires that the laboratory escape idea be rigorously investigated before being ruled out. China should allow for a dispassionate examination of the data and allow scientists to do what they are trained to do.' To fresh eyes, the most surprising name on the letter was Dr Ralph Baric's – not only the world's leading SARS-like virus researcher, but also a collaborator with Dr Shi of the WIV. Dr Baric had told the People's Pharmacy website in May 2020 that: 'The main problems that the Institute of Virology has is that the outbreak occurred in close proximity to that Institute.

That Institute has in essence the best collection of virologists in the world that have gone out and sought out, and isolated, and sampled bat species throughout Southeast Asia. So they have a very large collection of viruses in their laboratory. And so it's – you know – proximity is a problem. It's a problem.' He told the Italian television programme *Presa Diretta* in mid-2020: 'You can leave no trace that it was made in a laboratory ... if you're asking about intent or whether the virus existed before-hand, it would only be within the records of the institute of virology in Wuhan.'

In interviews after the *Science* letter was published, some of the signatories expressed concern that 'science is not living up to what it can be' (Dr Relman), that 'in many questions of science, it turns out the right answer is we don't know the right answer' (Dr Bloom), that 'it would be a troublesome precedent' not to evaluate the possibility of laboratory escape rigorously (Dr Sarah Cobey of the University of Chicago), that the collect-ing of bat viruses creates an opening 'for new viruses to get close to humans' (Dr Michael Worobey at the University of Arizona), and that an investigation of plausible hypotheses was necessary considering that eighty thousand animal samples in China had been tested with no sign of the virus or its precursor (Dr Akiko Iwasaki of Yale University). Writing for *MIT Technology Review*, the journalist Rowan Jacobsen was able to solicit a rare response from Dr Shi, who said that the letter's suspicions were misplaced and would damage the world's abil-ity to respond to pandemics. 'It's definitely not acceptable,' Dr Shi said of the group's call to see her lab's records. 'Who can provide an evidence that does not exist?'

The *Science* letter caught the attention of the media. Not only was there widespread coverage but the tone of much of the coverage had noticeably changed. The *New York Times* altered its description of the laboratory-leak theory in a tab heading of a previous online article from 'debunked' to 'unproven'. The fact-checking site PolitiFact removed a claim of 'debunked conspiracy theory' about a television interview on a laboratory leak, explaining 'that assertion is now more widely disputed'. Vox quietly changed a line on its website from 'The emergence of the virus in the same city as China's only level 4 biosafety lab, it turns out, is pure coincidence' to 'The emergence of the virus in the same city as China's only level 4 biosafety lab, it turns out, *appears to be* pure coincidence.' The *Washington Post* carried an editorial arguing that 'if the laboratory leak theory is wrong, China could easily clarify the situation by being more open and transparent. Instead, it acts as if there is something to hide.' Long articles by journalists Nathan Robinson in *Current Affairs* and Donald McNeil Jr in *Medium* both made the argument that the political left had been too quick to dismiss the possibility of a laboratory leak. 'I have often warned that liberals and leftists should be careful not to assume conservatives are always wrong about facts,' wrote Robinson, 'because sometimes they aren't.' The idea had been 'tarred by the fact that everyone backing it seemed to hate not just Democrats and the Chinese Communist Party, but even the Chinese themselves,' wrote McNeil. 'But now, 17 months later, China is persistently acting like a nation hiding something.'

The upshot of the WHO's inquiry into the origin of the virus was therefore not to provide an answer, but to galvanise some of the world's top scientists and journalists into reopening the question.

12.

Spillover

'If it looks like a duck, and quacks like a duck,
we have at least to consider the possibility that
we have a small aquatic bird of the family
Anatidae on our hands.'

DOUGLAS ADAMS

At the heart of this story lies a very simple fact. Somehow, an ancestral version of the SARS-CoV-2 virus came from a bat and ultimately infected people. That basic truth remains undeniable. Even if you think it was subject to genetic manipulation or some kind of human help along the way, this creature would still be fundamentally a product of evolution, a natural being, a wild organism. The majority of its genome is intact, original and natural, even if you put a question mark over a few small parts of it. It is one of many SARS-like viruses, and SARS-like viruses were invented by Mother Nature, not by people. So the theory that the pandemic began as a natural spillover was from the start, and remains to this day, highly plausible. It is the null hypothesis, the default assumption. It deserves its day in court. This chapter is the case for its defence.

The case begins something like this. Ladies and gentlemen of the jury, somewhere in a southern Chinese province, not necessarily in Mojiang County, there are horseshoe bats carrying viruses with genetic components reaching 98–99 per cent similarity to those in SARS-CoV-2. This must be true. The same has already been found for the 2003 SARS virus, and the scientists who discovered these bat coronaviruses concluded that recombination among these precursors resulted in the direct ancestor of the 2003 SARS virus. That we have not found these precursors for SARS-CoV-2 is hardly surprising. It is like searching for a needle in a haystack consisting of vast populations of bats ephemerally infected with viruses that are continuously recombining and evolving. The bat hunters described in this book have done a remarkable job in tracking down many sarbecoviruses in horseshoe bats, tying them to SARS, understanding how they are related to one another and how they evolve. Even a generation ago, little of this knowledge-building would have been possible. We would have been guessing where the virus came from, stumbling around in the scientific dark. It is the fact that we know so much and have at our fingertips so much greatly enhanced scientific capability this time around that enables us to be suspicious of gaps in knowledge and inaccessible evidence.

To blame those diligent virus hunters for the pandemic is like blaming a bystander or even a first responder for an accident. The main case against them is that they were present at the scene. But of course they were present at the scene! They were trying to help; they were trying to prevent a pandemic. They said it again and again. A pandemic caused by a novel SARS-like virus was always on the horizon. The very fact that they were there, sampling thousands of bats in caves all over southern China, is not proof they caused the pandemic, but it is part of the evidence that there was already a real threat of a natural pandemic. They warned repeatedly that bat viruses were 'poised for emergence', in the words of one of their papers. And emergence is what has happened. They are entitled to say, 'I told you so.'

Next, remember SARS. It is a close cousin of SARS-CoV-2 and it emerged. It spilled over. We still do not quite understand how or where. Bats may have infected civets which infected people. Or they may have infected people who infected civets. But it is universally accepted that there was no helping hand from science – there was no SARS research lab in the middle of Guangdong; indeed, researchers had to traverse mountains and rivers in search of viruses carrying the ancestral building blocks of the 2003 SARS virus. And the pattern of early cases, with food handlers and chefs among the first to be infected, pointed to infection through the food chain. A significant portion of animal traders in Guangdong had antibodies against SARS virus despite not having been previously diagnosed, speaking to the frequent exposure of animal traders in the province to SARS-like viruses. If SARS virus could spill over naturally, so could SARS-CoV-2. SADS, the pig diarrhoea disease of 2016, is another example: a direct infection of pigs by bats.

Or take Ebola. As we were writing these words, Ebola broke out again in Guinea. Nobody was claiming that this is because bat virus-hunting scientists were doing risky experiments. It was because, as in 2014, a person has been exposed to bats in their natural environment. Ebola has emerged about two dozen times in the past forty-five years in central Africa. Each time it has been a natural spillover.

Or take MERS or Nipah or Hendra: new viruses that spilled over from bats within living memory. They needed no help from scientists in laboratories (Marburg virus is an exception). Why, in a world where this happens again and again, should we suddenly abandon this simplest of explanations for something more complicated and novel?

Then consider recombination. The SARS-CoV-2 virus appears to have the backbone of one virus but parts of the spike gene of another. Which is what we can expect of these coronaviruses. We knew before the pandemic that recombination was their thing, especially in and around the spike gene. The long search for the progenitor of SARS eventually concluded that all the parts of the SARS virus were present in *R. sinicus* bats in the Shitou cave, but not in one virus and not in one bat. It was a jigsaw of interchangeable parts, continuously switching in and out among closely related viruses.

Then there is the evidence of seroprevalence: around the bat-filled caves of Yunnan, a small percentage of people have antibodies to SARS. There is undoubtedly occasional natural spillover going on. If that's true, why does the spillover of SARS-CoV-2 need to involve a scientist? True, it is a long way from Mojiang to Wuhan, but scientists were not the only ones making that journey. With bullet trains and freeways, rural people in China are on the move as never before. As the pandemic has shown us, a virus can and has travelled thousands of miles in an infected person. And a virus like SARS-CoV-2, which doesn't manifest severe symptoms in the majority of its victims, is likely to have remained undetected until a superspreading event in a densely populated major transit hub such as Wuhan city.

Sure, nobody appears to have got infected along the way between southern China and Wuhan before the first cases were detected in Wuhan. But that is not impossible. Suppose again the index patient was an illegal wildlife smuggler heading to Wuhan as surreptitiously as possible, interacting with few people along the way, driving his truck at night and not yet very infectious. He might not give his virus to anybody along the way.

A sensible attorney is prepared to concede some things to the other side so as not to spoil the case by arguing implausible things. In this case we must concede that the frozen-product or cold-chain hypothesis is nonsense. It provides a fanciful possibility that the virus came from abroad, and it was unfortunately endorsed as more likely than a laboratory leak by the China-WHO team. In mitigation, as we saw, one of the team members, Dr Fabian Leendertz, said they agreed to include the frozen-food theory merely 'to respect, a bit, the findings' of Chinese scientists. It is just not possible to devise a version of that story that makes sense without seeing infections elsewhere, especially at the site where the frozen product was prepared for export. The 'Popsicle Origins' hypothesis just passes the buck, posing the question of how the virus got on the frozen food in the first place. Yes, frozen products might occasionally carry the virus from place to place during the pandemic, but there is no solid evidence that a person has ever caught SARS-CoV-2 from a frozen product and this hypothesis cannot explain the origin of the pandemic. Even the China-WHO team leader, Dr Peter Ben Embarek, conceded that 'there were no widespread outbreaks of COVID-19 in food factories around the world' back in 2019 and that imported goods were 'not a possible route of introduction'.

Now comes the tricky part for the lawyer in court. We must explain away the peculiar behaviour of the Wuhan scientists, their apparent efforts to conceal the links between viruses in their collection and a mineshaft in Mojiang where three men met their deaths from a mysterious pneumonia. We must explain the refusal to divulge the sequences of the eight closely related SARS viruses from the mine more than a year after SARS-CoV-2 was first sequenced. Easy: the scientists were afraid of being blamed so they did what they could to prevent a narrative developing that would cast shade on them. In a secretive and autocratic system they were afraid to be open. Similarly, at the level of the local Wuhan government and even

at the level of the Chinese government, censoring information
and deflecting blame is to be expected – regardless of whether
the virus emerged naturally or via a laboratory leak. This had
already been demonstrated in the first SARS outbreak of 2003.
You may not like the lack of transparency, and we don't, but it
is hardly proof of guilt. In some ways, they have been remarkably open. 'Could they have come from our lab?' was Dr Shi
Zhengli's first thought when she heard about the first novel
coronavirus cases in Wuhan. Would she really have said that if
she was trying to cover up a link to her lab?

The lawyer also calls expert witnesses to the stand. Prominent
virologists have argued that SARS-CoV-2 is the seventh coronavirus known to infect humans – nothing unnatural about it.
The virus is also known to infect a wide range of species and
bind effectively to the ACE2 protein in different host species. It
can thus be termed a 'generalist' virus, not one that is specifically optimised for human beings alone – although humans are
near the top of its list. Furthermore, there is a simple reason to
doubt that this virus was engineered in a laboratory. SARS-
CoV-2's genomic sequence does not match any known virus
reported by any laboratory around the world. The closest virus
genome, RaTG13, is only a 96 per cent match – too distant to
be used as a template. There are also no reports of novel SARS-
like viruses being serially passaged in human cell culture or in
humanised animals. The mechanism by which the spike
protein's receptor-binding domain attaches to the human ACE2
receptor depends on certain amino acids in the sequence. A
team of genetic engineers would surely have borrowed a known
recipe. Yet, despite all the Chinese research theses being
unearthed by sleuths, none have revealed any sequences that
could point to a SARS-CoV-2 precursor being studied in a laboratory before the 2019 outbreak. Until actual evidence surfaces
to show that a SARS-CoV-2 precursor was in the possession of

scientists before the outbreak, the onus is on those who argue for laboratory origins to substantiate their speculations.

In addition, since the emergence of SARS-CoV-2, a growing number of its close relatives have been discovered in Southeast Asia and even Japan, within a broad range of *Rhinolophus* bat species, suggesting that these viruses are widespread. The most probable origin scenario is that SARS-CoV-2 evolved naturally, either in an animal host population (likely a species sharing a similar ACE2 with human beings) before spillover into humans, or in a remote human population before detection in Wuhan in late 2019. Consider that many of the early cases of Covid-19 were linked to a market in Wuhan so it remains possible that an animal at the market transmitted the virus into people. The belated publication in the middle of 2021 of a paper describing the sale in Wuhan markets of all sorts of live mammals must to some extent increase the possibility of an animal in the market serving as the intermediate host of SARS-CoV-2. Although no

Wuhan's Huanan seafood market before its closure on 31 December 2019.

animals or animal samples from Wuhan markets or elsewhere in China have been found to be the original source of the virus, it is possible that wildlife vendors might have quickly hidden their infected animals upon hearing news of the outbreak. Remember that the Huanan seafood market was thoroughly disinfected in such a hurry that the animal source might have been missed or even destroyed in the process. It remains entirely possible that the animal source will be found as Chinese scientists continue to test more farms and animal populations.

Ladies and gentlemen of the jury, the attorney goes on, you will hear from the other side that the furin cleavage site is deeply suspicious and puts scientists right at the scene of the crime. But why? Lots of coronaviruses have such a feature naturally. True, none of the sarbecoviruses have one, but so what? It could easily emerge by recombination, mutation or a mixture of the two. That specific region of the spike is prone to mutation. Viruses related to SARS-CoV-2 have been revealed to have possible insertions where the SARS-CoV-2 furin cleavage site is located; these could be intermediates to a functional furin cleavage site. There is a real example of such a site appearing as if by magic in an experiment involving a bovine coronavirus: it was already there as a rare variant and selection in the laboratory brought it to the fore. Something similar may have happened in the case of SARS-CoV-2 but in nature rather than inside a lab.

Then there is the lack of direct evidence of a laboratory leak. Where is the infected laboratory worker, the index case? Where is the record of an accident? If there was a leak, why, in the months since the pandemic began, has no scientist come forward with a confession or an eyewitness account? The Chinese regime may be authoritarian, but it is not omnipotent. The lack of a whistleblower with genuine evidence of an escape of the virus into the community must count for something.

Do not forget that there is no evidence at all that the SARS-CoV-2 virus was the subject of experiments in any laboratory. The WIV had published scores of papers on the various viruses it had manipulated and tested in its laboratories. Never once in this flood of papers did it mention a virus that could be a precursor to SARS-CoV-2. If the jury is to believe that such work was nonetheless happening, it would have to conclude that the pandemic began during a relatively brief window in which experiments had taken place but nothing had been submitted for publication.

Finally, there is the motivation of those pursuing the laboratory-leak hypothesis. Many of them have been passionate opponents of experiments on viruses with pandemic potential – some even since before the gain-of-function moratorium of 2014. They are bound to think that this is what they have been warning against. Others are in the arms control business and have devoted their lives to warning against the dangers of bioweapons. This pandemic is a godsend to such people and their budgets, so they are bound to recruit it to their cause. Others are fervent opponents of genetic engineering and this episode fits their world view. Others still are critics of the Chinese Communist Party and will readily believe any theory that blames the Chinese state. Still others, especially in journalism, are motivated by sensationalism. They want this to be a big story and a human-made leak is a much bigger story than a bat that shat on a passer-by. Finally, there are some who simply succumb to the all too human tendency to think that where there is pain there must be blame. The intentional stance, philosophers call it: the tendency to infer intention into even random events. If we think half seriously (and we sometimes do) that even the weather is vindictive when it chooses to rain on the day of a wedding, how much easier is it to think that a pandemic must be human-made? These witnesses are not dispassionate and sceptical assessors of the evidence but ready believers in whatever version of a laboratory origin scenario best fits their prejudices. In the fine old human tradition of confirmation bias, to which anyone is susceptible, they will seek out the evidence that is compatible with their theory and ignore what does not fit.

In conclusion, ladies and gentlemen of the jury, says the attorney, there is no evidence of a laboratory leak, and the default assumption must remain that this pandemic began with a natural event, as so many other epidemics have done in the past.

13.

Accident

'But I've never seen the Icarus story as a lesson
about the limitations of humans. I see it as a
lesson about the limitations of wax as an
adhesive.'

RANDALL MUNROE

It is now the turn of the attorney for the other side to make the
case that this pandemic derives from an accidental laboratory
leak. We begin by making clear that we are not alleging malfeasance, only a mistake. The hypothesis that the Chinese
government deliberately engineered a virus and released it to
cause harm in the world is not the case that is being argued.

Rather, the allegation we take seriously is that scientists at
the Wuhan Institute of Virology and, possibly, other
laboratories in the city, at the Center for Disease Control or the
Huazhong Agricultural University, were doing exactly what
they said they were doing: namely, studying viruses from bats
and other wildlife with a view to predicting pandemics and
eventually developing therapeutics or vaccines. But something
went wrong and a virus leaked, either infecting a person who
worked in the field, in a laboratory, or elsewhere in the city if

lab waste had not been properly decontaminated before being discarded.

The first point is that such research accidents, while rare, are sufficiently frequent that they can be expected to happen. Lottery wins are extremely rare, but they happen every week. If a lab works with tens of thousands of diverse zoonotic virus samples over a period of several years – especially at lower biosafety levels – it is only a question of when, and not if, a biosafety incident occurs. Unfortunately, there are no enforceable international biosafety and biosecurity standards. Many accidental releases of SARS, anthrax, smallpox, foot-and-mouth, Marburg virus, and other pathogens have occurred and continue to occur in even the most secure and well-run laboratories. Remember the SARS laboratory leaks in 2003 and 2004. One researcher had embarked on an international trip, and another researcher had taken multiple long-distance train rides after getting infected with SARS virus in the laboratory. For the former, it was pure luck that he had not developed symptoms and become infectious while in international transit. For the latter, close to a thousand people were quarantined and her laboratory-acquired infection had only been discovered a month after the fact.

Next consider that there is no direct evidence of a natural origin of the Covid-19 pandemic. As each month goes by, it becomes more and more extraordinary that the proponents of a natural origin fail to find any of the sort of evidence that very soon came to light in the case of the 2003 SARS epidemic. In that episode it quickly became apparent that food handlers were over-represented in the early cases, which led to the testing and discovery of infected animals in markets. Animal traders in Guangdong were found to have antibodies against SARS viruses despite not having been previously diagnosed with SARS or displaying symptoms – meaning that these groups of people had been regularly exposed to SARS-like viruses prevalent in the animal trading circuit. Guangdong was a hotspot for SARS viruses to transmit into humans. True, it then took months to find bats carrying similar viruses and then years to find the original bat reservoir of very closely related viruses. But the market evidence was strong and came early. Although the technology at that time was limited compared with today's, the intermediate hosts of the SARS virus were identified within weeks of first sequencing the genome of the virus. And when SARS spilled over once again at the end of 2003, Chinese doctors and scientists were even swifter at finding infected civets at the restaurant workplace of an index patient – on the very day of the patient's SARS diagnosis, the civets were sampled. No time was wasted.

This time, despite testing markets, farms, and eighty thousand animal samples spanning dozens of species across China, no similar evidence has emerged for SARS-CoV-2. Hundreds of samples taken from the animal carcasses at the Huanan seafood market all tested negative for any trace of the virus. A considerable fraction of the early Covid-19 cases in Wuhan also had no exposure to the market. Damningly, one of the early lineages of SARS-CoV-2 was not detected at or associated with cases from the market – pointing to the more likely scenario that a person had brought one of the early variants of the virus into the market where a superspreading event occurred in a poorly ventilated and crowded venue.

In stark contrast to the 2003 SARS virus, there has been no explanation of how it came to be so adept at spreading among humans. Both the 2003 SARS and SARS-CoV-2 are generalist viruses and can infect a wide range of animal species, but the stepwise adaptation of the 2003 SARS virus to its new human host in the early months of that outbreak is missing altogether in the case of SARS-CoV-2. When it was first detected in

Wuhan, it had apparently stabilised genetically – there was just a single mutation, D614G, in the early cases that could be said to slightly enhance its transmissibility among humans. More variants came along later, of course, as the virus became immensely widespread, increasing the opportunities for mutation in millions of victims. If the 2003 SARS virus had been allowed to spread to millions of people, we would have certainly seen the evolution of variants – even after the virus had picked up dozens of helpful adaptive mutations in the first months of its exploration in human beings. In the case of SARS-CoV-2, the Alpha variant has been hypothesised to be the product of long-term infection of an immunocompromised individual. The Delta variant can re-infect survivors of Covid-19 and emerged in a part of the world where Covid-19 was running rampant. In the absence of any intermediate host or remote human populations found to carry SARS-CoV-2, where did the virus perfect its ability to infect and transmit among humans? Could it have acquired its skill in a lab, dwelling in the cell cultures derived from the human respiratory tract, or even in humanised mice?

Ask yourself this: why did the Chinese authorities refuse to release to the WHO investigators the raw data about the early Covid-19 cases in Wuhan? If they wanted to quash rumours of a laboratory leak, the simplest thing to do would have been to share unredacted hospital records from Wuhan with outside investigators, alongside information on the locations and professions of the first cases. If these showed no connection to any virus laboratory in the city, it would argue against a laboratory leak. Yet no such details have been disclosed.

Ladies and gentlemen of the jury, says the attorney, let me pose a very simple question: why Wuhan? Of all the gin joints in all the towns in all the world, she walks into mine, says Humphrey Bogart's Rick in the film *Casablanca* when Ingrid

Dr Peter Daszak and Dr Thea Fischer, of the WHO team tasked with
investigating the origins of Covid-19, arriving at the Wuhan Institute
of Virology, 3 February 2021.

Bergman's Ilsa turns up in his bar. But if you were escaping
Nazi-occupied Europe, Casablanca was a well-known rat run
with regular flights to Lisbon, and Rick's Bar was supposedly
the main place to go in the town to buy travel papers, so it's not
quite as improbable as he makes it sound. It is surely far
stranger that a bat sarbecovirus should walk into the very city
with the largest collection of bat sarbecoviruses in the world
and the most active research programme studying such viruses
– and a city that is well over a thousand kilometres from the
region where closely related viruses are found naturally.

Scientists from Wuhan regularly made the long trip to southern China and trekked through forested terrain to reach remote
caves in order to collect their thousands upon thousands of
samples for study back in the metropolitan city of Wuhan. They
did not always wear full protective gear as they handled thousands of bats, swabbing the animals' anuses and noses and

generally getting far closer to them than local people ever would. It is highly unlikely that any other route to Wuhan was so conveniently available to such viruses. Even among the villagers living near caves in which SARS-like viruses had been found in bats, only 2.7 per cent had antibodies against SARS. Of the six individuals who were seropositive in that study, only one had left the province in the past year, and two had not even left the village. This is how remote these villages are. A bat virus with a lust for travel might wait decades to get a chance to visit nearby Kunming, let alone distant Wuhan. From the scientists' own extensive sampling of bat viruses across multiple Chinese provinces over the years, a simple truth emerges: within China, the particular lineage of bat sarbecoviruses that SARS-CoV-2 belongs to was only ever found in Yunnan province – certainly nowhere close to Wuhan.

We now know, as we did not in February 2020, that the WIV was in possession of a batch of very closely related viruses collected from an abandoned copper mine in Mojiang where workers had sickened with a SARS-like illness in 2012. We also now know that the WIV had not been forthcoming with vital information about these viruses, including what had been done with them in the laboratory. We know that for months the fact that more than one such virus had been collected from the mine went undisclosed. If not for the chance event that a large batch of sequences was uploaded with a scientific paper to an international genomic database before the pandemic began, the existence of eight of these viruses might never have been uncovered. And if not for the efforts of the sleuths and scientists painstakingly sifting through the raw data and supplementary materials of scientific papers, comparing sequences and compiling spreadsheets, the link between the Mojiang miners and the SARS-CoV-2-like viruses under study at the WIV might have never been brought to light.

The search for a virus that is a closer match elsewhere than Mojiang has been a failure. Some proponents of the natural-origin theory greeted with enthusiasm the discovery of viruses related to SARS-CoV-2 in Thailand, Cambodia and Japan, but none was as close a match to SARS-CoV-2 as the WIV's RaTG13 from the mine. Every new discovery of a less closely related virus serves only to underline the significance of the viruses brought from the Mojiang mine.

Then there is the almost ridiculous secrecy that surrounds the mine itself. To this day Dr Shi and her colleagues have refused to confirm the exact location, though it is no longer a secret thanks to the various theses and the work of sleuths. The spot remains strictly off limits to foreigners and is heavily guarded. International journalists trying to visit the Mojiang mine have been tailed, obstructed by supposedly broken-down vehicles, and detained by the police. This is not an approach that instils confidence in the innocence and irrelevance of the site.

Pull back from the detail and focus on the simple fact that the WIV had a database of at least fifteen thousand bat samples taken mainly from southern China. With more than twenty-two thousand entries in total, that database included the dates and locations of samples and the descriptions of viruses found in them. It became inaccessible to users outside of the WIV in September 2019 and was taken down altogether sometime in early 2020. To dispel rumours that SARS-CoV-2 was derived from this collection of virus sequences or samples, the WIV could have easily shared this database with other scientists. The excuse that there had been 'hacking attempts' does not make much sense, because that would not prevent the sharing of the data with other scientists. Yet, well over a year later, no one has reported having a copy or access to this database. What's the point of collecting viruses if you hide the data when a pandemic actually occurs?

The next piece of evidence against the laboratory scientists is the long and detailed record of their research. In paper after paper, they laid bare a record of experiments on sarbecoviruses and other coronaviruses that were ingenious, comprehensive and successful. They did not just bring viruses from caves and mines in southern China to the laboratory for storage; they sequenced their genomes, made infectious clones of them, rescued live viruses from culture, passaged them through a range of laboratory-made cell lines of different animal species and cell types, synthesised and altered their genomes to insert specific sequences, hybridised genomes to combine parts of one virus's spike gene with backbones from another virus, and used these viruses to infect human respiratory tract cells and human-ised mice genetically engineered to have human ACE2 in them. This type of research carries a risk of, unintentionally, generat-ing a more virulent or infectious version of a virus or selecting for bat viruses that are efficient at infecting human tissues and humanised animals. Their purpose was admirable: to under-stand the risk that each newly discovered virus posed and perhaps one day to devise a vaccine against all SARS-like viruses. But they also brought a possibility of starting a pandemic.

Nor can we consider fatal the argument made by proponents of a natural origin that the virus has not been reported in the scientific literature and shows no traces of genetic engineering. Scientists know that it often takes years to build up a solid scientific publication. Discoveries are not reported in real time. Samples are stored for years, awaiting analysis. We also know from the *Washington Post* that there were secret projects at the WIV and some reports or theses could be sealed for up to two decades. The WIV database is a case in point: to our knowl-edge, the existence of many of the entries has not yet been reported in formal scientific papers. The 630 viral samples reported in the Latinne et al. paper in 2020 had all been collected from 2010 to 2015 yet nobody thinks the WIV stopped collecting and isolating viruses in 2015. Besides, RaTG13 itself was collected in 2013, sequenced fully in 2018, but not published until 2020 when it became imperative to clarify its similarity to SARS-CoV-2. The other eight viruses collected in 2015 from the Mojiang mine were undisclosed until 2019 and their sequence was only released in mid-2021.

In any case, a laboratory leak does not require the pathogen in question to have been genetically modified. Many laboratory escapes have involved viruses collected from naturally infected animals or human beings. But if an engineered virus does escape, there might be no way to distinguish it from a natural mutant. Today's technology allows the seamless construction of entire virus genomes. We know that the WIV was collecting thousands of natural virus specimens, sampling from thousands of animals and people living in rural areas. We know that they were using and improving on these seamless techniques for switching parts in and out of virus genomes. And we know that they were testing the novel SARS-like viruses, natural or chimeric, in human cells and in humanised mice and civets. These conditions would have selected for a largely natural virus that could effectively use the human ACE2 receptor and perhaps spent time inside animals with functioning immune systems. These laboratory processes would not have left an obvious mark of laboratory origin on the virus.

The significance of the S1/S2 furin cleavage site in the spike gene of SARS-CoV-2 is unclear. On the one hand, such sites appear in many viruses, including other coronaviruses. On the other hand, it is striking that no other sarbecovirus has one, including all of SARS-CoV-2's closest relatives and the dozens of other SARS-like viruses collected from bats. To us, however, the most shocking aspect of the furin cleavage site is Dr Shi's failure to mention it in early papers first describing the novel coronavirus. The furin cleavage site stands out like a sore thumb and is a large part of the reason this virus is so infectious and virulent. In one recent collaboration between Dr Shi's group and scientists in America, a furin cleavage site had been engineered into the spike of a bat MERS-like coronavirus. So there is no doubt that Dr Shi and her colleagues understood the functional significance of this feature in coronaviruses. Yet the *Nature* paper co-authored by seasoned coronavirus specialists diligently described other minor features of the spike gene sequence but missed the show-stopping furin cleavage site. It was left to other teams of scientists in China, Canada and France to say 'Hey, look – this is the first sarbecovirus with a furin cleavage site, which might explain why it is so highly infectious.' Across town from the WIV, scientists at another Wuhan research institute, the Huazhong Agricultural University, had also recently collaborated with Dutch scientists to insert such a site into the spike of a live porcine epidemic diarrhoea coronavirus. It is a remarkable coincidence at the very least that the first sarbecovirus with a furin cleavage site showed up in one of the few gin joints in all the world where the knowledge and capabilities exist for inserting one into a coronavirus.

One final argument for the laboratory leak is that more virologists have shifted their view and now either think it is possible or even likely. Given the potential risk to the reputation of science as a whole and virology in particular, this cannot have been easy for them. Dr Ian Lipkin of Columbia University was a co-author on the 'Proximal Origin' paper that had stated in March 2020: 'We do not believe that any type of laboratory-based scenario is plausible.' However, in May 2021, Dr Lipkin said of the experiments done at the lower BSL-2 biosafety level at the WIV: 'That's screwed up. It shouldn't have happened. People should not be looking at bat viruses in BSL-2 labs. My view has changed.' Dr Bernard Roizman, a virologist at the University of Chicago, who signed the *Lancet* letter drafted by Dr Peter Daszak in February 2020, went further. In May 2021, he said: 'I'm convinced that what happened is that the virus was brought to a lab, they started to work with it … and some sloppy individual brought it out … they can't admit they did something so stupid.' Another author of the *Lancet* letter, Dr Charles Calisher, a Colorado State University virologist, also told ABC News that 'it is more likely that it came out of that lab'. And a third author of the letter, Dr Peter Palese, told the *New York Post* that 'A lot of disturbing information has surfaced since the *Lancet* letter I signed, so I want to see answers covering all questions.' Dr Gary Whittaker, a specialist in influenza and coronaviruses, joined Dr David Baltimore in calling the furin cleavage site in SARS-CoV-2 'extremely unusual, leading to the smoking gun hypothesis of manipulation'.

It must be conceded that many of those making the charge that SARS-CoV-2 leaked from a laboratory are not agreed as to which of several versions of the leak they find most plausible: whether SARS-CoV-2 is a natural virus that was stored there and leaked, like the SARS leaks in Beijing in 2004; or a virus that had been passaged in human cell cultures to the point where it evolved into an efficient human pathogen; or whether SARS-CoV-2 is an engineered chimeric virus. All are possible and until there is better evidence none of these can be ruled out.

Ladies and gentlemen of the jury, the attorney concludes, there is a stark absence of evidence for a zoonotic spillover at the start of this pandemic. In contrast, the proximity of the outbreak to the WIV – the largest collector of SARS-related coronaviruses in the world, where scientists were creating chimeric viruses and experimenting with close relatives of SARS-CoV-2 – makes a compelling case for a laboratory-based origin of the virus.

14.

The origin of Covid-19

'We live in a scientific age, yet we assume that
knowledge of science is the prerogative of only
a small number of human beings, isolated and
priestlike in their laboratories. This is not true.
The materials of science are the materials of
life itself. Science is part of the reality of living;
it is the way, the how and the why for
everything in our experience.'

RACHEL CARSON

There is a neat phrase used by the philosopher Daniel Dennett: burden-tennis. In a 1988 review of a book by Jerry Fodor he wrote: 'The book is a tireless exercise of that philosopher's pastime, burden-tennis. Burden, burden, who has the burden of proof now?' In many scientific debates, one side of the argument insists it is up to the other to prove its case, and vice versa. The onus is on you to show that I am wrong, says the professor who claims the moon is made of cheese, not on me to show that I am right. No, it is on you to show that I am wrong, says the professor who claims the moon is made of yoghurt, thwacking the burden of proof back over the philosophical net.

Much of the debate about the origin of the virus assumes that the laboratory-leak theory must prove itself. Natural spill-over, by contrast, is the default assumption, which does not have to prove anything. The ball is in its opponent's court. This holds any laboratory-leak theory to a higher standard of proof than any natural theory. The influential 'Proximal Origin' paper of March 2020 by Dr Kristian Andersen and his colleagues took this line, focusing on the lack of evidence for an accident rather than the evidence for a natural origin. A furin cleavage site could have arisen naturally, so it is up to those who think it was inserted to show otherwise – and so on. So too did the China-WHO joint study team in its February 2021 press conference, at which the team said that since they had not seen evidence for a leak, they therefore did not see the need to investigate the possibility further.

But why should the burden of proof be with those who posit a laboratory leak? True, there has been no major pandemic caused by a laboratory leak, so it would be a first. But we have entered an era when scientists are collecting, sequencing and manipulating viruses at unprecedented scale. There have already been numerous near misses around the world. Given the powerful circumstantial evidence that Wuhan was not a particularly likely place for a natural epidemic of a SARS-like virus to begin, but an obvious one for a laboratory-leaked one to start, it is surely reasonable to expect both hypotheses to be put to a similarly rigorous test.

We think that even if the burden of proof was on the laboratory leak initially, it has since shifted. That the closest relative of SARS-CoV-2, RaTG13, got to Wuhan via scientists shifts the burden of proof. So does the obfuscation and misdirection in the story of the Mojiang miners. So does the existence of the other eight SARS-CoV-2-like viruses from the mine. So does the missing database of more than twenty-two thousand entries. So

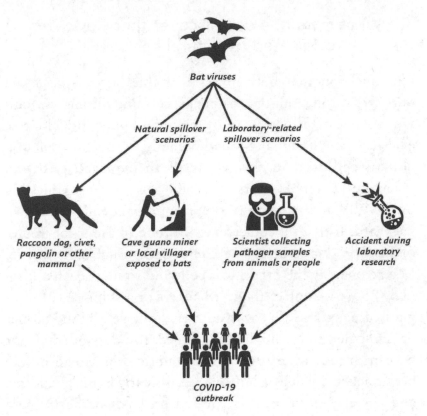

Four plausible paths that the virus could have taken from bats to people.

does the long record of virus collection, coronavirus genetic engineering and animal infection experiments in Wuhan. So does the track record of scientists inserting furin cleavage sites into viruses and the failure to find a furin cleavage site in another SARS-like virus despite widespread searches in China and other countries. And so does the failure of the market spill-over hypothesis to find any definitive evidence – no intermediate host, no immediate precursor to SARS-CoV-2, no animal samples found to be positive for the virus. These factors may not have proved the laboratory-leak hypothesis, but they have very clearly undermined the position of the natural spillover scenario as the default hypothesis.

What if we one day discover that Covid-19 originated by natural spillover?

One question that both authors of this book have asked ourselves throughout the writing process and during our individual investigations is what would happen if definitive evidence were eventually found pointing in the direction of a natural spillover. In the words of the journalist Rowan Jacobsen, 'Despite 15 years of coronavirus hunting and testing by the WIV, it was helpless to prevent a pandemic in its own backyard. If that's a coincidence, it's one of the great ironies of history.' Nonetheless, it could be a cruel coincidence. Maybe one day the origin will be found to have been a farm of SARS-CoV-2-infected ferret-badgers or a shipment of traumatised pangolins that was quickly destroyed by the farmer or smuggler to avoid being blamed for a pandemic. Or perhaps it may emerge that the zoonotic origins had in fact been tracked down by Chinese scientists early in the pandemic, and covertly destroyed or covered up so that the Chinese government could try to sell us the frozen foods not-from-China origins hypothesis. In this case, we would have maligned the very scientists who are working painstakingly to prevent pandemics.

It is a sobering thought. Until that happens, though, the current circumstances and the sparse evidence available demand that natural and laboratory-based origins are both treated as likely. Even if advocates for an investigation into the laboratory-leak hypothesis are eventually found to be wrong, we still think that it is imperative that a true investigation take place for the many reasons discussed in our book. The warnings about gain-of-function research, the worries about bringing viruses from the field into urban laboratories where accidents do happen, the anxieties about disturbing bats and other wild animal populations in the name of science – these are all still valid even if SARS-CoV-2 did emerge without the help of research activities.

And what if we discover that Covid-19 originated from a laboratory leak?

The possibility of SARS-CoV-2 having come from a lab has already stirred up chaos in the media and ugly spats between world leaders and governments. Indeed, it would have been much easier on scientists, science writers, journalists and the public if we could be sure that SARS-CoV-2 had spilled over into humans free of any laboratory involvement. If it is proven one day that SARS-CoV-2 somehow escaped from a laboratory, science's reputation could suffer a huge blow. Research that was previously thought to be life-saving and pandemic-preventing could suddenly stand exposed as a source of pandemics. The almost two decades of research that followed the SARS epidemic failed to keep its promises of preventing the next pandemic and yielding broad-spectrum SARS vaccines or therapeutics in time for the Covid-19 outbreak. (Although, in June 2021, Dr Baric's group published a proof-of-concept for a chimeric spike mRNA vaccine that can protect against various SARS-like virus infections in mice.) At best it left humanity with the agonising dilemma of Cassandra, the daughter of King Priam of Troy, who prophesied doom but went unheeded. At worst it was a Faustian bargain.

Virologists may find their grants terminated or at least diminished, their experiments banned or heavily regulated, and their reputations tarnished. Some might argue that science is generally regulated more tightly and conducted to higher standards outside China but that will not cut much ice with the general public especially considering the early vehement denials of the possibility of a laboratory leak by outspoken western scientists and science journalists. This backlash will also likely expand beyond the field of virology to impact other hotly controversial areas in science. Frankly, this reassessment will happen to some small degree even if laboratory science is fully exonerated in this case. Scientists, politicians, the media and the public have seen that even if it did not happen on this occasion, a dangerous virus could one day escape from a laboratory experiment. Experiments to alter viruses and make them potentially more dangerous, even unintentionally, need to be subject to a new round of review and debate.

Being fervent champions of science ourselves, we share the worries that scientists have. It has been demonstrated again and again – not least with the near miraculous development of novel, efficacious vaccines in 2020 – that science is vital to the future of humanity. For this precise reason, we knew that it was and is paramount for scientists to play leading roles in the investigation of the origin of the pandemic. This concern underlies our disappointment at the way some scientists closed minds and closed ranks against the lab-leak hypothesis so early in the pandemic, or at least showed far too little scepticism and curiosity on this issue. The world will have to negotiate a tougher and more transparent regime for regulating research on pathogens. And this cannot be done without scientists, alongside non-scientist stakeholders. The founding contributions from the internet sleuths and outsiders (several are scientists in their own right) cannot be denied, and should be celebrated, but it is scientists who must prove ourselves capable of credibly and transparently tracking the origins of SARS-CoV-2 – no matter where it leads us.

Setting a precedent

Nor do we think that finding the answer would be as damaging – to science and to international relations – as failing to find the answer. As Dr David Relman of Stanford University put it in November 2020: 'A more complete understanding of the origins of COVID-19 clearly serves the interests of every person in every country on this planet. It will limit further recriminations and diminish the likelihood of conflict; it will lead to more effective responses to this pandemic, as well as efforts to anticipate and prevent the next one.'

There is also a far more serious implication for scientists and humanity if world governments and organisations like the WHO ignore or dismiss the laboratory-leak hypothesis. Regimes around the world that are carrying out military-civilian, dual-use pathogen research, and terrorists who are also considering the use of bioweapons, are paying attention to what has happened. Not only will they have noticed the vast scale of disruption caused by an epidemic; they will also have noticed how easily the Chinese authorities dismissed a laboratory leak and neutered an international investigation, with the willing help of many scientific experts worldwide. Nefarious actors may have learned that they can easily get away with the creation and release of dangerous pathogens – with an unpredictably large impact on their target populations. This is true whether or not a laboratory leak was actually responsible in the case of Covid-19.

Fortunately, the mood is changing. While we were researching this book, more and more scientists began shifting their positions to agree that a laboratory-leak hypothesis must be properly investigated at the very least to set a precedent. Despite representing a wide spectrum of opinions on the likelihood of a laboratory leak as the cause of the pandemic, many scientists have united to write open letters calling for a credible and rigorous investigation of both natural and lab-based origin hypotheses. Indeed, the growing calls for an investigation have stimulated renewed conversation around the types of pathogen research being conducted around the world and what scientists can do to make this work safer.

In February 1975, a group of scientists gathered near a beach in California, at Asilomar near Monterey, to agree to voluntary guidelines to ensure that recombinant DNA technology would be used safely. They agreed to draw up safe-containment procedures in the laboratory, to limit the experiments to strains of bacteria that were unable to live outside the laboratory, and to avoid working on highly pathogenic organisms. Designed to reassure the public that science was being transparent – not least in the febrile atmosphere that followed the Watergate scandal – the conference set the tone of the debate for years to come. For almost fifty years thereafter biotechnology was indeed reassuringly safe. But it has long since abandoned the cautious limits set at Asilomar. As this book has documented, pathogenic viruses have indeed been the subject of frequent recombination or engineering experiments. As the pandemic ends, a new Asilomar is vitally needed.

Even if we never discover the origins of Covid-19, engaging in the process of investigation will have yielded life-saving outcomes if the world adopts new policies and mitigation strategies to prevent outbreaks that result from natural spillovers or laboratory accidents. A growing number of voices have already started to say that, regardless of the origin of Covid-19, actions must be taken to prevent laboratory-based outbreaks and regulate high-risk pathogen research. It remains to be seen whether there is enough momentum to drive the organisation of a new international moratorium and biosecurity treaty focused on pathogen research.

The beginnings of a true investigation into the origins of Covid-19

On 26 May 2021, the US President Joe Biden issued a statement describing the findings of the US intelligence community and calling for a redoubling of its efforts to collect and analyse information that could determine the origin of Covid-19. The statement revealed for the first time that the intelligence community had 'coalesced around two likely scenarios' – that of the virus emerging 'from human contact with an infected animal or from a laboratory accident' – but did not have adequate information to assess which was more likely. Since that announcement, bipartisan hearings and Senate briefings in the US have been convened to hear from scientists and other relevant experts what is currently known about the origin of the virus and what an international response or investigation should look like. Following one Senate briefing, Kansas senator Roger Marshall said, 'It would be utterly irresponsible to suffer through the worst pandemic in a century and not have the origins fully investigated.' And New York senator Kirsten Gillibrand said, 'We owe it to the American people to do everything we can to prevent future pandemics.'

We end this book with a tribute to those who made this change possible, especially the sleuths of Drastic, who laboured away for many months, mostly unrewarded with anything other than ridicule and insult. They dug deeper and darker tunnels than any scientist, into subterranean ores of data, in search of nuggets of truth about what happened in Wuhan and Yunnan in the years before 2020. People like the Seeker, Rossana Segreto, Francisco de Ribera, Yuri Deigin, Charles Small, Brian Reed, Billy Bostickson, Monali Rahalkar, Gilles Demaneuf, Moreno Colaiacovo and Babarlelephant – as well as others whose contributions we have not found the space to describe. Twitter was their only forum. Had they not persevered in their own time and at their own expense when all the well-funded forces of science, journalism, business and politics took little interest, we might never have had the chance to find as much truth as we have. That would have been a devastating insult to the millions who lost their lives to the terrible pandemic of Covid-19.

We also pay our respects to the whistleblowers in China who first sounded the alarm and hastened the global release of information regarding the novel outbreak in Wuhan. Some of them gave their lives and well-being to do so and are heroes in the truest meaning of the word. The impact of their bravery cannot be overstated. How many people would willingly jeopardise their safety and that of their loved ones to tell the world what little they know? The fate of the investigation into the origin of the pandemic still depends on courageous souls in China. There is hope that some of the few people who hold pieces of the puzzle within China's borders will one day, even if decades later, be able to illuminate the story of how SARS-CoV-2 found its way into Wuhan. We have seen time and again that the truth often comes out many years later.

The reader may want to know what the authors of this book think happened. Of course, we do not know for sure. There is still too much uncertainty and too much that is concealed. In this book we have tried to lay out the evidence and follow it wherever it leads, but it has not led us to a definitive conclusion. At the start of the pandemic neither of us thought the virus came from a laboratory leak. We both thought that a natural event would soon emerge as by far the most likely explanation. In the first month of 2020, the story of Covid-19 seemed to follow the recipe of the first SARS outbreak, where a chain of infection from animals to people quickly became apparent. However, we now think it very possible that the pandemic did result from the work of scientists either when collecting samples in the field or when working with those samples in a laboratory. The lab-leak hypothesis transformed from a conspiracy theory to just barely a possibility, to a plausible hypothesis worthy of a credible investigation. The world deserves to know the truth.

Epilogue
Truth will out

'The world deserves answers.'

PRESIDENT JOE BIDEN

On 27 August 2021, the US government's intelligence agencies published a brief, declassified summary of the results of their inquiry into the origin of the virus. The agencies were agreed that the virus was not a bioweapon and they thought with 'low confidence' that it was 'probably' not the product of genetic engineering. They also thought that Chinese officials had no foreknowledge of the pandemic but complained that China 'continues to hinder the global investigation, resist sharing information and blame other countries'. Beyond that they could not reach agreement. Four 'elements' of the intelligence community concluded with 'low confidence' that the source of the pandemic was natural exposure to an animal infected with it or a close progenitor virus. Three could not decide. Only one agency had reached a conclusion with 'moderate confidence' and that was that the first infection came about as a result of 'a laboratory-associated incident, probably involving experimentation, animal handling, or sampling by the Wuhan Institute of Virology'. The inquiry had therefore solved nothing, added little but reinforced the need to take both natural and lab-origin hypotheses seriously, to investigate them properly and to act as if both could happen again. A year before it would have been cataclysmic news that the US government thought that there was a fairly level playing field between a natural spillover and a laboratory leak. However, by August 2021, much of the American public and the international media had already shifted to regard both hypotheses as plausible – a laboratory origin was no longer extremely unlikely or the conspiracy theory it had been condemned as by numerous scientists and experts in 2020.

Remember that for the first year of the pandemic, as the death toll mounted, the debate about the origin of the virus had stayed on a single track. With rare exceptions, the natural spillover hypothesis was assumed by scientists, governments, officials, broadcasters and journalists to be by far the most likely explanation of how the virus first infected a human being. Taking their cue from Dr Peter Daszak's *Lancet* letter in February 2020 and Dr Kristian Andersen's *Nature Medicine* article in March that year, many public figures of influence in the west assumed that the idea of a laboratory leak was a fantasy of a few far-right politicians. Chinese sources were even more emphatic. The scientific establishment barely wavered, repeatedly telling journalists that the laboratory-leak hypothesis was a conspiracy theory. Few reporters or politicians knew enough biology to come to a different conclusion and most were content to take it on trust. Science journalists, unlike political or business journalists, are generally not in the habit of challenging their sources.

True, there were uncomfortable gaps in the story and they were growing wider, not narrower. For anybody following closely, the anomalies were piling up, as was the evidence that quite a lot of information was being held back.

Yet few of these stories broke through into the mainstream media, which showed magnificent incuriosity – with some notable exceptions. Within the World Health Organization and its advisors, only Jamie Metzl publicly expressed scepticism. Yuri Deigin's essay in *Medium* in April 2020, 'Lab-Made? SARS-CoV-2 Genealogy Through the Lens of Gain-of-Function Research', caught the attention of plenty of online readers, but was otherwise ignored. Milton Leitenberg's essay in June 2020 in the *Bulletin of the Atomic Scientists*, 'Did the SARS-CoV-2 Virus Arise from a Bat Coronavirus Research Program in a Chinese Laboratory? Very Possibly', had limited impact. Rowan Jacobsen's profile of Alina in *Boston Magazine* in September 2020, 'Could Covid-19 Have Escaped from a Lab?' led to few follow-ups. Nicholson Baker's long essay in *New York Magazine* in January 2021, 'The Lab-Leak Hypothesis', made only a few more waves. Our own articles in the *Wall Street Journal* in January and the *Telegraph* in March caused barely a ripple. Essays exploring gain-of-function research, the leakiness of laboratories or the bat virus-hunting expeditions of the EcoHealth Alliance appeared, but mostly in obscure, low-traffic outlets. Their authors were mostly unpaid mavericks who cared little for their own reputations. So the 'scientific consensus' held.

Only on Twitter did a lively debate continue. The Drastic group not only explored every angle, but their numbers grew, and a handful of senior scientists joined their conversations – especially in France, where the 'Paris group' crystallised thanks to the dedicated efforts of Dr Virginie Courtier-Orgogozo of the Institut Jacques Monod in Paris. The farcical WHO press conference in Beijing in February convinced almost nobody that frozen food was a more likely source of the pandemic than a laboratory experiment. Indeed, the press conference was almost certainly counterproductive for the Chinese regime, as the responses from western governments showed. Yet still the western media and scientific leadership largely steered clear of giving the laboratory-leak hypothesis any oxygen.

Part of the problem was the confusion of two different hypotheses: that the virus was deliberately engineered as a bioweapon or vaccine; or that it was a natural (or at least mostly natural) virus that accidentally escaped by infecting somebody during fieldwork or an experiment. Neither Dr Daszak nor Dr Andersen made such a clear distinction in their publications, using arguments mostly against the first idea, but implying that this ruled out both types of laboratory-leak scenarios. Certainly, that was the way many journalists read them. Moreover, many of the essays on the other side of the debate fell into the same trap, focusing mainly on the evidence for deliberate manipulation of the virus, rather than accidental release. Deigin's argument focused on the furin cleavage site. Steven Quay's lengthy manuscript on zenodo in January 2021, 'A Bayesian Analysis Concludes Beyond a Reasonable Doubt that SARS-CoV-2 Is Not a Natural Zoonosis But Instead Is Laboratory Derived', suggested a vaccine trial that went wrong. Two independent scientists, Angus Dalgleish and Birger Sorensen, struggled to get scientific journals to publish their analysis that the virus was certainly human-made.

Thus it was easy to polarise the debate and refract it through a political lens. The *Washington Post* accused Senator Tom Cotton of 'fanning the embers of a conspiracy theory that has been repeatedly debunked by experts', when he called for an open-minded investigation of all possibilities; the *New York Times* said the Wuhan laboratory had been 'the focus of unfounded conspiracy theories promoted by the Trump admin-istration', as if an accident could be a conspiracy; and National Public Radio reported that 'scientists debunk lab accident theory' when actually they had (barely) rebutted the genetically engineered theory.

Quite suddenly in May 2021, the dam crumbled. It is not clear what the trigger was. There were already rumblings and cracks in the concrete before the letter in *Science* that eighteen scientists signed in mid-May. Nicholas Wade's long essay in *Medium*, 'Origin of Covid – Following the Clues', went more viral than Deigin's or Baker's had. The *Wall Street Journal* detailed an intelligence report describing three WIV researchers needing hospital treatment in November 2019. By the start of June, just as we were completing this book, the world had changed. Dr Anthony Fauci altered his stance, conceding that he was 'not convinced' of a natural origin. A flood of coverage of the laboratory-leak hypothesis ensued in newspapers, on television, on the radio and online. We found ourselves in constant demand to comment, with even outlets such as National Public Radio for this first time showing an interest in the origin of the virus. Suddenly every aspect of the story came under scrutiny in the media. The gain-of-function controversies, and whether Dr Fauci and the NIH had knowingly funded such research in Wuhan, took centre stage in America.

The censorship that had marked the coverage now unravelled – rapidly. Facebook reversed its policy of censoring or labelling as misinformation any post that discussed a laboratory origin of Covid-19. Vox quietly edited past articles that had dismissed a lab origin out of hand. Fact checkers used by the media re-edited their articles or retracted their claims: 'We are removing this fact-check from our database pending a more thorough review,' wrote PolitiFact concerning its criticism of Li Meng Yan's work. Ian Birrell of the *Mail on Sunday* wrote an excoriating article detailing the role of the media in refusing to consider, let alone publish, anything that cast doubt on the natural spillover hypothesis: 'The failures of both new and traditional media, coalescing around the comfort blanket of the scientific establishment and their loathing of Trump, should provoke serious reflection.' Even the *New York Times* finally caved in and published a deep dive into the possibility of a laboratory-based origin of Covid-19 by Zeynep Tufekci. She wrote: 'Now, for the second time in 50 years, there are questions about whether we are dealing with a pandemic caused by scientific research.' (The first pandemic refers to the 1977 H1N1 influenza outbreak that was deemed to be the result of flawed vaccine trials.)

By August, even the Chinese state-controlled media appeared to have given up trying to persuade the world of a natural spill-over explanation and started to claim that a laboratory leak was indeed likely, only that it happened at the University of North Carolina. On 9 August 2021, *Global Times* published an article arguing that 'frequent accidents, attributed to lax safety procedures' at Dr Ralph Baric's lab in Chapel Hill pointed to that location as the likely source of the pandemic: 'With a mixed record on safety, the US' ambiguous, double-standard attitude toward the Covid-19 lab leak theory has led many in the public to become increasingly suspicious: it keeps smearing Chinese labs for "leaking the virus," while attempting to cover up its domestic situation.' How a leak in North Carolina might have caused a pandemic in China it did not explain.

The attempts by the Chinese government to deflect blame had by now descended into farce. On 10 August the Swiss embassy in Beijing took to Twitter to point out that a scientist named Wilson Edwards, supposedly based in Switzerland and widely amplified in Chinese state media since 24 July when he had joined Facebook to post a complaint about the origin tracing of Covid-19 and American pressure on the WHO, did not exist. Thus rumbled, 'Wilson Edwards' disappeared from Facebook soon afterwards.

At home in China, the actions of the Chinese state were far from farcical for brave whistleblowers who had tried to bring light to the nature of the virus as it had just emerged in Wuhan.

Two young activists, Chen Mai and Cai Wei, were found guilty by the Chaoyang District People's Court on 13 August of 'picking quarrels and stirring up trouble'. They were sentenced to prison for a year and three months each, though released for time served. Their crime? Creating a repository on the website GitHub of articles that had appeared online in China and which included reports, interviews and personal stories about the early weeks of the pandemic in Wuhan.

A citizen journalist, Zhang Zhan, was sentenced to four years in prison for daring to report on the Chinese government's shortcomings in handling the Wuhan outbreak. Hearing of her hunger strike and rapidly declining health, human rights activists warned that she might die in prison, following in the footsteps of so many other Chinese dissidents. Chinese reporters who spoke to Peter Hessler from the *New Yorker* remarked that the Chinese government-run media preferred success stories rather than the suffering of Wuhan residents. Scientists and officials were tight-lipped for good reason. One admitted that 'he knew that he would be judged by history' and would consider sharing his story if the climate were to change years later. A worker at a research institution wept and would not answer questions from the reporter but confessed that he had been recording details in his diary. As Hessler speculated hopefully, 'nowadays there are so many ways to preserve information. In time, we will learn more, but the delay is important to the Communist Party. It handles history the same way that it handles the pandemic – a period of isolation is crucial.'

It was therefore unsurprising when President Biden commented in August: 'Critical information about the origins of this pandemic exists in the People's Republic of China, yet from the beginning, government officials in China have worked to prevent international investigators and members of the global public health community from accessing it.' This confirmation of a cover-up by the Chinese government alongside a lack of new information surrounding the origin of Covid-19 was far from satisfying. The *New York Times* reported that 'current and former officials said the FBI believed that the virus was created in the lab'. The *Wall Street Journal* remarked that 'This might be a moment when quietly leaking the IC report's classified details can help public knowledge and keep the pressure on China.' Behind the inconclusive summary of the inquiry rumours continue to circulate. The three WIV workers who fell ill in November 2019 were said to have worked in the bat coronavirus laboratory and to have symptoms that suggested Covid-19: ground-glass opacities in the lungs and a loss of smell.

More voices are calling for a systematic and transparent review of the crucial information that exists outside of China. The EcoHealth Alliance, proud of having closely collaborated with the WIV on the surveillance of bat viruses for many years, must possess some data or correspondence that could be helpful in evaluating the likelihood of a laboratory origin. Assertions such as those by the president of the EcoHealth Alliance, Dr Daszak, that he had seen the WIV's pathogen database and that it contained no information of relevance are simply not acceptable in light of the extreme stakes. As late as September 2021, new and potentially concerning information from parties inside the US continued to be released to the public without any sense of urgency. At the end of July, the NIH had written a letter in response to a senator, revealing that during the time it had funded the EcoHealth Alliance, more chimeric coronavirus work had been proposed – this time involving MERS-CoV and related novel bat coronaviruses – to evaluate their behaviour in cells and animal infection models at the WIV.

The latest clues have indeed emerged outside China. On 6 September 2021, Labor Day in the United States, the *Intercept* published more than nine hundred pages of freshly released documents describing the NIH-funded research proposals and project updates by the EcoHealth Alliance and the WIV. The *Intercept* had to litigate to get these documents, which they had requested one year before under the Freedom of Information Act. The contents would undoubtedly have alarmed the public if released in early 2020. Yet by September 2021, many of the key pieces of information had already been scavenged by internet sleuths and scientists carefully poring over the dozens of research publications by the WIV. The documents confirmed: that the work on SARS-like coronaviruses had been conducted at BSL-2 and -3; that the WIV had at their fingertips a wide variety of boutique cell lines obtained from bats, primates, humans and other animals, which were used for the isolation and study of novel viruses that other virology laboratories do not possess; that they had purposefully sampled thousands of animals and human beings in rural China and in live animal markets that they expected to be at high risk of exposure to SARS-like viruses; that there were protocols for sampling bats, civets, raccoon dogs and other potential intermediate hosts of SARS-like CoVs, with a focus on avoiding being bitten by the animals while collecting samples; that the WIV brought these samples from spillover hotspots in southern China – places where 'wildlife trade, hunting, and bat guano collection is common' – back to Wuhan for in-depth study and the attempted isolation or synthesis of novel viruses. The scientists acknowledged that 'fieldwork involves the highest risk of exposure to SARS or other CoVs, while working in caves with high bat density overhead and the potential for fecal dust to be inhaled. There is also some risk of exposure to pathogens or physical injury while handling bats, civets, rodents or other animals, their blood samples or their excreta.' One important revelation was that the WIV had kept records of people who tested positive for signs of infection by animal viruses, and also tracked research-related incidents such as being bitten by bats or market animals. Does the EcoHealth Alliance have access to these medical records and data on animal bites suffered by researchers?

In addition, the documents showed that the WIV had shared with their US collaborators that by mid-2018, they had obtained results from testing chimeric SARS-like coronaviruses in humanised mice at BSL-3, potentially on the Wuhan University campus instead of at the WIV. In one case, a chimeric SARS-like virus, not previously disclosed, was found to replicate more efficiently, by two to three orders of magnitude, in the lungs of humanised mice than the non-chimeric version of the virus, and without causing observable weight loss (a way to gauge the severity of disease in mice). Other chimeric SARS-like viruses were also shown to grow much better in humanised mouse lungs compared to the non-chimeric virus, while retaining the ability to sicken the mice – causing up to 20 per cent body weight loss by the sixth day post-infection.

These preliminary findings from 2018 spoke to how unpredictable the results of generating chimeric viruses could be. You might get a virus that is less deadly but more transmissible, or a virus that is both more deadly and more transmissible. It is unclear if more humanised mouse experiments had been conducted after these striking findings and if other naturally isolated or chimeric SARS-like viruses had been tested in this animal model and perhaps later in civets. To make things worse, the WIV is known to have conducted classified research and has an institutional policy on protecting the confidentiality of work relating to state secrets and research secrets.

Just one week later, what purported to be another relevant research proposal by Dr Peter Daszak of the EcoHealth Alliance was shared with Drastic by an anonymous source. This proposal requesting $14.2 million in funding was submitted in March 2018 to the US Department of Defense's R&D agency, the Defense Advanced Research Projects Agency (DARPA). Its subcontractors included Drs Ralph Baric, Linfa Wang and Shi Zhengli. The application was ultimately unsuccessful, but its contents laid bare their extensive roadmap for collecting and experimenting with SARS-like viruses with pandemic potential. One of the criticisms of the proposal by the DARPA programme manager was that, although the approach potentially involved gain-of-function research, it did 'not mention or assess potential risks of Gain of Function (GoF) research and DURC [Dual Use Research of Concern]'.

The proposal was to test the spillover potential of viruses that it described as 'a clear-and-present danger to our military and to global health security' using a range of infection experiments including in captive bats in the lab, humanised mice and 'batenized' mice, that is, mice with bat immune cells. The early 2018 document described plans for a point that Dr Daszak had denied in 2020, namely that wild-caught *Rhinolophus* bats were to be kept and experimented with at the WIV. Further, it was revealed that by that time, the EcoHealth Alliance and its collaborators had already found more than 180 unique SARS-like viruses across approximately ten thousand samples.

Most importantly, the proposal contained the first written statements by the EcoHealth Alliance that it and its collaborators had plans to insert cleavage sites into engineered sarbecoviruses: 'We will analyze all SARSr-CoV S gene sequences for appropriately conserved proteolytic cleavage sites in S2 and for the presence of potential furin cleavage sites ... we will introduce appropriate human-specific cleavage sites and evaluate growth potential in Vero [monkey kidney] cells and HAE [human airway epithelial] cultures.' If functional cleavage sites were discovered in rare SARS-like viruses detected in their sequencing data, they planned to introduce these into virus strains in the lab. In other words, this early 2018 proposal told us that the EcoHealth Alliance and WIV were in possession of a growing semi-private collection of SARS-like viruses; they had intended to expand their recombinant virus infection experiments across a range of cells and animals; they had also delineated a workflow for identifying novel cleavage sites and inserting these into novel spikes and novel SARS-like viruses in the lab.

These revelations show that all is still to play for. Yet this is not a game. Our preference throughout was for a balanced debate that led to the truth, not for a victory for one side or the other. The world now faces the strong possibility that scientific research, intended to avert a pandemic, instead started one; that all that collecting of viruses and sampling of bats in remote caves – and then hiding the specimens in secret databases – had put humanity in harm's way. That two decades of research on the genomes of sarbecoviruses had not produced a vaccine but a plague. And that from the start there had been an active campaign by Chinese scientists and government officials, aided and abetted by far too many scientists, scientific journal editors and journalists in the west, to prevent the world finding out what happened.

If another pandemic of ambiguous origin occurs in the next decade – call it SARS-CoV-3, MERS-CoV-2, influenza, Nipah – then unless we learn key lessons from this pandemic, we will make the same mistakes. The world shows little sign yet of either finally shutting down the wildlife trade or addressing the risks of the burgeoning pathogen research worldwide, let alone both. There is little progress, if any, of governments designing a better system to encourage whistleblowing or transparency in science, public health and global pathogen surveillance. Rather the reverse: honest discussions among leaders and scientists appear to be happening increasingly on burner phones and in secure email channels. We can but hope like Shakespeare's Launcelot in *The Merchant of Venice*, that 'at the length truth will out'.

Timeline

2012

Six men are hospitalised in April and May after working in a bat-infested mine near Tongguan in Mojiang County, Yunnan. Infection by a bat virus is suspected and top laboratories including that of Dr Shi Zhengli of the Wuhan Institute of Virology (WIV) start to search the mine for viruses

2013

WIV scientists find RaTG13, a sarbecovirus (SARS-like coronavirus), in a sample taken from the Mojiang mine to Wuhan

2014

US moratorium on gain-of-function research

Beijing scientists reveal they searched the Mojiang mine but didn't find the virus that sickened the miners

Scientists from the lab of the then deputy director of the Chinese CDC, Dr George Gao, search the Mojiang mine for viruses

2015

The WIV finds eight other sarbecoviruses in the Mojiang mine

Dr Shi of the WIV co-authors a paper reporting the manipulation of a furin cleavage site in MERS and MERS-like spike genes

Dr Shi co-authors a paper with Dr Ralph Baric creating a chimeric sarbecovirus with a spike gene from a bat sarbecovirus

2016

Dr Shi publishes a small fragment of RaTG13 under the name BtCoV/4991

2017

Dr Shi reports the creation of eight recombinant bat sarbecoviruses

Moratorium on gain-of-function research ends; framework on work with enhanced potential pandemic pathogens is introduced

2018

RaTG13's full genome is sequenced

The WIV files patent for bat breeding cages

2019

The WIV takes its extensive pathogen database offline on 12 September; as far as we can tell, they were studying nine of the closest relatives to SARS-CoV-2 known to scientists at the time in the laboratory – all collected from the Mojiang mine

Wuhan Military World Games committee holds an exercise in how to respond to a novel coronavirus outbreak (September)

The WIV files patent for finger-wound device

WIV study of people living in sarbecovirus spillover zone in southern China says that bat coronavirus spillover is rare

Dr Peter Daszak describes progress with finding 100 novel sarbecoviruses; some can infect human cells and humanised mice in the laboratory

Covid-19/SARS-CoV-2 is first detected in Wuhan in late 2019

Detailed timeline in 2019–20

17 Nov
Fifty-five-year-old man in Hubei province is allegedly the earliest case to be retrospectively diagnosed

1 Dec
First clinically diagnosed case according to *Lancet* study

8 Dec
First clinically diagnosed case according to China-WHO joint study on the origin of Covid-19

24 Dec
Patient sample from Wuhan Central Hospital is sent for testing

26 Dec
The sample tests positive for a coronavirus

27 Dec
First SARS-CoV-2 genome is obtained and allegedly shared with the WIV

30 Dec
Doctors at Wuhan Central Hospital, Dr Ai Fen and Dr Li Wenliang, blow the whistle on novel SARS-related coronavirus infections

This information is leaked online and reaches the Chinese CDC director, Dr George Gao

Dr Shi is called to return to Wuhan from a Shanghai conference in order to look into the novel coronavirus outbreak; she wonders if it came from her laboratory because she 'never expected this kind of thing to happen in Wuhan, in central China'.

The WIV's pathogen database fact sheet is updated

ProMED sends a global alert regarding the novel coronavirus outbreak

31 Dec

Huanan seafood market is disinfected while hundreds of samples are taken from surfaces and animal carcasses or products

1 Jan

The authorities crack down on doctors and scientists sharing information about the outbreak in the media or online

3 Jan

Dr Li Wenliang is interrogated by police and made to sign a confession to wrongdoing

Chinese authorities order all patient samples to be destroyed or transferred to designated institutions

Dr Zhang Yongzhen in Shanghai receives Wuhan patient samples

The SARS-CoV-2 genome is shared with companies to develop diagnostics

4 Jan

The Chinese vaccine developer Sinopharm starts to manufacture a vaccine for SARS-CoV-2

5 Jan

Dr Zhang obtains the SARS-CoV-2 genome and uploads it to the international GenBank database under embargo; he tells Wuhan Central Hospital that they are dealing with a dangerous pathogen

7 Jan

Dr Zhang submits a paper to *Nature* describing the SARS-CoV-2 virus and its genome

11 Jan

As pressure grows, Dr Zhang shares the genome with his co-author Dr Edward Holmes who posts the genome online

13 Jan

Taiwan sends experts to Wuhan to check if human-to-human transmission is occurring

First Covid-19 case outside China: a traveller from Wuhan to Thailand

14 Jan

World Health Organization (WHO) announces that the Chinese authorities have found 'no clear evidence of human-to-human transmission of the novel #coronavirus (2019-nCoV) identified in #Wuhan, #China'

18 Jan

Sir Jeremy Farrar informs the WHO of data pointing to human-to-human transmission

Wuhan mayor hosts record-breaking banquet in the city

20 Jan

Chinese government confirms human-to-human transmission of the virus

Dr Shi submits her first two papers on the novel coronavirus to prestigious scientific journals; neither point out the furin cleavage site

22 Jan

Chinese CDC says the virus likely came from wild animals illegally sold at Huanan seafood market

Pangolin coronavirus data is released by Chinese scientists on the international NCBI database following a 2019 paper

23 Jan

Wuhan is locked down

Dr Shi preprints a seminal paper on the novel coronavirus and RaTG13

WHO votes not to declare an international emergency

31 Jan

China reports to the OIE World Animal Health Information System that no animal samples from the Huanan seafood market have tested positive for the virus

1 Feb

Top scientists and virologists in the West hold conference call to discuss the origin of the virus, including the possibility of a lab origin; several participants consider a lab origin at least as likely as a natural origin

6 Feb

Wuhan scientists post an article online suggesting a laboratory origin (the article is quickly withdrawn)

Dr Peter Daszak drafts a letter to condemn conspiracy theories suggesting anything other than a natural origin and recruits signatories

7 Feb
Dr Li Wenliang dies from Covid-19
Chinese scientists announce discovery of a pangolin virus that is allegedly a 99 per cent match to SARS-CoV-2

18–20 Feb
Four preprints are posted online describing the same pangolin coronavirus that has a similar spike receptor-binding domain to SARS-CoV-2, but is only 90 per cent similar overall

24 Feb
Chinese CDC informs its offices that information regarding the novel coronavirus must not be shared publicly

7 Mar
Dr Daszak's letter co-signed by twenty-seven prominent scientists is published by the *Lancet*

11 Mar
WHO declares a pandemic

17 Mar
Dr Kristian Andersen and colleagues publish 'Proximal Origin' correspondence in *Nature Medicine*: 'We do not believe that any type of laboratory-based scenario is plausible'

2 May
Alina's preprint on the adaptation of SARS-CoV-2 for humans is posted online

18 May

The Seeker posts the medical thesis describing the Mojiang miners under Alina's Twitter thread on her preprint

25 May

Dr Gao of the Chinese CDC announces that Huanan seafood market is a victim and that the virus had existed long before the market

29 May

The Seeker posts a doctoral thesis from Dr Gao's lab describing the Mojiang miners and providing GPS coordinates to the mine

31 May

The WIV and EcoHealth Alliance publish Latinne et al. preprint with data on viruses collected up till 2015

7 Jul

Alina and Dr Shing Hei Zhan post a preprint revealing scientific issues with the pangolin papers

31 Jul

The WHO agrees terms of reference for its joint study with the Chinese government on the origins of the virus

2 Aug

Francisco de Ribera and 'Babar' conclude that eight viruses from the Latinne et al. manuscript were from the Mojiang mine

11 Nov

An editor's note is added to a pangolin paper in *Nature* (though no correction has been made to date)

17 Nov
An addendum is added by Dr Shi to her *Nature* paper acknowl-
edging connection to the Mojiang mine and reporting additional
sarbecoviruses found in the mine

Detailed timeline in 2021

14 Jan
WHO-recruited international experts arrive in Wuhan

15 Jan
US State Department issues a fact sheet pointing to possibility
of a laboratory origin

9 Feb
China-WHO joint study press conference in Wuhan

4 Mar
Open letter by international experts and scientists, including
Alina, calls for a full investigation including that of a lab origin

30 Mar
China-WHO joint study publishes its report

14 May
Letter in *Science* co-signed by prominent scientists, including
Alina, calls for a credible investigation into both natural and
laboratory origin hypotheses

26 May
US President Joe Biden announces ninety-day inquiry by the
intelligence community into the origin of Covid-19

22 Jun

Dr Jesse Bloom unearths early Wuhan Covid-19 data deleted from the international NCBI database in June 2020; the data exonerate the Huanan seafood market

22 Jul

Chinese government announces that it will not cooperate with second phase of WHO's Covid-19 origins investigation

3 Aug

Chinese authorities share draft changes to the Regulation on the Administration of Laboratory Animals for public consultation, including the text 'Animals used in experiments cannot go back on the market'

6 Aug

Scientist in Beijing is revealed to have caught Covid-19 in a laboratory in early 2020

9 Aug

Chinese state-owned *Global Times* publishes allegations that laboratories at the University of North Carolina could have leaked the virus

27 Aug

Biden administration releases inconclusive report from US intelligence agencies about the origin of the virus

6 Sep

Documents released to the *Intercept* by National Institutes of Health reveal new information about the WIV's creation of chimeric viruses with increased pathogenicity

Acknowledgements

Both authors had an enormous amount of help, support, encouragement and advice from a range of people: confidential sources, senior scientists, junior researchers, intelligence officials, anonymous investigators, journalists, politicians, businesspeople, colleagues, literary agents, publishers, friends and family members. Because it would be unwise and unhelpful, possibly even dangerous, to name some of them, we have taken the unusual decision to name none of them. We hope those who would not mind being mentioned can forgive us, but we sincerely thank all of those who helped us so much with this extraordinary project, often going out of their way to do so.

Dr Shi Zhengli, Dr George Gao and Dr Peter Daszak, among others, were given the opportunity to respond to a series of questions relating to their portrayal in this book but chose not to respond to the authors in the requested time period.

Notes

1. The copper mine

13 *the BBC's John Sudworth and colleagues attempted to drive to the mine* John Sudworth. Twitter. 25 May 2021. https://twitter.com/TheJohnSudworth/status/1397223374745706510 John Sudworth. 'Covid: Wuhan Scientist Would "Welcome" Visit Probing Lab Leak Theory'. BBC. 21 December 2020. https://www.bbc.co.uk/news/world-asia-china-55364445

14 *John Sudworth was only one of many thwarted reporters* Dake Kang, Maria Cheng and Sam McNeil. 'China Clamps Down in Hidden Hunt for Coronavirus Origins'. Associated Press. 30 December 2020. https://apnews.com/article/united-nations-coronavirus-pandemic-china-only-on-ap-bats-24fbadc58cee3a40bca2ddf7a14d2955

14 *A team from NBC's* Today *show were told that wild elephants* Today. 2 April 2021, video report. https://twitter.com/TODAYshow/status/1377957516907311104

14 *In 2021, a team of undercover French journalists got close* Envoye Speciale. 11 March 2021, video report. https://twitter.com/franciscodeasis/status/1370183826731888641

14 Wall Street Journal *reported that its journalist did manage to get very close to the mine* Jeremy Page, Betsy McKay and Drew Hinshaw. 'The Wuhan Lab Leak Question: A Disused Chinese Mine Takes Center Stage'. *Wall Street Journal*. 24 May 2021. https://www.wsj.com/articles/wuhan-lab-leak-question-chinese-mine-covid-pandemic-11621871125

15 *So it is to satellite images that we turn* Brian Reed. Twitter. 4 February 2021. https://twitter.com/Drinkwater5Reed/status/1357213217085403139. Francisco de Ribera. Twitter. 6

April 2021. https://twitter.com/franciscodeasis/status/
1379386245911093250

15 *Two books unearthed by Reed Mojiang Hani Autonomous
    County History* 2002. *Geographical names of Mojiang County
    Autonomous County, Yunnan Province* 1985. Available at https://
    pastebin.com/u/franciscodeasis/1/ZE8kL74V. See also:
    Memorabilia of the Party History 1959. http://www.peds.gov.cn/
    mj/zx_nr.asp?id=3694

16 *A study of bat guano in a Thai cave* Wacharapluesadee, S.,
    Sintunawa, C., Kaewpom, T., Khongnomnan, K., Olival, K. J.,
    Epstein, J. H., Rodpan, A., Sangsri, P., Intarut, N., Chindamporn,
    A., Suksawa, K., & Hemachudha, T. (2013). 'Group C
    Betacoronavirus in Bat Guano Fertilizer, Thailand'. *Emerging
    Infectious Diseases*, 19(8), 1349–1351. https://doi.org/10.3201/
    eid1908.130119

18 *the 2013 medical thesis* https://www.documentcloud.org/
    documents/6981198-Analysis-of-Six-Patients-With-Unknown-
    Viruses.html

19 *They identified a new* Henipavirus-*like paramyxovirus* Wu, Z.,
    Yang, L., Yang, F., Ren, X., Jiang, J., Dong, J., Sun, L., Zhu, Y.,
    Zhou, H., & Jin, Q. (2014). 'Novel Henipa-like virus, Mojiang
    Paramyxovirus, in rats, China, 2012'. *Emerging Infectious
    Diseases*, 20(6), 1064–1066. https://doi.org/10.3201/
    eid2006.131022. Richard Stone. 'A New Killer Virus in China?'.
    Science. 20 March 2014. https://www.sciencemag.org/
    news/2014/03/new-killer-virus-china

19 *virus was incapable of replication in monkey, human or hamster
    cells* Rissanen, I., Ahmed, A. A., Azarm, K., Beaty, S., Hong, P.,
    Nambulli, S., Duprex, W. P., Lee, B., & Bowden, T. A. (2017).
    'Idiosyncratic Mòjiāng Virus Attachment Glycoprotein Directs a
    Host-Cell Entry Pathway Distinct from Genetically Related
    Henipaviruses'. *Nature Communications*, 8, 16060. https://doi.
    org/10.1038/ncomms16060

20 *They published a tiny part of its genomic sequence in 2016* Ge, X.
    Y., Wang, N., Zhang, W., Hu, B., Li, B., Zhang, Y. Z., Zhou, J. H.,
    Luo, C. M., Yang, X. L., Wu, L. J., Wang, B., Zhang, Y., Li, Z. X.,
    & Shi, Z. L. (2016). 'Coexistence of Multiple Coronaviruses in
    Several Bat Colonies in an Abandoned Mineshaft'. *Virologica
    Sinica*, 31(1), 31–40. https://doi.org/10.1007/s12250-016-3713-9

22 *Dr Yingle's paper was published in a prominent journal* Chen, L.,
    Liu, W., Zhang, Q., Xu, K., Ye, G., Wu, W., Sun, Z., Liu, F., Wu,

K., Zhong, B., Mei, Y., Zhang, W., Chen, Y., Li, Y., Shi, M., Lan, K., & Liu, Y. (2020). 'RNA Based mNGS Approach Identifies a Novel Human Coronavirus from Two Individual Pneumonia Cases in 2019 Wuhan Outbreak'. *Emerging Microbes & Infections*, 9(1), 313–319. https://doi.org/10.1080/22221751.2020.1725399

22 *on 3 February a paper had been published in the prestigious journal* Nature Zhou, P., Yang, X. L., Wang, X. G., Hu, B., Zhang, L., Zhang, W., Si, H. R., Zhu, Y., Li, B., Huang, C. L., Chen, H. D., Chen, J., Luo, Y., Guo, H., Jiang, R. D., Liu, M. Q., Chen, Y., Shen, X. R., Wang, X., Zheng, X. S., … Shi, Z. L. (2020). 'A Pneumonia Outbreak Associated with a New Coronavirus of Probable Bat Origin'. *Nature*, 579(7798), 270–273. https://doi.org/10.1038/s41586-020-2012-7

23 *published an article on his Virology Blog* Vincent Racaniello. 'Pangolins and the Origin of SARS-CoV-2 Coronavirus'. Virology.ws. 20 February 2020. https://www.virology.ws/2020/02/20/pangolins-and-the-origin-of-sars-cov-2-coronavirus/

25 *Daszak claimed in an interview* George Arbuthnott, Jonathan Calvert and Philip Sherwell. 'Revealed: Seven Year Coronavirus Trail from Mine Deaths to a Wuhan Lab'. *Sunday Times*. 4 July 2020. https://www.thetimes.co.uk/article/seven-year-covid-trail-revealed-l5vxt7jqp

26 *these online sleuths have filled the gap* Moreno Colaiacovo. 'The Origin of SARS-CoV-2 Is a Riddle: Meet the Twitter Detectives Who Aim to Solve It'. *Medium*. 3 December 2020. https://mygenomix.medium.com/the-origin-of-sars-cov-2-is-a-riddle-meet-the-twitter-detectives-who-aim-to-solve-it-5050216fd279

27 *calling itself the 'Decentralized Radical Autonomous Search Team Investigating COVID-19'* https://drasticresearch.org/the team/

30 *published an addendum* Zhou, P., Yang, X. L., Wang, X. G., Hu, B., Zhang, L., Zhang, W., Si, H. R., Zhu, Y., Li, B., Huang, C. L., Chen, H. D., Chen, J., Luo, Y., Guo, H., Jiang, R. D., Liu, M. Q., Chen, Y., Shen, X. R., Wang, X., Zheng, X. S., … Shi, Z. L. (2020). 'Addendum: A Pneumonia Outbreak Associated with a New Coronavirus of Probable Bat Origin'. *Nature*, 588(7836), E6. https://doi.org/10.1038/s41586-020-2951-z

32 *it told members of the WHO-China global study of origins of SARS-CoV-2* Joint WHO-China Study. WHO-Convened Global

Study of Origins of SARS-CoV-2: China Part. Joint Report –
ANNEXES. World Health Organization. 14 January–10 February
2021. https://www.who.int/docs/default-source/coronaviruse/
who-convened-global-study-of-origins-of-sars-cov-2-china-part-
annexes.pdf

2. Viruses

35 *Tyrrell had sent a sample to Almeida* Almeida J. (2008). June
Almeida (née Hart). *BMJ: British Medical Journal*, *336*(7659),
1511. https://doi.org/10.1136/bmj.a434

35 *She and seven colleagues wrote to the journal* Nature 'Virology:
Coronaviruses'. (1968). *Nature*, *220*(5168), 650. https://doi.
org/10.1038/220650b0

35 *Within the beta-coronaviruses lies a 'species'* Linked source was
incorrect

36 *Viruses are all around us all the time* Reche, I., D'Orta, G.,
Mladenov, N., Winget, D. M., & Suttle, C. A. (2018). 'Deposition
Rates of Viruses and Bacteria Above the Atmospheric Boundary
Layer'. *ISME journal*, *12*(4), 1154–1162. https://doi.org/10.1038/
s41396-017-0042-4

36 *can also infect animals such as cats, tigers, lions, minks* Smriti
Mallapaty. 'The Search for Animals Harbouring Coronavirus –
And Why It Matters'. *Nature*. 2 March 2021. https://www.nature.
com/articles/d41586-021-00531-z

36 *weighs about thirty quintillionths of an ounce* Bar-On, Y. M.,
Flamholz, A., Phillips, R., & Milo, R. (2020). 'SARS-CoV-2
(COVID-19) by the Numbers'. *eLife*, *9*, e57309. https://doi.
org/10.7554/eLife.57309

36 *could fit inside a single soda can* Christian Yates. 'Why All the
World's Coronavirus Would Fit in a Can of Cola'. *The
Conversation*. 10 February 2021. https://www.bbc.com/future/
article/20210210-why-the-entire-coronavirus-would-fit-in-a-can-
of-coca-cola

36 *The infectious diseases that have ravaged human populations*
Michael S. Rosenwald. 'History's Deadliest Pandemics, from
Ancient Rome to Modern America'. *Washington Post*. 7 April
2020. https://www.washingtonpost.com/graphics/2020/local/
retropolis/coronavirus-deadliest-pandemics/

37 *Bacteria were known from the late 1600s* Aria Nouri. 'The
Discovery of Bacteria'. AAAS. 5 July 2011. https://www.aaas.org/
discovery-bacteria

37 *the criteria known as Koch's postulates* Grimes, D. J. (2008). 'Koch's Postulates – Then and Now'. *Microbe, 1*(5): 223–228. DOI:10.1128/microbe.1.223.1 http://www.antimicrobe.org/history/Microbe-Grimes-Kochs%20Postulates-2006.pdf

37 *by inoculating healthy plants with sap from diseased plants* Artenstein A. W. (2012). 'The Discovery of Viruses: Advancing Science and Medicine by Challenging Dogma'. *International Journal of Infectious Diseases: Official Publication of the International Society for Infectious Diseases, 16*(7), e470–e473. https://doi.org/10.1016/j.ijid.2012.03.005

37 *Martinus Beijerinck coined the word 'virus'* Theresa Machemer. 'How a Few Sick Tobacco Plants Led Scientists to Unravel the Truth About Viruses'. *Smithsonian Magazine*. 24 March 2020. https://www.smithsonianmag.com/science-nature/what-are-viruses-history-tobacco-mosaic-disease-180974480/

40 *when the virus makes it into the lungs* David Cyranoski. 'Profile of a Killer: The Complex Biology Powering the Coronavirus Pandemic'. *Nature*. 4 May 2020. https://www.nature.com/articles/d41586-020-01315-7

43 *In 2005, Dr Marc van Ranst at Leuven University in Belgium examined parts of the genomes of the two viruses* Vijgen, L., Keyaerts, E., Moës, E., Thoelen, I., Wollants, E., Lemey, P., Vandamme, A. M., & Van Ranst, M. (2005). 'Complete Genomic Sequence of Human Coronavirus OC43: Molecular Clock Analysis Suggests a Relatively Recent Zoonotic Coronavirus Transmission Event'. *Journal of Virology, 79*(3), 1595–1604. https://doi.org/10.1128/JVI.79.3.1595-1604.2005

43 *the coincidence of the date of the pandemic and the likely date of OC43's arrival in the human species* 'Q&A: Why History Suggests Covid-19 Is Here to Stay'. *Biocev*. 14 April 2021. https://www.biocev.eu/en/about/projects/micobion.2/q-a-why-history-suggests-covid-19-is-here-to-stay.271

44 *The outbreak began in central Asia, in the independent city state of Bukhara in May 1889* Charles River Editors. *The 1889–1890 Flu Pandemic: The History of the 19th Century's Last Major Global Outbreak* (pp. 13–14). Kindle Edition

45 *A key insight came from the evolutionary biologist Dr Paul Ewald* Ewald P. W. (2011). 'Evolution of Virulence, Environmental Change, and the Threat Posed by Emerging and Chronic Diseases'. *Ecological Research, 26*(6), 1017–1026. https://doi.org/10.1007/s11284-011-0874-8

3. The Wuhan whistleblowers

49 *At least two western infectious diseases experts* James Rainey and Kiera Feldman. 'Impeachment. Primaries. Kobe. Coronavirus Rushed in While Our Focus Was Elsewhere'. *Los Angeles Times*. 12 April 2020. https://www.latimes.com/california/story/2020-04-12/coronavirus-attention-impeachment-primaries-kobe; see also: Vincent Racaniello and Ian Lipkin. 'TWiV Special: Conversation with a COVID-19 Patient, Ian Lipkin'. *This Week in Virology*. 28 March 2020. https://www.microbe.tv/twiv/twiv-special-lipkin/; Nicholas A. Christakis @NAChristakis Twitter. 31 August 2021. https://twitter.com/NAChristakis/status/1432875764895133710; 'Virologists on the Coronavirus Outbreak'. VPRO Documentary. 27 March 2020. https://www.youtube.com/watch?v=5DeUBLeMnjY

49 *Wuhan Central Hospital took a sample from a patient's lung on 24 December 2019* Jane McMullen. 'China's COVID Secrets'. PBS *Frontline*. 2 February 2021. https://www.pbs.org/wgbh/frontline/film/chinas-covid-secrets/transcript/

49 *the first person to sound the alarm on the emerging coronavirus* Li, X., Cui, W., & Zhang, F. (2020). 'Who Was the First Doctor to Report the COVID-19 Outbreak in Wuhan, China?' *Journal of Nuclear Medicine: Official Publication, Society of Nuclear Medicine*, 61(6), 782–783. https://doi.org/10.2967/jnumed.120.247262; Billie Thomson. 'The Doctor Who "Discovered the Coronavirus": Wuhan Medic Raised the Alarm to Authorities in December After Noticing "Mysterious Pneumonia" That Could Spread Between People'. *Daily Mail*. 29 April 2020. https://www.dailymail.co.uk/news/article-8268509/Meet-doctor-discovered-coronavirus.html

50 *Across town, on 27 December, Dr Ai Fen* Lily Kuo. 'Coronavirus: Wuhan Doctor Speaks Out Against Authorities'. *Guardian*. 11 March 2020. https://www.theguardian.com/world/2020/mar/11/coronavirus-wuhan-doctor-ai-fen-speaks-out-against-authorities

51 *Don't circulate the message outside this group* 'New Corona Pneumonia "Whistleblower" Li Wenliang: Reality Is More Important Than Redress'. Caixin. 31 January 2020. http://china.caixin.com/2020-01-31/101509761.html

51 *who contacted Dr Marjorie Pollack, the deputy editor of ProMED-mail* Jane McMullen. 'Covid-19: Five Days That Shaped

The Outbreak'. BBC. 26 January 2021. https://www.bbc.com/news/world-55756452

51 *As the world learned of the outbreak on the last day of 2019* 'Chinese Officials Investigate Cause of Pneumonia Outbreak in Wuhan'. Reuters. 31 December 2019. https://www.reuters.com/article/us-china-health-pneumonia/chinese-officials-investigate-cause-of-pneumonia-outbreak-in-wuhan-idUSKBN1YZ0GP

51 *the Huanan seafood market was being shuttered by the authorities* Jeremy Page and Natasha Khan. 'On the Ground in Wuhan, Signs of China Stalling Probe of Coronavirus Origins'. *Wall Street Journal*. 12 May 2020. https://www.wsj.com/articles/china-stalls-global-search-for-coronavirus-origins-wuhan-markets-investigation-11589300842

52 *Dr Ai was rebuked by her hospital* Lily Kuo. 'Coronavirus: Wuhan Doctor Speaks Out Against Authorities'. *Guardian*. 11 March 2020. https://www.theguardian.com/world/2020/mar/11/coronavirus-wuhan-doctor-ai-fen-speaks-out-against-authorities

52 *Dr Li developed Covid-19 symptoms* Stephanie Hegarty. 'The Chinese Doctor Who Tried to Warn Others About Coronavirus'. BBC. 6 February 2020. https://www.bbc.com/news/world-asia-china-51364382

52 *left behind a pregnant wife and a young child* 'Li Wenliang: Widow of Chinese Coronavirus Doctor Gives Birth to Son'. BBC. 12 June 2020. https://www.bbc.com/news/world-asia-53021852

52 *the government's management of the outbreak and Dr Li's whistleblowing* Nie, J. B., & Elliott, C. (2020). 'Humiliating Whistle-Blowers: Li Wenliang, the Response to Covid-19, and the Call for a Decent Society'. *Journal of Bioethical Inquiry*, 17(4), 543–547. https://doi.org/10.1007/s11673-020-09990-x

52 *the Chinese authorities had obtained the genome sequence* 'Exclusive | Coronavirus Gene Sequencing Traceability: When Did the Alarm Sound?' Caixin. 26 February 2020. https://web.archive.org/web/20200227094018/http://china.caixin.com/2020-02-26/101520972.html; see also Jeremy Farrar (2021). *Spike: The Virus vs. The People*. Profile.

53 *A day after Dr Li signed the confession letter* John Sudworth. 'Wuhan: City of Silence'. BBC. 7 July 2021. https://www.bbc.co.uk/news/extra/ewsu2giezk/city-of-silence-china-wuhan

53 *Sinopharm kicked into high gear manufacturing a vaccine* Dake Kang. 'China Testing Blunders Stemmed from Secret Deals with

Firms'. Associated Press. 3 December 2020. https://apnews.com/
article/china-virus-testing-secret-deals-firms-312f4a953e0
264a3645219a08c62a0ad; Norman. Twitter. 29 November
2020. https://twitter.com/norman7177/status/133309
6602765062144

53 *According to Chinese Human Rights Defenders* 'China: Protect
Human Rights While Combatting Coronavirus Outbreak. Chinese
Human Rights Defenders'. 31 January 2020. https://www.nchrd.
org/2020/01/china-protect-human-rights-while-combatting-
coronavirus-outbreak/

53 *Taiwanese health authorities were carefully monitoring Chinese
social media* 'The Facts Regarding Taiwan's Email to Alert WHO
to Possible Danger of COVID-19'. Taiwan Centers for Disease
Control. 11 April 2020. https://www.cdc.gov.tw/Category/
ListContent/sOn2_m9QgxKqhZ7omgiz1A?uaid=PAD-
lbwDHeN_bLa-viBOuw

53 *The Taiwanese government sent Professor Chuang Yin-ching* Tom
Cheshire. 'COVID-19 One Year On: The Taiwanese Scientist Who
Tried to Warn the World of Coronavirus'. Sky News. 7 December
2020. https://news.sky.com/story/covid-19-one-year-on-the-
taiwanese-scientist-who-tried-to-warn-the-world-of-
coronavirus-12152979

54 *Taiwan has still not been invited to join as a member of the
WHO* 'Why Taiwan Has Become a Problem for WHO'.
BBC. 30 March 2020. https://www.bbc.com/news/world-
asia-52088167

54 *the WHO had been informed by Sir Jeremy Farrar* Jeremy Farrar
(2021). *Spike: The Virus vs. The People.*

54 *Separate groups of scientists analysing genetic data* van Dorp, L.,
Acman, M., Richard, D., Shaw, L. P., Ford, C. E., Ormond, L.,
Owen, C. J., Pang, J., Tan, C., Boshier, F., Ortiz, A. T., & Balloux,
F. (2020). 'Emergence of Genomic Diversity and Recurrent
Mutations in SARS-CoV-2'. *Infection, Genetics and Evolution:
Journal of Molecular Epidemiology and Evolutionary Genetics in
Infectious Diseases*, 83, 104351. https://doi.org/10.1016/j.
meegid.2020.104351; see also Kumar, S., Tao, Q., Weaver, S.,
Sanderford, M., Caraballo-Ortiz, M. A., Sharma, S., Pond, S., &
Miura, S. (2021). 'An Evolutionary Portrait of the Progenitor
SARS-CoV-2 and Its Dominant Offshoots in COVID-19
Pandemic'. *bioRxiv: The Preprint Server for Biology*,
2020.09.24.311845. https://doi.org/10.1101/2020.09.24.311845

54 *According to government data seen by the* South China Morning Post *in March 2020* Josephine Ma. 'Coronavirus: China's First Confirmed Covid-19 Case Traced Back to November 17'. *South China Morning Post*. 13 March 2020. https://www.scmp.com/news/china/society/article/3074991/coronavirus-chinas-first-confirmed-covid-19-case-traced-back

55 *The hospitals and laboratories that possessed early samples from December 2019 were instructed by the authorities to destroy them* Sophia Ankel. 'China Confirms US Accusations That It Destroyed Early Samples of the Novel Coronavirus, But Says It Was Done for "Biosafety Reasons"'. *Business Insider*. 16 May 2020. https://www.businessinsider.com/china-confirms-that-it-destroyed-early-samples-of-new-coronavirus-2020-5

55 *a later study by the University of Kent in Canterbury* Roberts, D. L., Rossman, J. S., & Jarić, I. (2021). 'Dating First Cases of COVID-19'. *PLoS Pathogens*, 17(6), e1009620. https://doi.org/10.1371/journal.ppat.1009620

55 *nine thousand athletes from a hundred countries participated* Diane Francis. 'Canadian Forces Have Right to Know If They Got COVID at the 2019 Military World Games in Wuhan'. *Financial Post*. 25 June 2021. https://financialpost.com/diane-francis/diane-francis-canadian-forces-have-right-to-know-if-they-got-covid-at-the-2019-military-world-games-in-wuhan

55 *Xinhua reported that on 18 September* 'Xinhua: On Sep 18, 2019, Wuhan Held an Emergency Response Drill for Novel Coronavirus'. Xinhua. 26 September 2019. http://chinascope.org/archives/21681? doing_wp_cron=1613400458.4895019531250000000000

55 *'investigative thread languishing due to neglect'* Josh Rogin. Twitter. 26 June 2021. https://twitter.com/joshrogin/status/1408812303345192968

55 *more than a hundred people could well have already contracted the disease* Julia Hollingsworth and Yong Xiong. 'The Truthtellers'. CNN. February 2021. https://www.cnn.com/interactive/2021/02/asia/china-wuhan-covid-truthtellers-intl-hnk-dst/

56 *a study by Wuhan and other Chinese scientists published on 24 January 2020* Chen, N., Zhou, M., Dong, X., Qu, J., Gong, F., Han, Y., Qiu, Y., Wang, J., Liu, Y., Wei, Y., Xia, J., Yu, T., Zhang, X., & Zhang, L. (2020). 'Epidemiological and Clinical Characteristics of 99 Cases of 2019 Novel Coronavirus

Pneumonia in Wuhan, China: A Descriptive Study'. *Lancet (London, England)*, 395(10223), 507–513. https://doi. org/10.1016/S0140-6736(20)30211-7

56 *Dr Jesse Bloom of the Fred Hutchinson Cancer Research Center made a new breakthrough* Jesse D. Bloom. 'Recovery of Deleted Deep Sequencing Data Sheds More Light on the Early Wuhan SARS-CoV-2 Epidemic'. *bioRxiv: The Preprint Server for Biology* 2021.06.18.449051; doi: https://doi.org/10.1101/2021.06. 18.449051

58 *the vice minister of China's National Health Commission tried to clarify* Carl Zimmer. 'Those Virus Sequences That Were Suddenly Deleted? They're Back' *New York Times*. 30 July 2021. https:// www.nytimes.com/2021/07/30/science/coronavirus-sequences-lab-leak.html

59 *he set about sequencing the genome of the virus* Charlie Campbell. 'Exclusive: The Chinese Scientist Who Sequenced the First COVID-19 Genome Speaks Out About the Controversies Surrounding His Work'. *Time*. 24 August 2020. https://time. com/5882918/zhang-yongzhen-interview-china-coronavirus-genome/; 'China's Response to the New Coronavirus Gains International Recognition'. 17 January 2020. http://www.rmzxb. com.cn/c/2020-01-17/2508144.shtml

60 *According to an investigation by the Associated Press* Dake Kang. 'China Testing Blunders Stemmed from Secret Deals with Firms'. Associated Press. 3 December 2020

60 *Dr Nick Loman from the University of Birmingham* @ pathogenomenick Twitter response to Alina Chan. 5 February 2021. https://twitter.com/pathogenomenick/ status/1357808339321827337

60 *Jeremy Farrar tweeted* @JeremyFarrar Tweet. 10 January 2020. https://twitter.com/JeremyFarrar/status/1215647022893670401; see also Jeremy Farrar (2021). *Spike: The Virus vs. The People*.

61 *published the novel coronavirus genome on a site called Virological.org* Edward Holmes. 'Novel 2019 Coronavirus Genome'. Virological. 10 January 2020. https://virological.org/t/ novel-2019-coronavirus-genome/319

61 *According to accounts by Dr Holmes* Kate Aubusson. 'Virus Rebel Professor Edward Holmes Named NSW Scientist of the Year'. *Sydney Morning Herald*. 27 October 2020. https://www. smh.com.au/national/nsw/virus-rebel-professor-edward-holmes-named-nsw-scientist-of-the-year-20201026-p568qj.html

61 *Later, Dr Zhang connected with GenBank to release the genome from embargo* Wuhan seafood market pneumonia virus isolate Wuhan-Hu-1, complete genome. GenBank: MN908947.1. https://www.ncbi.nlm.nih.gov/nuccore/MN908947.1

62 *the Chinese National Health Commission had told the WIV's director general, Dr Wang Yanyi* Nick Schifrin. 'How Virus Research Has Become a Point of Tension for the US and China'. PBS *Newshour*. 22 May 2020. https://www.pbs.org/newshour/show/how-virus-research-has-become-a-point-of-tension-for-the-u-s-and-china

62 *Dr Zhang's laboratory was closed for 'rectification' on 12 January* Pinghui Zhuang. 'Chinese Laboratory That First Shared Coronavirus Genome with World Ordered to Close for "Rectification", Hindering Its Covid-19 Research'. *South China Morning Post*. 28 February 2020. https://www.scmp.com/news/china/society/article/3052966/chinese-laboratory-first-shared-coronavirus-genome-world-ordered

62 *Documents leaked to the Associated Press* Dake Kang, Maria Cheng and Sam McNeil. 'China Clamps Down in Hidden Hunt for Coronavirus Origins'. Associated Press. 30 December 2020. https://apnews.com/article/united-nations-coronavirus-pandemic-china-only-on-ap-bats-24fbadc58cee3a40bca2ddf7a14d2955

62 *CNN reported that staff from the Chinese education ministry verified the directive* Nectar Gan, Caitlin Hu and Ivan Watson. 'Beijing Tightens Grip Over Coronavirus Research, Amid US-China Row on Virus Origin'. CNN. 16 April 2020. https://www.cnn.com/2020/04/12/asia/china-coronavirus-research-restrictions-intl-hnk/index.html

63 *Wuhan's mayor Zhou Xianwang had hosted a record-breaking banquet* James Griffiths. 'Is Wuhan's Mayor Being Set Up to Be the Fall Guy for the Virus Outbreak?' CNN. 29 January 2020. https://www.cnn.com/asia/live-news/coronavirus-outbreak-01-29-20-intl-hnk/h_6d8cf9d5c0b2cf01447dd24325ed6dd3; James Kynge, Sun Yu and Tom Hancock. 'Coronavirus: The Cost of China's Public Health Cover-Up'. *Financial Times*. 6 February 2020. https://www.ft.com/content/fa83463a-4737-11ea-aeb3-955839e06441; 'Timeline: China's COVID-19 Outbreak and Lockdown of Wuhan'. Associated Press. 22 January 2021. https://apnews.com/article/pandemics-wuhan-china-coronavirus-pandemic-e6147ec0ff88affb99c811149424239d

63 *the first case of Covid-19 outside China was confirmed in Thailand* Lisa Schnirring. 'Report: Thailand's Coronavirus Patient Didn't Visit Outbreak Market'. Center for Infectious Disease Research and Policy. 14 January 2020. https://www.cidrap.umn. edu/news-perspective/2020/01/report-thailands-coronavirus-patient-didnt-visit-outbreak-market

63 *The first case in Japan was confirmed on 15 January* Lisa Schnirring. 'Japan Has 1st Novel Coronavirus Case; China Reports Another Death'. Center for Infectious Disease Research and Policy. 16 January 2020. https://www.cidrap.umn.edu/news-perspective/2020/01/japan-has-1st-novel-coronavirus-case-china-reports-another-death

64 *Covid-19 cases started to bubble up around the world* USA: https://www.nejm.org/doi/full/10.1056/NEJMoa2001191. France: https://www.ncbi.nlm.nih.gov/pmc/articles/PMC7029452/. Canada: https://www.canadianhealthcarenetwork.ca/covid-19-a-canadian-timeline. Australia: https://www.abc.net.au/news/2020-01-25/first-confirmed-coronavirus-case-australian-as-china-toll-rises/11900428. Germany: https://www.thelancet.com/journals/laninf/article/PIIS1473-3099(20)30314-5/fulltext. India: https://www.ncbi.nlm.nih.gov/pmc/articles/PMC7530459/. Italy: https://www.corriere.it/cronache/20_gennaio_31/virus-primi-due-casi-italia-due-cinesi-marito-moglie-italia-dieci-giorni-e365df1c-43b3-11ea-bdc8-faf1f56f19b7.shtml. Sweden: https://www.bloomberg.com/news/articles/2020-01-31/sweden-reports-first-case-of-confirmed-coronavirus. UK: https://www.bbc.co.uk/news/uk-england-55622386

64 *Dr Botao Xiao and Lei Xiao, husband and wife scientists in Wuhan, published an online article* Botao Xiao and Lei Xiao. 'The Possible Origins of 2019-nCoV Coronavirus'. ResearchGate preprint. 6 February 2020. Available at: https://img-prod.tgcom24.mediaset.it/images/2020/02/16/114720192-5eb8307f-017c-4075-a697-348628da0204.pdf

66 *Dr Botao Xiao wrote in an email to the* Wall Street Journal *on 26 February* James T. Areddy. 'Coronavirus Epidemic Draws Scrutiny to Labs Handling Deadly Pathogens'. *Wall Street Journal*. 5 March 2020. https://www.wsj.com/articles/coronavirus-epidemic-draws-scrutiny-to-labs-handling-deadly-pathogens-11583349777

66 *The interview with Dr Shi was published in* Scientific American Jane Qiu. 'How China's "Bat Woman" Hunted Down Viruses from SARS to the New Coronavirus'. *Scientific American*. 11

March 2020. https://www.scientificamerican.com/article/
how-chinas-bat-woman-hunted-down-viruses-from-sars-to-the-
new-coronavirus1/

4. The seafood market

68 *Dr Gao was just as certain that the market was not the source of
the outbreak* Aristos Georgiou. 'Wuhan Seafood Market Was a
"Victim" of Coronavirus, Says Director of China's CDC'.
Newsweek. 27 May 2020. https://www.newsweek.com/wuhan-
seafood-market-victim-coronavirus-china-cdc-1506766

69 *the Paris-based World Organisation for Animal Health (OIE) was
briefed by the Chinese government* Jeremy Page and Natasha
Khan. 'On the Ground in Wuhan, Signs of China Stalling Probe of
Coronavirus Origins'. *Wall Street Journal*. 12 May 2020. https://
www.wsj.com/articles/china-stalls-global-search-for-coronavirus-
origins-wuhan-markets-investigation-11589300842

70 *On 21 February 2003, Dr Liu Jianlun, a sixty-four-year-old
medical professor from Guangzhou* Update 95 – SARS:
Chronology of a Serial Killer. World Health Organization. https://
www.who.int/csr/don/2003_07_04/en/ (Website removed.)

70 *In Guangdong, domestic cat meat is a delicacy* Josephine Ma.
'Guangdong Gourmets Eat 10,000 Cats a Day'. *South China
Morning Post*. 4 December 2002. https://www.scmp.com/
article/399602/guangdong-gourmets-eat-10000-cats-day

70 *domestic cats had been determined* Martina, B. E., Haagmans, B.
L., Kuiken, T., Fouchier, R. A., Rimmelzwaan, G. F., Van
Amerongen, G., Peiris, J. S., Lim, W., & Osterhaus, A. D. (2003).
'Virology: SARS Virus Infection of Cats and Ferrets'. *Nature*,
425(6961), 915. https://doi.org/10.1038/425915a

71 *SARS seems to have erupted several times across seven
municipalities* Xu, R. H., He, J. F., Evans, M. R., Peng, G. W.,
Field, H. E., Yu, D. W., Lee, C. K., Luo, H. M., Lin, W. S., Lin, P.,
Li, L. H., Liang, W. J., Lin, J. Y., & Schnur, A. (2004).
'Epidemiologic Clues to SARS Origin in China'. *Emerging
Infectious Diseases*, 10(6), 1030–1037. https://doi.org/10.3201/
eid1006.030852

71 *the WHO was severely critical of the Chinese government for its
lack of openness* Thomas Crampton. 'WHO Criticizes China Over
Handling of Mystery Disease'. *New York Times*. 7 April 2003.
https://www.nytimes.com/2003/04/07/international/asia/
who-criticizes-china-over-handling-of-mystery-disease.html

72 *The patients in ambulances were driven around the city while the WHO team visited the hospital* John Pomfret. 'Beijing Told Doctors to Hide SARS Victims'. *Washington Post*. 20 April 2003. https://www.washingtonpost.com/archive/politics/2003/04/20/beijing-told-doctors-to-hide-sars-victims/3eb7d1aa-d2ff-477b-bc15-d0164377b123/

73 *it was not until 1989 that scientists discovered a very closely related virus in wild chimps* Zoltan Fehervari. 'Origin Story'. *Nature*. 28 November 2018. https://www.nature.com/articles/d42859-018-00008-6

74 *a team of scientists from Hong Kong visited a live animal market in Shenzhen* Guan, Y., Zheng, B. J., He, Y. Q., Liu, X. L., Zhuang, Z. X., Cheung, C. L., Luo, S. W., Li, P. H., Zhang, L. J., Guan, Y. J., Butt, K. M., Wong, K. L., Chan, K. W., Lim, W., Shortridge, K. F., Yuen, K. Y., Peiris, J. S., & Poon, L. L. (2003). 'Isolation and Characterization of Viruses Related to the SARS Coronavirus from Animals in Southern China'. *Science (New York, N.Y.)*, *302*(5643), 276–278. https://doi.org/10.1126/science.1087139

75 *Similar studies were conducted in the same month of May by Guangdong scientists* Centers for Disease Control and Prevention (CDC) (2003). 'Prevalence of IgG Antibody to SARS-Associated Coronavirus in Animal Traders – Guangdong Province, China, 2003'. *MMWR. Morbidity and Mortality Weekly Report*, *52*(41), 986–987

75 *The highest antibody prevalence (72.7 per cent) was found in those who primarily traded civets* Tu, C., Crameri, G., Kong, X., Chen, J., Sun, Y., Yu, M., Xiang, H., Xia, X., Liu, S., Ren, T., Yu, Y., Eaton, B. T., Xuan, H., & Wang, L. F. (2004). 'Antibodies to SARS Coronavirus in Civets'. *Emerging Infectious Diseases*, *10*(12), 2244–2248. https://doi.org/10.3201/eid1012.040520

76 *The waitress was diagnosed with possible SARS on 2 January 2004* Wang, M., Yan, M., Xu, H., Liang, W., Kan, B., Zheng, B., Chen, H., Zheng, H., Xu, Y., Zhang, E., Wang, H., Ye, J., Li, G., Li, M., Cui, Z., Liu, Y. F., Guo, R. T., Liu, X. N., Zhan, L. H., Zhou, D. H., … Xu, J. (2005). 'SARS-CoV Infection in a Restaurant from Palm Civet'. *Emerging Infectious Diseases*, *11*(12), 1860–1865. https://doi.org/10.3201/eid1112.041293

77 *ten thousand civets were exterminated* Philip P. Pan. 'Fearing SARS, China Begins Mass Killing of Civet Cats'. *Washington Post*. 7 January 2004. https://www.washingtonpost.com/archive/

politics/2004/01/07/fearing-sars-china-begins-mass-killing-of-civet-cats/a4955b83-76f1-415d-9797-5a37a788b1c4/

77   *a team of Hong Kong scientists captured 127 bats, 60 rodents and 20 monkeys from 11 locations in the wild in Hong Kong* Lau, S. K., Woo, P. C., Li, K. S., Huang, Y., Tsoi, H. W., Wong, B. H., Wong, S. S., Leung, S. Y., Chan, K. H., & Yuen, K. Y. (2005). 'Severe Acute Respiratory Syndrome Coronavirus-Like Virus in Chinese Horseshoe Bats'. *Proceedings of the National Academy of Sciences of the United States of America*, 102(39), 14040–14045. https://doi.org/10.1073/pnas.0506735102

78   *Dr Zaki took a sample from the throat* Islam Hussein. 'The Story of the First MERS Patient'. *Nature Middle East*. 2 June 2014. https://www.natureasia.com/en/nmiddleeast/article/10.1038/nmiddleeast.2014.134

79   *MERS cases have been sporadic* 'Middle East Respiratory Syndrome'. World Health Organization. http://www.emro.who.int/health-topics/mers-cov/mers-outbreaks.html

79   *Some MERS-like coronaviruses have been identified in bats* Goldstein, S. A., & Weiss, S. R. (2017). 'Origins and Pathogenesis of Middle East Respiratory Syndrome-Associated Coronavirus: Recent Advances'. *F1000Research*, 6, 1628. https://doi.org/10.12688/f1000research.11827.1

80   *In November 2019, according to the* Los Angeles Times Alice Su. 'Why China's Wildlife Ban Is Not Enough to Stop Another Virus Outbreak'. *Los Angeles Times*. 2 April 2020. https://www.latimes.com/world-nation/story/2020-04-02/why-china-wildlife-ban-not-enough-stop-coronavirus-outbreak

80   *including species such as the yellow breasted bunting* Agence France Presse. 'Yellow-Breasted Buntings "Being Eaten to Extinction by China"'. *Guardian*. 9 June 2015. https://www.theguardian.com/environment/2015/jun/09/buntings-are-being-eaten-to-extinction-by-china-study-says

81   *under Chinese President Xi Jinping, TCM has been championed* 'Xi Jinping Stresses Role of Traditional Chinese Medicine for "Healthy China"'. CGTN. 25 October 2019. https://news.cgtn.com/news/2019-10-25/Xi-stresses-role-of-traditional-Chinese-medicine-for-Healthy-China--L4OQBRhOGA/index.html

81   *Beijing health authorities publicly solicited submissions on a draft regulation that could criminalise anyone who slandered TCM* Helen Davidson. 'Beijing Draws Up Plans to Outlaw Criticism of Traditional Chinese Medicine'. *Guardian*. 3 June 2020. https://

www.theguardian.com/world/2020/jun/03/beijing-draws-up-plans-to-outlaw-criticism-of-traditional-chinese-medicine

81 *However, the smaller bats are used in TCM* Wassenaar, T. M., & Zou, Y. (2020). '2019_nCoV/SARS-CoV-2: Rapid Classification of Betacoronaviruses and Identification of Traditional Chinese Medicine as Potential Origin of Zoonotic Coronaviruses'. *Letters in Applied Microbiology*, 70(5), 342–348. https://doi.org/10.1111/lam.13285

82 *Dr Shi was equally certain* Jon Cohen. 'Trump "Owes Us an Apology". Chinese Scientist at the Center of COVID-19 Origin Theories Speaks Out'. *Science*. 24 July 2020. https://www.sciencemag.org/news/2020/07/trump-owes-us-apology-chinese-scientist-center-covid-19-origin-theories-speaks-out

82 *At first suspicion fell on snakes* Ji, W., Wang, W., Zhao, X., Zai, J., & Li, X. (2020). 'Cross-Species Transmission of the Newly Identified Coronavirus 2019-nCoV'. *Journal of Medical Virology*, 92(4), 433–440. https://doi.org/10.1002/jmv.25682

83 *approximately nine thousand tonnes of snakes were sold in Chinese markets each year* Gene Marks. 'Risky Business: China's Snake Farmers Cash In on Global Venom Market'. *Guardian*. 12 July 2018. https://www.theguardian.com/business/2018/jul/12/snake-farm-village-china-zisiqiao-venom-market-medicine

83 *the lucrative business in Zisiqiao was shuttered* Henry Holloway. 'Chinese "Snake Village" That Factory Farmed Reptiles for Food Shuts Down Over Coronavirus After Killer Bug Link'. *Sun*. 8 April 2020. https://www.thesun.co.uk/news/11355807/china-snake-village-coronavirus-meat/

83 *in a paper published in* Clinical Microbiology Reviews *and concluded starkly – and presciently* Cheng, V. C., Lau, S. K., Woo, P. C., & Yuen, K. Y. (2007). 'Severe Acute Respiratory Syndrome Coronavirus as an Agent of Emerging and Reemerging Infection'. *Clinical Microbiology Reviews*, 20(4), 660–694. https://doi.org/10.1128/CMR.00023-07

84 *Two of the authors repeated the warning in February 2019* Wong, A., Li, X., Lau, S., & Woo, P. (2019). 'Global Epidemiology of Bat Coronaviruses'. *Viruses*, 11(2), 174. https://doi.org/10.3390/v11020174

86 *The Chinese CDC's map only emerged in a leak to the* South China Morning Post John Power and Simone McCarthy. 'WHO's Coronavirus Detectives Look to Wuhan Market as Undisclosed

Map Surfaces'. *South China Morning Post*. 15 December 2020. https://www.scmp.com/week-asia/health-environment/article/3113952/whos-coronavirus-detectives-look-wuhan-market; for map, see: https://cdn.i-scmp.com/sites/default/files/d8/images/methode/2020/12/15/8163e590-3b81-11eb-9b80-f4f1a4017c77_image_hires_125131.png

87 *This China-WHO report revealed a different picture* Origins of SARS-CoV-2 Virus. World Health Organization. 30 March 2021. https://www.who.int/health-topics/coronavirus/origins-of-the-virus

5. The pangolin papers

89 *On 7 February 2020, a press release appeared on the website of South China Agricultural University* Li Yan. 'Pangolins a Potential Intermediate Host of Novel Coronavirus: Study'. Xinhua. 7 February 2020. http://www.ecns.cn/news/2020-02-07/detail-ifztmcih6512454.shtml

91 *a group of journalists calling themselves the Global Environmental Reporting Collective put out a report on the pangolin trade called 'Trafficked to Extinction'* Global Environmental Reporting Collective. 'Trafficked to Extinction'. The Pangolin Reports. September 2019. https://globalstory.pangolinreports.com

92 *In March 2019, the Guangdong and Guangxi Anti-Smuggling Bureaus, working with more than three hundred police forces in several countries* Choo, S. W., Zhou, J., Tian, X., Zhang, S., Qiang, S., O'Brien, S. J., Tan, K. Y., Platto, S., Koepfli, K. P., Antunes, A., Sitam, F. T. (2020). 'Are Pangolins Scapegoats of the COVID-19 Outbreak-CoV Transmission and Pathology Evidence?' *Conservation Letters*, 13(6): e12754. https://doi.org/10.1111/conl.12754

93 *When pangolin genomes were first fully sequenced in 2016* Choo, S. W., Rayko, M., Tan, T. K., Hari, R., Komissarov, A., Wee, W. Y., Yurchenko, A. A., Kliver, S., Tamazian, G., Antunes, A., Wilson, R. K., Warren, W. C., Koepfli, K. P., Minx, P., Krasheninnikova, K., Kotze, A., Dalton, D. L., Vermaak, E., Paterson, I. C., Dobrynin, P., ... Wong, G. J. (2016). 'Pangolin Genomes and the Evolution of Mammalian Scales and Immunity'. *Genome Research*, 26(10), 1312–1322. https://doi.org/10.1101/gr.203521.115

93 *In an October 2019 paper in the journal* Viruses Liu, P., Chen, W., & Chen, J. P. (2019). 'Viral Metagenomics Revealed Sendai Virus

and Coronavirus Infection of Malayan Pangolins (*Manis javanica*)'. *Viruses*, *11*(11), 979. https://doi.org/10.3390/v11110979

94   *a burst of scientific preprints followed on 18 and 20 February* Lam, T. T., Jia, N., Zhang, Y. W., Shum, M. H., Jiang, J. F., Zhu, H. C., Tong, Y. G., Shi, Y. X., Ni, X. B., Liao, Y. S., Li, W. J., Jiang, B. G., Wei, W., Yuan, T. T., Zheng, K., Cui, X. M., Li, J., Pei, G. Q., Qiang, X., Cheung, W. Y., … Cao, W. C. (2020). 'Identifying SARS-CoV-2-related Coronaviruses in Malayan Pangolins'. *Nature*, *583*(7815), 282–285. https://doi.org/10.1038/s41586-020-2169-0; Xiao, K., Zhai, J., Feng, Y., Zhou, N., Zhang, X., Zou, J. J., Li, N., Guo, Y., Li, X., Shen, X., Zhang, Z., Shu, F., Huang, W., Li, Y., Zhang, Z., Chen, R. A., Wu, Y. J., Peng, S. M., Huang, M., Xie, W. J., … Shen, Y. (2020). 'Isolation of SARS-CoV-2-related Coronavirus from Malayan Pangolins'. *Nature*, *583*(7815), 286–289. https://doi.org/10.1038/s41586-020-2313-x; Liu, P., Jiang, J. Z., Wan, X. F., Hua, Y., Li, L., Zhou, J., Wang, X., Hou, F., Chen, J., Zou, J., & Chen, J. (2020). 'Are Pangolins the Intermediate Host of the 2019 Novel Coronavirus (SARS-CoV-2)?' *PLoS Pathogens*, *16*(5), e1008421. https://doi.org/10.1371/journal.ppat.1008421; Zhang, T., Wu, Q., & Zhang, Z. (2020). 'Probable Pangolin Origin of SARS-CoV-2 Associated with the COVID-19 Outbreak'. *Current Biology : CB*, *30*(7), 1346–1351.e2. https://doi.org/10.1016/j.cub.2020.03.022

95   *this superspreader event in Boston would eventually lead to an estimated three hundred thousand cases worldwide* Lara Salahi and Kaitlin McKinley Becker. '1 Year Later: The "Superspreader" Conference That Sparked Boston's COVID Outbreak'. NBC. 26 February 2021. https://www.nbcboston.com/news/coronavirus/1-year-later-the-superspreader-conference-that-sparked-bostons-coronavirus-outbreak/2314011/

95   *In late March, top experts who had been anticipating and watching for potentially dangerous changes* Pien Huang. 'The Coronavirus Is Mutating. But That May Not Be A Problem For Humans'. NPR Goats and Soda. 25 March 2020. https://www.npr.org/sections/goatsandsoda/2020/03/25/820998549/the-coronavirus-is-mutating-but-that-may-not-be-a-problem-for-humans

98   *Ian Birrell, writing for the* Mail on Sunday *in London* Ian Birrell. 'Landmark Study: Virus Didn't Come from Animals in Wuhan Market'. *Mail on Sunday*. 17 May 2020. http://www.ianbirrell.

com/landmark-study-virus-didnt-come-from-animals-in-wuhan-market/

101 *In the same month, the US Right to Know organisation published emails* https://usrtk.org/wp-content/uploads/2020/11/Pangolin_Papers_Perlman_Emails.pdf

102 *We know from internal Chinese government documents, seen by the Associated Press* Dake Kang, Maria Cheng and Sam McNeil. 'China Clamps Down in Hidden Hunt for Coronavirus Origins'. Associated Press. 30 December 2020. https://apnews.com/article/united-nations-coronavirus-pandemic-china-only-on-ap-bats-24fbadc58cee3a40bca2ddf7a14d2955

102 *Scientists working with the China Biodiversity Conservation and Green Development Foundation* Choo, S. W., Zhou, J., Tian, X., Zhang, S., Qiang, S., O'Brien, S. J., Tan, K. Y., Platto, S., Koepfli, K. P., Antunes, A., Sitam, F. T. (2020). 'Are Pangolins Scapegoats of the COVID-19 Outbreak-CoV Transmission and Pathology Evidence?' *Conservation Letters*, *13*(6): e12754. https://doi.org/10.1111/conl.12754

103 *In June 2020, Dr Peter Daszak and his collaborators reported an analysis of 334 smuggled pangolins* Lee, J., Hughes, T., Lee, M. H., Field, H., Rovie-Ryan, J. J., Sitam, F. T., Sipangkui, S., Nathan, S., Ramirez, D., Kumar, S. V., Lasimbang, H., Epstein, J. H., & Daszak, P. (2020). 'No Evidence of Coronaviruses or Other Potentially Zoonotic Viruses in Sunda Pangolins (Manis javanica) Entering the Wildlife Trade via Malaysia'. *EcoHealth*, *17*(3), 406–418. https://doi.org/10.1007/s10393-020-01503-x

105 *On 15 January 2021, the US State Department released a fact sheet* US Department of State. Fact Sheet: Activity at the Wuhan Institute of Virology. 15 January 2021. https://2017-2021.state.gov/fact-sheet-activity-at-the-wuhan-institute-of-virology/index.html

107 *China announced a ban on the trade of wild animals* James Gorman. 'China's Ban on Wildlife Trade a Big Step, But Has Loopholes, Conservationists Say'. *New York Times*. 27 February 2020. https://www.nytimes.com/2020/02/27/science/coronavirus-pangolin-wildlife-ban-china.html

107 *In October 2020, the Environmental Investigation Agency, an international watchdog, revealed that eBay and Taobao* Jonathan Watts. 'China Still Allowing Use of Pangolin Scales in Traditional Medicine'. *Guardian*. 12 October 2020. https://www.theguardian.com/environment/2020/oct/13/china-still-allowing-use-of-pangolin-scales-in-traditional-medicine

107 *ACTAsia, part of the Animal Care Trust non-profit organisation* 'China's Revised Law Paves the Way for Wildlife to Become Farm Animals'. ACTAsia. 23 November 2020. https://www.actasia.org/news/chinas-revised-law-paves-the-way-for-wildlife-to-become-farm-animals/

108 *the Environmental Investigation Agency has pointed out could be a cover* Environmental Investigation Agency. 'Commercial Pangolin Farming Will Very Likely Have a Negative Impact on Pangolin Conservation'. 30 July 2019. https://eia-international.org/news/commercial-pangolin-farming-will-very-likely-have-a-negative-impact-on-pangolin-conservation/

108 *In April 2021, Reuters reported that pangolins continued to be trafficked* David Stanway. 'As WHO Highlights COVID Animal Origins, China Wildlife Crackdown Needs More Teeth – Experts'. 1 April 2021. https://www.reuters.com/article/us-health-coronavirus-china-wildlife-idUSKBN2BO6Y3

6. Bats and the virus hunters

109 *SARS, SARS-CoV-2 and other SARS-like coronaviruses predominantly thrive in just one genus of bat* Hu, B., Ge, X., Wang, L. F., & Shi, Z. (2015). 'Bat Origin of Human Coronaviruses'. *Virology Journal*, 12, 221. https://doi.org/10.1186/s12985-015-0422-1

110 *China is home to many horseshoe bat species* http://www.bio.bris.ac.uk/research/bats/China%20bats/Rhinolophidae.htm

111 *In the last few decades, no other group of animals has proved as prolific as bats* Irving, A. T., Ahn, M., Goh, G., Anderson, D. E., & Wang, L. F. (2021). 'Lessons from the Host Defences of Bats, a Unique Viral Reservoir'. *Nature*, 589(7842), 363–370. https://doi.org/10.1038/s41586-020-03128-0

112 *Bats have existed for an estimated fifty million years and have evolved unique immune systems* Banerjee, A., Baker, M. L., Kulcsar, K., Misra, V., Plowright, R., & Mossman, K. (2020). 'Novel Insights Into Immune Systems of Bats'. *Frontiers in Immunology*, 11, 26. https://doi.org/10.3389/fimmu.2020.00026

113 *In 2018, Dr Peng Zhou, Dr Shi Zhengli and colleagues published the results of an experiment* Xie, J., Li, Y., Shen, X., Goh, G., Zhu, Y., Cui, J., Wang, L. F., Shi, Z. L., & Zhou, P. (2018). 'Dampened STING-Dependent Interferon Activation in Bats'. *Cell Host & Microbe*, 23(3), 297–301.e4. https://doi.org/10.1016/j.chom.2018.01.006

114 *a paper by an Australian team in 2018* Banerjee, A., Misra, V., Schountz, T., & Baker, M. L. (2018). 'Tools to Study Pathogen-Host Interactions in Bats'. *Virus Research*, *248*, 5–12. https://doi.org/10.1016/j.virusres.2018.02.013

114 *The team included Dr Linfa Wang* Jane Qiu. 'How China's "Bat Woman" Hunted Down Viruses from SARS to the New Coronavirus'. *Scientific American*. 11 March 2020. https://www.scientificamerican.com/article/how-chinas-bat-woman-hunted-down-viruses-from-sars-to-the-new-coronavirus1/

115 *The group went on to publish a review of bats and SARS* Wang, L. F., Shi, Z., Zhang, S., Field, H., Daszak, P., & Eaton, B. T. (2006). 'Review of Bats and SARS'. *Emerging Infectious Diseases*, *12*(12), 1834–1840. https://doi.org/10.3201/eid1212.060401

115 *Over the next decade and more, many other influential papers from this research consortium describing novel viruses isolated from bats would follow* Li, W., Shi, Z., Yu, M., Ren, W., Smith, C., Epstein, J. H., Wang, H., Crameri, G., Hu, Z., Zhang, H., Zhang, J., McEachern, J., Field, H., Daszak, P., Eaton, B. T., Zhang, S., & Wang, L. F. (2005). 'Bats Are Natural Reservoirs of SARS-Like Coronaviruses'. *Science (New York, N.Y.)*, *310*(5748), 676–679. https://doi.org/10.1126/science.1118391; Kai Kupferschmidt. 'This Biologist Helped Trace SARS to Bats. Now, He's Working to Uncover the Origins of COVID-19'. *Science*. 30 September 2020. https://www.sciencemag.org/news/2020/09/biologist-helped-trace-sars-bats-now-hes-working-uncover-origins-covid-19

115 *the EcoHealth Alliance non-profit organisation* https://www.ecohealthalliance.org/about

116 *the United States Agency for International Development's (USAID) Emerging Pandemic Threats (EPT) programme began in 2009* https://ohi.vetmed.ucdavis.edu/programs-projects/predict-project; See also: https://ohi.vetmed.ucdavis.edu/sites/g/files/dgvnsk5251/files/inline files/PREDICT%20LEGACY%20-%20FINAL%20FOR%20WEB_0.pdf

117 *By 2018, the EcoHealth Alliance had grown its income to almost $17 million a year* EcoHealth Alliance Fiscal Year 2018 Annual Report https://www.ecohealthalliance.org/wp-content/uploads/2019/09/EHA-FY18-Annual-Report.pdf

117 *One journalist, Sam Husseini, found his emails and voicemails ignored* Sam Husseini. 'Peter Daszak's EcoHealth Alliance Has Hidden Almost $40 Million in Pentagon Funding and Militarized Pandemic Science'. *Independent Science News*. 16 December

2020. https://www.independentsciencenews.org/news/peter-daszaks-ecohealth-alliance-has-hidden-almost-40-million-in-pentagon-funding/

118   *the Shitou cave near Kunming* Jane Qiu. 'How China's "Bat Woman" Hunted Down Viruses from SARS to the New Coronavirus'. *Scientific American*. 11 March 2020. https://www.scientificamerican.com/article/how-chinas-bat-woman-hunted-down-viruses-from-sars-to-the-new-coronavirus1/

118   *a paper in* Nature *on 30 October 2013* Ge, X. Y., Li, J. L., Yang, X. L., Chmura, A. A., Zhu, G., Epstein, J. H., Mazet, J. K., Hu, B., Zhang, W., Peng, C., Zhang, Y. J., Luo, C. M., Tan, B., Wang, N., Zhu, Y., Crameri, G., Zhang, S. Y., Wang, L. F., Daszak, P., & Shi, Z. L. (2013). 'Isolation and Characterization of a Bat SARS-like Coronavirus That Uses the ACE2 Receptor'. *Nature*, *503*(7477), 535–538. https://doi.org/10.1038/nature12711

119   *alpha-coronaviruses in pigs* Zhou, P., Fan, H., Lan, T., Yang, X. L., Shi, W. F., Zhang, W., Zhu, Y., Zhang, Y. W., Xie, Q. M., Mani, S., Zheng, X. S., Li, B., Li, J. M., Guo, H., Pei, G. Q., An, X. P., Chen, J. W., Zhou, L., Mai, K. J., Wu, Z. X., ... Ma, J. Y. (2018). 'Fatal Swine Acute Diarrhoea Syndrome Caused By an HKU2-Related Coronavirus of Bat Origin'. *Nature*, *556*(7700), 255–258. https://doi.org/10.1038/s41586-018-0010-9

121   *'This is the first documented spillover of a bat coronavirus that caused severe diseases in domestic animals'* Cui, J., Li, F., & Shi, Z. L. (2019). 'Origin and Evolution of Pathogenic Coronaviruses'. *Nature Reviews. Microbiology*, *17*(3), 181–192. https://doi.org/10.1038/s41579-018-0118-9

121   *In October 2015, they took blood from 139 women and 79 men living in four villages* Wang, N., Li, S. Y., Yang, X. L., Huang, H. M., Zhang, Y. J., Guo, H., Luo, C. M., Miller, M., Zhu, G., Chmura, A. A., Hagan, E., Zhou, J. H., Zhang, Y. Z., Wang, L. F., Daszak, P., & Shi, Z. L. (2018). 'Serological Evidence of Bat SARS-Related Coronavirus Infection in Humans, China'. *Virologica Sinica*, *33*(1), 104–107. https://doi.org/10.1007/s12250-018-0012-7

122   *Over the next two years the team repeated the survey on a wider scale* Li, H., Mendelsohn, E., Zong, C., Zhang, W., Hagan, E., Wang, N., Li, S., Yan, H., Huang, H., Zhu, G., Ross, N., Chmura, A., Terry, P., Fielder, M., Miller, M., Shi, Z., & Daszak, P. (2019). 'Human-Animal Interactions and Bat Coronavirus Spillover Potential Among Rural Residents in Southern China'. *Biosafety*

and Health, 1(2), 84–90. https://doi.org/10.1016/j.
bsheal.2019.10.004

123 *In April 2020, in an interview with Vox* Eliza Barclay. 'Why These
Scientists Still Doubt the Coronavirus Leaked From a Chinese
Lab'. Vox. 29 April 2020. https://www.vox.com/2020/4/23/
21226484/wuhan-lab-coronavirus-china

123 *Dr Tian Junhua, an expert on pathogens* 'Entering the Mountain
to Catch Bats in the Middle of the Night, Wuhan Experts Catch
Tens of Thousands of "Worms" to Study the Virus'. *Changjiang
Times*. 3 May 2017. http://www.changjiangtimes.
com/2017/05/567037.html; Guo, W. P., Lin, X. D., Wang, W.,
Tian, J. H., Cong, M. L., Zhang, H. L., Wang, M. R., Zhou, R. H.,
Wang, J. B., Li, M. H., Xu, J., Holmes, E. C., & Zhang, Y. Z.
(2013). 'Phylogeny and Origins of Hantaviruses Harbored By
Bats, Insectivores, and Rodents'. *PLoS Pathogens*, 9(2), e1003159.
https://doi.org/10.1371/journal.ppat.1003159; see also https://
www.researchgate.net/profile/Junhua-Tian-2

124 *he claimed to have trapped nearly ten thousand bats* 'The Post-80s
"Disease Control" Guy Lay Down to Watch Cockroaches at
Night in Order to Carry Out Research and Catch Tens of
Thousands of Insects'. *Wuhan Evening News*. 3 May 2017. http://
www.xinhuanet.com/local/2017-05/03/c_1120909064_2.htm

124 *Dr Tian is a co-author with Dr Zhang Yongzhen in Shanghai on
the* Nature *paper* Wu, F., Zhao, S., Yu, B., Chen, Y. M., Wang, W.,
Song, Z. G., Hu, Y., Tao, Z. W., Tian, J. H., Pei, Y. Y., Yuan, M. L.,
Zhang, Y. L., Dai, F. H., Liu, Y., Wang, Q. M., Zheng, J. J., Xu, L.,
Holmes, E. C., & Zhang, Y. Z. (2020). 'A New Coronavirus
Associated with Human Respiratory Disease in China'. *Nature*,
579(7798), 265–269. https://doi.org/10.1038/s41586-020-2008-3

124 *The withdrawn paper (speculating on possible lab origins of
Covid-19) by two Wuhan scientists* Botao Xiao and Lei Xiao.
'The Possible Origins of 2019-nCoV Coronavirus'. ResearchGate
preprint. 6 February 2020. Available at: https://img-prod
tgcom24.mediaset.it/images/2020/02/16/114720192-5eb8307f-
017c-4075-a697-348628da0204.pdf; Guo, W. P., Lin, X. D.,
Wang, W., Tian, J. H., Cong, M. L., Zhang, H. L., Wang, M. R.,
Zhou, R. H., Wang, J. B., Li, M. H., Xu, J., Holmes, E. C., &
Zhang, Y. Z. (2013). 'Phylogeny and Origins of Hantaviruses
Harbored By Bats, Insectivores, and Rodents'. *PLoS
Pathogens*, 9(2), e1003159. https://doi.org/10.1371/journal.
ppat.1003159

125 *In 2015, the researchers got a step closer* Yang, X. L., Hu, B., Wang, B., Wang, M. N., Zhang, Q., Zhang, W., Wu, L. J., Ge, X. Y., Zhang, Y. Z., Daszak, P., Wang, L. F., & Shi, Z. L. (2015). 'Isolation and Characterization of a Novel Bat Coronavirus Closely Related to the Direct Progenitor of Severe Acute Respiratory Syndrome Coronavirus'. *Journal of Virology*, 90(6), 3253–3256. https://doi.org/10.1128/JVI.02582-15

126 *From 2012 to 2015, a study led by Dr Zhang Yongzhen sampled 1,067 bats* Lin, X. D., Wang, W., Hao, Z. Y., Wang, Z. X., Guo, W. P., Guan, X. Q., Wang, M. R., Wang, H. W., Zhou, R. H., Li, M. H., Tang, G. P., Wu, J., Holmes, E. C., & Zhang, Y. Z. (2017). 'Extensive Diversity of Coronaviruses in Bats from China'. *Virology*, 507, 1–10. https://doi.org/10.1016/j.virol.2017.03.019

126 *a team of biologists from the Third Military Medical University in Chongqing* Hu, D., Zhu, C., Ai, L., He, T., Wang, Y., Ye, F., Yang, L., Ding, C., Zhu, X., Lv, R., Zhu, J., Hassan, B., Feng, Y., Tan, W., & Wang, C. (2018). 'Genomic Characterization and Infectivity of a Novel SARS-Like Coronavirus in Chinese Bats'. *Emerging Microbes & Infections*, 7(1), 154. https://doi.org/10.1038/s41426-018-0155-5

127 *In a 2017 interview with Xinhua News Agency* Keoni Everington. 'Video Shows Wuhan Lab Scientists Admit to Being Bitten by Bats'. *Taiwan News*. 15 January 2021. https://www.taiwannews.com.tw/en/news/4102619

127 *In his book* Spillover, *David Quammen describes squirming deep into a narrow cave* Quammen, David. *Spillover* (p. 200). Random House. Kindle Edition

128 *Did the WIV ever keep bats in the laboratory in Wuhan?* Glen Owen and Jake Ryan. 'Revealed: Secret Bat Cages at Wuhan Lab Where Researchers Planned to Breed Animals for Virus Experiments – Despite Denials of British Scientist on the WHO Team "Investigating" the Origins of Covid'. *Mail on Sunday*. 13 February 2021. https://www.dailymail.co.uk/news/article-9257413/Secret-bat-cages-Wuhan-lab-researchers-planned-breed-animals-virus-experiments.html

128 *Yet in 2009 a colleague of Dr Shi's gave an interview to* Science Times 'Shi Zhengli: A Female Scientist Accompanying the Virus'. *Science Times*. 10 March 2009. http://news.sciencenet.cn/sbhtmlnews/2009/3/216816.html

129 *WIV had filed a patent on bat breeding* Wuhan Institute of Virology of CAS. 'Artificial Breeding Method for Wild Bat of

Predatory Worm'. https://patents.google.com/patent/
CN112205352A/en

129   *Predict came to an end after two budget cycles costing $207
       million* Donald G. McNeil Jr. 'Scientists Were Hunting for the
       Next Ebola. Now the U.S. Has Cut Off Their Funding'. *New York
       Times.* 25 October 2019. https://www.nytimes.com/2019/10/25/
       health/predict-usaid-viruses.html

130   *in 2018, an even more ambitious global project called the Global
       Virome Project (GVP) had been launched* Glenn McDonald. 'The
       Global Virome Project Is Hunting Hundreds of Thousands of
       Deadly Viruses'. The Seeker. 3 January 2018. https://www.seeker.
       com/health/the-global-virome-project-is-hunting-hundreds-of-
       thousands-of-deadly-viruses

130   *Drs Edward Holmes, Andrew Rambaut and Kristian Andersen,
       published an article in* Nature Holmes, E. C., Rambaut, A., &
       Andersen, K. G. (2018). 'Pandemics: Spend on Surveillance, Not
       Prediction'. *Nature, 558*(7709), 180–182. https://doi.org/10.1038/
       d41586-018-05373-w

132   *Dr Daszak had told the* Economist *that approximately sixteen
       thousand bats had been sampled* 'The Hunt for the Origins of
       SARS-CoV-2 Will Look Beyond China'. *Economist.* 22 July
       2020. https://www.economist.com/science-and-technology/2020/
       07/22/the-hunt-for-the-origins-of-sars-cov-2-will-look-beyond-
       china

7. Laboratory leaks

134   *In the Singapore case* Lim, P. L., Kurup, A., Gopalakrishna, G.,
       Chan, K. P., Wong, C. W., Ng, L. C., Se-Thoe, S. Y., Oon, L., Bai,
       X., Stanton, L. W., Ruan, Y., Miller, L. D., Vega, V. B., James, L.,
       Ooi, P. L., Kai, C. S., Olsen, S. J., Ang, B., & Leo, Y. S. (2004).
       'Laboratory-Acquired Severe Acute Respiratory Syndrome'. *New
       England Journal of Medicine, 350*(17), 1740–1745. https://doi.
       org/10.1056/NEJMoa032565; Senior K. (2003). 'Recent
       Singapore SARS Case a Laboratory Accident'. *Lancet. Infectious
       Diseases, 3*(11), 679. https://doi.org/10.1016/S1473-
       3099(03)00815-6; See also 'Singapore Man Acquired SARS in
       Government Lab, Panel Says'. Center for Infectious Disease
       Research and Policy. 23 September 2003. https://www.cidrap.
       umn.edu/news-perspective/2003/09/singapore-man-acquired-sars-
       government-lab-panel-says; and Lawrence K. Altman. 'Lab
       Infection Blamed for Singapore SARS Case'. *New York Times.* 24

September 2003. https://www.nytimes.com/2003/09/24/world/lab-infection-blamed-for-singapore-sars-case.html

135 *In the Taiwan case* Orellana C. (2004). 'Laboratory-Acquired SARS Raises Worries on Biosafety'. *Lancet. Infectious Diseases*, 4(2), 64. https://doi.org/10.1016/s1473-3099(04)00911-9; see also Sars Case in Laboratory Worker in Taiwan, China. World Health Organization. 17 December 2003. https://www.who.int/csr/don/2003_09_24/en/ (Website removed.); 'Taiwan's New SARS Case Raises Questions About Sloppy Procedures'. Associated Press. 17 December 2003. https://usatoday30.usatoday.com/news/health/2003-12-17-singapore-sars_x.htm; Martin Furmanski. 'Threatened Pandemics and Laboratory Escapes: Self-Fulfilling Prophecies'. *Bulletin of the Atomic Scientists*. 31 March 2014. https://thebulletin.org/2014/03/threatened-pandemics-and-laboratory-escapes-self-fulfilling-prophecies/; 'Taiwanese SARS Researcher Infected'. Center for Infectious Disease Research and Policy. 17 December 2003. https://www.cidrap.umn.edu/news-perspective/2003/12/taiwanese-sars-researcher-infected; Wen-Chao Wu, Li-Li Lee and Ho-Sheng Wu. 'Audit Report for Laboratories of Biosafety Level 3 and Higher in Taiwan, 2007'. *Taiwan Epidemiology Bulletin*. 25 July 2008. https://www.cdc.gov.tw/En/File/Get/53fRU0drzfunw22tKaDplw; 'Taiwan Sars Man "Feared Shame"'. BBC. 19 December 2003. http://news.bbc.co.uk/2/hi/asia-pacific/3333775.stm; 'No SARS Reported in Taiwan Patient's Contacts'. Center for Infectious Disease Research and Policy. 22 December 2003. https://www.cidrap.umn.edu/news-perspective/2003/12/no-sars-reported-taiwan-patients-contacts; and Wen-Chao Wu, Li-Li Lee, Wei-Fong Chen, Shih-Yan Yang, Ho-Sheng Wu, Wen-Yi Shih and Steve Hsu-Sung Kuo. 'Development of Laboratory Biosafety Management: The Taiwan Experience'. *Applied Biosafety*. 12:18–25 2007. https://www.liebertpub.com/doi/pdfplus/10.1177/153567600701200104

136 *the Beijing lab leaks of SARS in early 2004* Parry J. (2004). 'Chinese Authorities on Alert as SARS Breaks Out Again'. *BMJ (Clinical Research ed.)*, 328(7447), 1034. https://doi.org/10.1136/bmj.328.7447.1034; see also Walgate, R. 'SARS Escaped Beijing Lab Twice'. *Genome Biol 4*, spotlight-20040427-03 (2004). https://doi.org/10.1186/gb-spotlight-20040427-03; Parry J. (2004). 'Breaches of Safety Regulations Are Probable Cause of Recent SARS Outbreak, WHO Says'. *BMJ (Clinical Research ed.)*,

328(7450), 1222. https://doi.org/10.1136/bmj.328.7450.1222-b; Liang, W. N., Zhao, T., Liu, Z. J., Guan, B. Y., He, X., Liu, M., Chen, Q., Liu, G. F., Wu, J., Huang, R. G., Xie, X. Q., & Wu, Z. L. (2006). 'Severe Acute Respiratory Syndrome – Retrospect and Lessons of 2004 Outbreak in China'. *Biomedical and Environmental Sciences : BES*, *19*(6), 445–451; WHO SARS Risk Assessment and Preparedness Framework. World Health Organization. October 2004. https://www.who.int/csr/resources/ publications/CDS_CSR_ARO_2004_2.pdf?ua=1; China's Latest SARS Outbreak Has Been Contained, But Biosafety Concerns Remain – Update 7. World Health Organization. 18 May 2004. (Website removed.) https://www.who.int/csr/don/2004_05_18a/ en/. https://www.who.int/csr/don/2004_04_22/en/; Jim Yardley and Lawrence K. Altman. 'China Is Scrambling to Curb SARS Cases After a Death'. *New York Times*. 24 April 2004. https:// www.nytimes.com/2004/04/24/world/china-is-scrambling-to-curb-sars-cases-after-a-death.html

138 *After this episode, the director of the Chinese CDC and several officials resigned* Lei Du and Martin Enserink. 'SARS Crisis Topples China Lab Chief'. *Science*. 2 July 2004. https://www. sciencemag.org/news/2004/07/sars-crisis-topples-china-lab-chief

140 *a previously undisclosed US intelligence report was leaked to the* Wall Street Journal Michael R. Gordon, Warren P. Strobel and Drew Hinshaw. 'Intelligence on Sick Staff at Wuhan Lab Fuels Debate on Covid-19 Origin'. *Wall Street Journal*. 23 May 2021. https://www.wsj.com/articles/intelligence-on-sick-staff-at-wuhan-lab-fuels-debate-on-covid-19-origin-11621796228

140 *According to a former State Department official, David Asher, the wife of one worker died* Ian Birrell. '"A Lab Leak Isn't 100% Certain But It Seems to Be the Only Logical Source of Covid": Washington Expert Who Led Inquiry into the Cause of the Virus Reveals Three Wuhan Lab Scientists Fell Ill in November 2019'. *Mail on Sunday*. 27 March 2021

140 *early in the pandemic a senior scientist at a Beijing research institute was infected with SARS-CoV-2* Sainath Suryanarayanan. 'Senior Chinese Scientist Acquired SARS-CoV-2 in Lab Infection Accident, Virologist Says'. *US Right to Know*. 5 August 2021. https://usrtk.org/biohazards/senior-chinese-scientist-acquired-sars-cov-2-in-lab-infection-accident/

141 *the* New York Times *reported a story of leopards* Chris Buckley. 'Was That a Giant Cat? Leopards Escape, and a Zoo Keeps Silent

(at First)'. *New York Times*. 10 May 2021. https://www.nytimes. com/2021/05/10/world/asia/china-zoo-leopards.html; see also 'Chinese Safari Park Concealed Leopards' Escape for 2 Weeks'. Associated Press. 10 May 2021. https://apnews.com/article/china-travel-lifestyle-2070394ea87488bd6fab17e39fe6d6f4

142 *In 2003, in Kunming, Yunnan, laboratory rats infected students with a form of hantavirus* Zhang, Y., Zhang, H., Dong, X., Yuan, J., Zhang, H., Yang, X., Zhou, P., Ge, X., Li, Y., Wang, L. F., & Shi, Z. (2010). 'Hantavirus Outbreak Associated with Laboratory Rats in Yunnan, China'. *Infection, Genetics and Evolution: Journal of Molecular Epidemiology and Evolutionary Genetics in Infectious Diseases*, 10(5), 638–644. https://doi.org/10.1016/j. meegid.2010.03.015

143 *In Lanzhou, the capital of Gansu province in the north-west of the country* '10,528 Residents Test Positive for Brucellosis in Lanzhou after Brucella Leakage in Local Factory'. *Global Times*. 3 December 2020. https://www.globaltimes.cn/content/1208864. shtml; see also 'Over 6,000 People in China's Lanzhou Test Positive for Brucellosis: State Media'. Reuters. 5 November 2020. https://www.reuters.com/article/us-health-brucellosis-china/over-6000-people-in-chinas-lanzhou-test-positive-for-brucellosis-state-media-idUSKBN27L1LA; Jessie Yeung and Eric Cheung. 'Bacterial Outbreak Infects Thousands After Factory Leak in China'. CNN. 17 September 2020. https://www.cnn.com/2020/09/17/asia/china-brucellosis-outbreak-intl-hnk/index.html; and 'Explainer: How Thousands in China Got Infected by Brucellosis in One Single Outbreak'. *Channel News Asia*. 6 November 2020. https://www. channelnewsasia.com/news/asia/china-brucellosis-disease-outbreak-lanzhou-13478438

144 *Biosecurity lapses happen in other countries* Martin Furmanski. 'Laboratory Escapes and "Self-Fulfilling Prophecy" Epidemics'. Armscontrolcenter.org. 17 February 2014. https:// armscontrolcenter.org/wp-content/uploads/2016/02/Escaped-Viruses-final-2-17-14-copy.pdf

144 *In 2019, research was shut down due to inadequate wastewater decontamination practices at laboratories at Fort Detrick, the United States' army biomedical and research facility* Tom Bowman. 'Fort Detrick: From Biowarfare to Biodefense'. NPR. 1 August 2008. https://www.npr.org/templates/story/story. php?storyId=93196647; see also Denise Grady. 'Deadly Germ Research Is Shut Down at Army Lab Over Safety Concerns'. *New*

York Times. 5 August 2019. https://www.nytimes.
com/2019/08/05/health/germs-fort-detrick-biohazard.html

144 *In March 2021, Alison Young, a journalist who has been tracking
     lab leaks since 2007, published a scorching article in* USA Today
     Alison Young. 'Could an Accident Have Caused COVID-19? Why
     the Wuhan Lab-Leak Theory Shouldn't Be Dismissed'. *USA
     Today*. 22 March 2021. https://www.usatoday.com/in-depth/
     opinion/2021/03/22/why-covid-lab-leak-theory-wuhan-shouldnt-
     dismissed-column/4765985001/; see also Alison Young.
     'Hundreds of Bioterror Lab Mishaps Cloaked in Secrecy'. *USA
     Today*. 17 August 2014. https://www.usatoday.com/story/news/
     nation/2014/08/17/reports-of-incidents-at-bioterror-select-agent-
     labs/14140483/

145 *Smallpox escaped from laboratories in the United Kingdom three
     separate times* Monica Rimmer. 'How Smallpox Claimed Its Final
     Victim'. BBC. 10 August 2018. https://www.bbc.com/news/
     uk-england-birmingham-45101091; see also 'Report of the
     Investigation into the Cause of the 1978 Birmingham Smallpox
     Occurrence'. 22 July 1980. https://assets.publishing.service.gov.
     uk/government/uploads/system/uploads/attachment_data/
     file/228654/0668.pdf

145 *Foot-and-mouth virus escaped from a laboratory at Pirbright*
     Peter Walker. 'Foot and Mouth Reports Blame Lab Site Drains'.
     Guardian. 7 September 2007. https://www.theguardian.com/
     uk/2007/sep/07/footandmouth.immigrationpolicy

145 *Marburg was first discovered after it infected laboratory workers
     in Germany* 'Outbreaks Chronology: Marburg Hemorrhagic
     Fever'. Centers for Disease Control and Prevention. https://www.
     cdc.gov/vhf/marburg/outbreaks/chronology.html

146 *Anthrax, a bacterium, escaped from a biological-warfare
     laboratory in the city of Sverdlosk* 'The 1979 Anthrax Leak'. PBS
     Frontline. https://www.pbs.org/wgbh/pages/frontline/shows/
     plague/sverdlovsk/. https://www.pbs.org/wgbh/pages/frontline/
     shows/plague/sverdlovsk/alibekov.html; https://www.pbs.org/
     wgbh/pages/frontline/shows/plague/interviews/alibekov.html

147 *anthrax spores were sent through the mail* 'Timeline: How The
     Anthrax Terror Unfolded'. National Public Radio. 15 February
     2011. https://www.npr.org/2011/02/15/93170200/timeline-how-
     the-anthrax-terror-unfolded; see also Stephen Engelberg. 'New
     Evidence Adds Doubt to FBI's Case Against Anthrax Suspect'.
     ProPublica. 10 October 2011. https://www.propublica.org/article/

new-evidence-disputes-case-against-bruce-e-ivins; and Stephen Engelberg, Greg Gordon, Jim Gilmore and Mike Wiser. 'Did Bruce Ivins Hide Attack Anthrax from the FBI?' PBS *Frontline*. 10 October 2011. https://www.pbs.org/wgbh/frontline/article/did-bruce-ivins-hide-attack-anthrax-from-the-fbi/

148 *In an article in 2015, Alison Young revealed* Alison Young. 'Inside America's Secretive Biolabs'. *USA Today*. 28 May 2015. https://www.usatoday.com/story/news/2015/05/28/biolabs-pathogens-location-incidents/26587505/

148 *An Influenza A (H1N1) strain of flu swept the world in 1977* Nakajima, K., Desselberger, U., & Palese, P. (1978). 'Recent Human Influenza A (H1N1) Viruses Are Closely Related Genetically to Strains Isolated in 1950'. *Nature*, 274(5669), 334–339. https://doi.org/10.1038/274334a0

149 *Scientists from the Chinese Academy of Medical Sciences published a paper in the* Bulletin of the World Health Organization *in 1978* Kung, H. C., Jen, K. F., Yuan, W. C., Tien, S. F., & Chu, C. M. (1978). 'Influenza in China in 1977: Recurrence of Influenzavirus A Subtype H1N1'. *Bulletin of the World Health Organization*, 56(6), 913–918

150 *Dr Michelle Rozo and Dr Gigi Kwik Gronvall at the UPMC Center for Health Security in Baltimore, Maryland, reanalysed genetic sequences from the 1977 isolates* Rozo, M., & Gronvall, G. K. (2015). 'The Reemergent 1977 H1N1 Strain and the Gain-of-Function Debate'. *mBio*, 6(4), e01013-15. https://doi.org/10.1128/mBio.01013-15

151 *Dr Martin Furmanski of the Center for Arms Control and Non-Proliferation in Washington, DC, disagreed* Furmanski M. (2015). 'The 1977 H1N1 Influenza Virus Reemergence Demonstrated Gain-of-Function Hazards'. *mBio*, 6(5), e01434-15. https://doi.org/10.1128/mBio.01434-15

152 *On 3 February 2020, an official letter seeking help in determining the origins of the novel coronavirus* OSTP Coronavirus Request to NASEM_02.06.2020.pdf. 3 February 2020. https://www.nationalacademies.org/documents/link/LD006CF17B5004C7B41F63CD7E0A0F4EDF738C451F32/fileview/D20D1390AB906330493E8A40B27E965A6761F4117EA2/OSTP%20Coronavirus%20Request%20to%20NASEM_02.06.2020. pdf?hide=thumbs+breadcrumbs+favs+props+ nextprev+sidebar+pin+actions&scheme=light&fitwidth

153 *a preprint titled 'Uncanny Similarity of Unique Inserts in the 2019-nCoV Spike Protein to HIV-1 gp120 and Gag'* Prashant Pradhan, Ashutosh Kumar Pandey, Akhilesh Mishra, Parul Gupta, Praveen Kumar Tripathi, Manoj Balakrishnan Menon, James Gomes, Perumal Vivekanandan, Bishwajit Kundu. 'Uncanny Similarity of Unique Inserts in the 2019-nCoV Spike Protein to HIV-1 gp120 and Gag'. *bioRxiv: The Preprint Server for Biology.* 2020.01.30.927871 [withdrawn]

153 *Dr Shi Zhengli from the WIV posted a furious response* Yang Rui, Feng Yuding, Zhao Jinchao. 'Exclusive | Shi Zhengli Responds to Questions. Experts Agree that the Novel Coronavirus Is Not Artificial'. Caixin. 5 February 2020. https://www.caixin. com/2020-02-05/101511817.html

154 *Email exchanges among the convened experts were later obtained via Freedom of Information* Sainath Suryanarayanan. 'New Emails Show Scientists' Deliberations On How To Discuss SARS-Cov-2 Origins'. US Right to Know. 14 December 2020. https:// usrtk.org/tag/covid-19/

155 *On 1 December, Dr Bedford reaffirmed his stance* @trvrb Twitter. 1 December 2020. https://twitter.com/trvrb/ status/1333885880054910976; see also @trvrb Twitter. 16 December 2020. https://twitter.com/trvrb/status/ 1339267459061227521

155 *From a separate batch of FOI'ed emails by US Right to Know* https://usrtk.org/wp-content/uploads/2020/11/Biohazard_FOIA_ Maryland_Emails_11.6.20.pdf

156 *the emails unearthed months later by US Right to Know came as a surprise* Sainath Suryanarayanan. 'EcoHealth Alliance Orchestrated Key Scientists' Statement on "Natural Origin" of SARS-CoV-2'. US Right to Know. 18 November 2020. https:// usrtk.org/biohazards-blog/ecohealth-alliance-orchestrated-key-scientists-statement-on-natural-origin-of-sars-cov-2/

157 *The reason for their absence was once again revealed by more emails obtained by US Right to Know* https://usrtk.org/ wp-content/uploads/2021/02/Baric_Daszak_email.pdf

157 *the* Lancet *published an addendum* Editors of the Lancet. 'Addendum: Competing Interests and the Origins of SARS-CoV-2'. *Lancet.* 21 June 2021. https://www.thelancet.com/ journals/lancet/article/PIIS0140-6736(21)01377-5/fulltext

158 *The Hubei Wildlife Rescue Centre* Michael Standaert. 'Wuhan Facilities Shed Light on China's Oversight on Wildlife Use'.

Aljazeera. 27 August 2021. https://www.aljazeera.com/
news/2021/8/27/wuhan-facilities-sheds-light-on-chinas-
problematic-oversight-of

158 *A short distance away is the Wuhan Institute of Biological
Products* http://en.sasac.gov.cn/2020/08/10/c_5352.htm

159 *As Dr Shi clarified to* Science *magazine in July 2020* Shi Zhengli.
'Reply to Science Magazine'. 24 July 2020. https://www.
sciencemag.org/sites/default/files/Shi%20Zhengli%20Q%
26A.pdf

160 *As Dr Ralph Baric told the* MIT Technology Review Rowan
Jacobsen. '"We Never Created a Supervirus." Ralph Baric
Explains Gain-of-Function Research'. *MIT Technology Review.*
26 July 2021. https://www.technologyreview.com/2021/07/26/
1030043/gain-of-function-research-coronavirus-ralph-baric-
vaccines/

160 *China long had ambitions to host a BSL-4 laboratory* David
Cyranoski. 'China to Permit Lab Poised to Study World's Most
Dangerous Pathogens'. *Scientific American.* 22 February 2017.
https://www.scientificamerican.com/article/china-to-permit-lab-
poised-to-study-worlds-most-dangerous-pathogens/

161 *A Paris correspondent of the* American Spectator, *Joseph Harriss,
wrote* Joseph A. Harriss. 'The Suspect French Lab in Wuhan'.
American Spectator. 4 May 2020. https://spectator.org/
the-suspect-french-lab-in-wuhan/

162 *In January 2021, the US State Department under the Trump
administration released a statement* Mike Pompeo. Ensuring a
Transparent, Thorough Investigation of Covid-19's Origin. US
Department of State. 15 January 2021. Available at: https://web.
archive.org/web/20210116020513/; https:/www.state.gov/
ensuring-a-transparent-thorough-investigation-of-covid-19s-
origin/

163 *read a joint memo from the two institutions* Le Duc, J. W., &
Yuan, Z. (2018). 'Network for Safe and Secure Labs'. *Science
(New York, N.Y.), 362*(6412), 267. https://doi.org/10.1126/
science.aav7120; see also 'The US Counselor visited Wuhan
Institute of Virology, CAS'. Wuhan Institute of Virology. 3 April
2018. http://english.whiov.cas.cn/Exchange2016/Foreign_
Visits/201804/t20180403_191334.html. Available here: https://
archive.is/6lc3C

163 *Some of these were leaked to Josh Rogin of the* Washington Post
in April 2020 Josh Rogin. 'State Department Cables Warned of

Safety Issues at Wuhan Lab Studying Bat Coronaviruses'. *Washington Post*. 14 April 2020. https://www.washingtonpost. com/opinions/2020/04/14/state-department-cables-warned-safety-issues-wuhan-lab-studying-bat-coronaviruses/

163 *One of the cables, drafted by two officials from the US embassy's environment, science and health sections and sent on 19 January 2018* https://foia.state.gov/Search/Results.aspx?searchText=& beginDate=&endDate=&published BeginDate=20200716& publishedEndDate=20200716& caseNumber=

164 *One of the authors of the cables told Rogin* Josh Rogin. 'In 2018, Diplomats Warned of Risky Coronavirus Experiments in a Wuhan Lab. No One Listened'. *Politico*. 8 March 2021. https://www. politico.com/news/magazine/2021/03/08/josh-rogin-chaos-under-heaven-wuhan-lab-book-excerpt-474322

165 *Dr Yuan Zhiming, chair of the institutional biosafety committee at the WIV, published an article* Yuan Zhiming. 'The Interacademy Partnership'. https://www.interacademies.org/ person/yuan-zhiming; see also Zhiming Y. (2019). 'Current Status and Future Challenges of High-Level Biosafety Laboratories in China'. *Journal of Biosafety and Biosecurity*, 1(2), 123–127. https://doi.org/10.1016/j.jobb.2019.09.005

165 *the Wuhan Center for Disease Control and Prevention* Wuhan Center for Disease Control and Prevention. 'Announcement of Single Source Procurement Method of Hazardous Chemical Waste Disposal Procurement Project in Laboratory of the Municipal Center for Disease Control and Prevention'. 27 June 2019. https:// web.archive.org/web/20200510182006/https://www.whcdc.org/ index.php/view/11147.html

165 *the Chinese Ministry of Science and Technology issued new biosafety rules* Caiyu Liu and Shumei Leng. 'Biosafety Guideline Issued to Fix Chronic Management Loopholes at Virus Labs'. *Global Times*. 16 February 2020. https://www.globaltimes.cn/ content/1179747.shtml; see also Kinling Lo. 'China Tightens Laboratory Animal Rules Amid Calls for Coronavirus Lab-Leak Probe'. *South China Morning Post*. 5 August 2021. https://www. scmp.com/news/china/science/article/3144005/china-tightens-lab-animal-rules-amid-calls-covid-19-laboratory

166 *a comprehensive 61.5 MB online database* https://archive.is/ jPPkB#selection-1359.74-1367.353; see also Anonymous OSINT Contributor, Billy Bostickson, Gilles Demaneuf. 'An Investigation into the WIV Databases That Were Taken Offline'. ResearchGate

preprint. February 2021. https://www.researchgate.net/publication/349073738_An_investigation_into_the_WIV_databases_that_were_taken_offline

166 *A summary paper by the WIV and the EcoHealth Alliance published in 2020 listed 630 novel coronaviruses* Latinne, A., Hu, B., Olival, K. J., Zhu, G., Zhang, L., Li, H., Chmura, A. A., Field, H. E., Zambrana-Torrelio, C., Epstein, J. H., Li, B., Zhang, W., Wang, L. F., Shi, Z. L., & Daszak, P. (2020). 'Origin and Cross-Species Transmission of Bat Coronaviruses in China'. *Nature Communications*, *11*(1), 4235. https://doi.org/10.1038/s41467-020-17687-3

166 *an interview in early December 2019* TWiV 615: Peter Daszak of EcoHealth Alliance. Video Report. 19 May 2020. https://www.youtube.com/watch?v=IdYDL_RK--w&t=1700s

167 *USAID discovered when it reviewed the work in 2017* 'USAID Emerging Pandemic Threats 2 Program Evaluation'. USAID. March 2018. https://pdf.usaid.gov/pdf_docs/PA00SW1M.pdf

167 *the WIV had received funding from the European Virus Archive Global (EVAg) project, which had a mission to 'collect, amplify, characterize, standardize and authenticate viruses to develop and maintain the largest active globally accessible virus archive'* Parliamentary questions. European Parliament. 26 January 2021. https://www.europarl.europa.eu/doceo/document/E-9-2020-005963-ASW_EN.html; see also https://www.european-virus-archive.com/faqs

167 *The discovery of how the database went offline in September 2019* https://archive.is/jE9iy; https://archive.is/9MNVZ; https://extendsclass.com/csv-editor.html#efcaafd; https://archive.is/Sz1dO#selection-1365.0-1373.96; https://archive.is/jPPkB#selection-1387.0-1395.28; https://archive.is/AGtFv; https://archive.is/JZrh6#selection-941.200-941.437. See also Stéphane Foucart. 'Les silences de la Chine, un virus repéré dès 2013, la fausse piste du pangolin … Enquête sur les origines du SARS-CoV-2'. *Le Monde*. 22 December 2020. https://www.lemonde.fr/sciences/article/2020/12/22/a-l-origine-de-la-pandemie-de-covid-19-un-virus-sars-cov-2-aux-sources-toujours-enigmatiques_6064168_1650684.html; and Miranda Devine. 'What Is China Covering Up About the Coronavirus?' *New York Post*. 6 May 2020. https://nypost.com/2020/05/06/what-is-china-covering-up-about-the-coronavirus-devine/

169 *Dr Shi told the BBC in December 2020* John Sudworth. 'Covid: Wuhan Scientist Would "Welcome" Visit Probing Lab Leak Theory'. BBC. 21 December 2020. https://www.bbc.co.uk/news/world-asia-china-55364445

170 *Billy Bostickson, even phrased a question for other scientists to ask* @BillyBostickson Twitter. 23 December 2020. https://twitter.com/BillyBostickson/status/1341624212612497409

170 *Dr Daszak was a participant in a public webinar* @McWLuke Twitter. 10 March 2021. https://twitter.com/McWLuke/status/1369686004924309504

170 *He told the* New York Times Jennifer Kahn. 'How Scientists Could Stop the Next Pandemic Before It Starts'. *New York Times.* 21 April 2020. https://www.nytimes.com/2020/04/21/magazine/pandemic-vaccine.html

8. Gain of function

172 *One of the most lethal strains, known as influenza A H5N1* https://www.who.int/home/cms-decommissioning (https://www.who.int/news-room/q-a-detail/influenza-h5n1)

172 *In May 2012, Dr Yoshihiro Kawaoka's research group* Imai, M., Watanabe, T., Hatta, M., Das, S. C., Ozawa, M., Shinya, K., Zhong, G., Hanson, A., Katsura, H., Watanabe, S., Li, C., Kawakami, E., Yamada, S., Kiso, M., Suzuki, Y., Maher, E. A., Neumann, G., & Kawaoka, Y. (2012). 'Experimental Adaptation of an Influenza H5 HA Confers Respiratory Droplet Transmission to a Reassortant H5 HA/H1N1 Virus in Ferrets'. *Nature,* 486(7403), 420–428. https://doi.org/10.1038/nature10831

173 *Dr Ron Fouchier at Erasmus University in Rotterdam published a similar paper* Herfst, S., Schrauwen, E. J., Linster, M., Chutinimitkul, S., de Wit, E., Munster, V. J., Sorrell, E. M., Bestebroer, T. M., Burke, D. F., Smith, D. J., Rimmelzwaan, G. F., Osterhaus, A. D., & Fouchier, R. A. (2012). 'Airborne Transmission of Influenza A/H5N1 Virus Between Ferrets'. *Science (New York, N.Y.),* 336(6088), 1534–1541. https://doi.org/10.1126/science.1213362

173 *Dr Fouchier told the* New York Times Denise Grady and Donald G. McNeil Jr. 'Debate Persists on Deadly Flu Made Airborne'. 26 December 2011. https://www.nytimes.com/2011/12/27/science/debate-persists-on-deadly-flu-made-airborne.html

176 *On 20 December 2011, the US National Science Advisory Board for Biosecurity* Press Statement on the NSABB Review of H5N1

Research. National Institutes of Health. 20 December 2011. https://www.nih.gov/news-events/news-releases/press-statement-nsabb-review-h5n1-research; see also Mikaela Conley. 'Dutch Scientist Agrees to Omit Published Details of Highly Contagious Bird Flu Findings'. ABC News. 21 December 2011. https://abcnews.go.com/Health/dutch-scientist-agrees-omit-details-killer-bird-flu/story?id=15204649

176 *Drs Francis Collins, Anthony Fauci and Gary Nabel from the NIH posited that 'new data provide valuable insights* Anthony Fauci, Gary Nabel and Francis Collins. 'A Flu Virus Risk Worth Taking'. *Washington Post*. 30 December 2011. https://www.washingtonpost.com/opinions/a-flu-virus-risk-worth-taking/2011/12/30/gIQAM9sNRP_story.html

177 *a* New York Times *editorial on 7 January 2012* 'An Engineered Doomsday'. *New York Times*. 7 January 2012. https://www.nytimes.com/2012/01/08/opinion/sunday/an-engineered-doomsday.html

177 *Dr Fouchier and Dr Kawaoka joined other scientists around the world in declaring a voluntary and temporary sixty-day moratorium on further experiments* Fouchier, R. A., García-Sastre, A., Kawaoka, Y., Barclay, W. S., Bouvier, N. M., Brown, I. H., Capua, I., Chen, H., Compans, R. W., Couch, R. B., Cox, N. J., Doherty, P. C., Donis, R. O., Feldmann, H., Guan, Y., Katz, J., Klenk, H. D., Kobinger, G., Liu, J., Liu, X., ... Webster, R. G. (2012). 'Pause on Avian Flu Transmission Research'. *Science (New York, N.Y.)*, *335*(6067), 400–401. https://doi.org/10.1126/science.335.6067.400

177 *a laboratory at the Harbin Veterinary Research Institute in China successfully mixed a duck isolate of H5N1 with the 2009 H1N1 flu virus* Zhang, Y., Zhang, Q., Kong, H., Jiang, Y., Gao, Y., Deng, G., Shi, J., Tian, G., Liu, L., Liu, J., Guan, Y., Bu, Z., & Chen, H. (2013). 'H5N1 Hybrid Viruses Bearing 2009/H1N1 Virus Genes Transmit in Guinea Pigs by Respiratory Droplet'. *Science (New York, N.Y.)*, *340*(6139), 1459–1463. https://doi.org/10.1126/science.1229455

178 *a run of three mistakes involving smallpox, anthrax and the bird flu virus in US laboratories* Jocelyn Kaiser. 'Lab Incidents Lead to Safety Crackdown at CDC'. *Science*. 11 July 2014. https://www.sciencemag.org/news/2014/07/lab-incidents-lead-safety-crackdown-cdc

179 *Dr Marc Lipsitch had organised the Cambridge Working Group*
     http://www.cambridgeworkinggroup.org/about

179 *Dr Steven Salzberg, an influenza researcher at Johns Hopkins*
     University http://www.cambridgeworkinggroup.org/documents/
     salzberg.pdf

179 *By October 2014, the White House Office of Science and*
     Technology Policy had announced a funding moratorium 'US
     Government Gain-of-Function Deliberative Process and Research
     Funding Pause on Selected Gain-of-Function Research Involving
     Influenza, MERS, and SARS Viruses'. 17 October 2014. https://
     www.phe.gov/s3/dualuse/documents/gain-of-function.pdf; see also
     Nell Greenfieldboyce. 'How a Tilt Toward Safety Stopped a
     Scientist's Virus Research'. NPR. 7 November 2014. https://www.
     npr.org/sections/health-shots/2014/11/07/361219361/how-a-tilt-
     toward-safety-stopped-a-scientists-virus-research; and
     'Gain-of-Function Deliberative Process Written Public Comments.
     Oct. 19, 2014 – Jun. 8, 2016'. https://osp.od.nih.gov/wp-content/
     uploads/2013/06/Gain_of_Function_Deliberative_Process_
     Written_Public_Comments.pdf

180 *reverse genetics* https://www.ncbi.nlm.nih.gov/books/
     NBK21843/

181 *In 1995, he published a book chapter about mouse hepatitis virus*
     Baric, R. S., Fu, K., Chen, W., & Yount, B. (1995). 'High
     Recombination and Mutation Rates in Mouse Hepatitis Virus
     Suggest That Coronaviruses May Be Potentially Important
     Emerging Viruses'. *Advances in Experimental Medicine and*
     Biology, 380, 571–576. https://doi.org/10.1007/978-1-4615-
     1899-0_91

181 *a 'no-see'm' method, patented in 2006* 'Methods and
     Compositions for Infectious cDNA of SARS Coronavirus'.
     JUSTIA Patents. 19 January 2006. https://patents.justia.com/
     patent/20060240530; see also Yount, B., Denison, M. R., Weiss, S.
     R., & Baric, R. S. (2002). 'Systematic Assembly of a Full-Length
     Infectious cDNA of Mouse Hepatitis Virus Strain A59'. *Journal of*
     Virology, 76(21), 11065–11078. https://doi.org/10.1128/
     jvi.76.21.11065-11078.2002

182 *used from at least 2016 by the WIV scientists in some of their*
     experiments Zeng, L. P., Gao, Y. T., Ge, X. Y., Zhang, Q., Peng, C.,
     Yang, X. L., Tan, B., Chen, J., Chmura, A. A., Daszak, P., & Shi,
     Z. L. (2016). 'Bat Severe Acute Respiratory Syndrome-Like
     Coronavirus WIV1 Encodes an Extra Accessory Protein, ORFX,

Involved in Modulation of the Host Immune Response'. *Journal of Virology*, *90*(14), 6573–6582. https://doi.org/10.1128/JVI.03079-15

182   *it took coronavirus experts in Europe and America* Thi Nhu Thao, T., Labroussaa, F., Ebert, N., V'kovski, P., Stalder, H., Portmann, J., Kelly, J., Steiner, S., Holwerda, M., Kratzel, A., Gultom, M., Schmied, K., Laloli, L., Hüsser, L., Wider, M., Pfaender, S., Hirt, D., Cippà, V., Crespo-Pomar, S., Schröder, S., ... Thiel, V. (2020). 'Rapid Reconstruction of SARS-CoV-2 Using a Synthetic Genomics Platform'. *Nature*, *582*(7813), 561–565. https://doi.org/10.1038/s41586-020-2294-9; see also Xie, X., Muruato, A., Lokugamage, K. G., Narayanan, K., Zhang, X., Zou, J., Liu, J., Schindewolf, C., Bopp, N. E., Aguilar, P. V., Plante, K. S., Weaver, S. C., Makino, S., LeDuc, J. W., Menachery, V. D., & Shi, P. Y. (2020). 'An Infectious cDNA Clone of SARS-CoV-2'. *Cell Host & Microbe*, *27*(5), 841–848.e3. https://doi.org/10.1016/j.chom.2020.04.004

182   *In 2007, Dr Shi published a paper announcing her arrival* Ren, W., Qu, X., Li, W., Han, Z., Yu, M., Zhou, P., Zhang, S. Y., Wang, L. F., Deng, H., & Shi, Z. (2008). 'Difference in Receptor Usage Between Severe Acute Respiratory Syndrome (SARS) Coronavirus and SARS-like Coronavirus of Bat Origin'. *Journal of Virology*, *82*(4), 1899–1907. https://doi.org/10.1128/JVI.01085-07

183   *researchers at Utrecht University in the Netherlands had done a similar reverse genetics experiment* Kuo, L., Godeke, G. J., Raamsman, M. J., Masters, P. S., & Rottier, P. J. (2000). 'Retargeting of Coronavirus By Substitution of the Spike Glycoprotein Ectodomain: Crossing the Host Cell Species Barrier'. *Journal of Virology*, *74*(3), 1393–1406. https://doi.org/10.1128/jvi.74.3.1393-1406.2000

183   *The Baric group published an even more ground-breaking experiment* Rockx, B., Sheahan, T., Donaldson, E., Harkema, J., Sims, A., Heise, M., Pickles, R., Cameron, M., Kelvin, D., & Baric, R. (2007). 'Synthetic Reconstruction of Zoonotic and Early Human Severe Acute Respiratory Syndrome Coronavirus Isolates That Produce Fatal Disease in Aged Mice'. *Journal of Virology*, *81*(14), 7410–7423. https://doi.org/10.1128/JVI.00505-07; see also Dyer O. (2003). 'Two Strains of the SARS Virus Sequenced'. *BMJ: British Medical Journal*, *326*(7397), 999

184   *Dr Raymond Pickles, had by now ingeniously managed to culture the cells that line the human respiratory tract* Sims, A. C., Baric, R.

S., Yount, B., Burkett, S. E., Collins, P. L., & Pickles, R. J. (2005). 'Severe Acute Respiratory Syndrome Coronavirus Infection of Human Ciliated Airway Epithelia: Role of Ciliated Cells in Viral Spread in the Conducting Airways of the Lungs'. *Journal of Virology*, 79(24), 15511–15524. https://doi.org/10.1128/JVI.79.24.15511-15524.2005

184 *the creation of 'humanised' mice to accurately and rapidly study diseases such as SARS* Jia, H., Yue, X., & Lazartigues, E. (2020). 'ACE2 Mouse Models: A Toolbox for Cardiovascular and Pulmonary Research'. *Nature Communications*, 11(1), 5165. https://doi.org/10.1038/s41467-020-18880-0; see also Rowan Jacobsen. 'Inside the Risky Bat-Virus Engineering That Links America to Wuhan'. *MIT Technology Review*. 29 June 2021. https://www.technologyreview.com/2021/06/29/1027290/gain-of-function-risky-bat-virus-engineering-links-america-to-wuhan/; Sun, S. H., Chen, Q., Gu, H. J., Yang, G., Wang, Y. X., Huang, X. Y., Liu, S. S., Zhang, N. N., Li, X. F., Xiong, R., Guo, Y., Deng, Y. Q., Huang, W. J., Liu, Q., Liu, Q. M., Shen, Y. L., Zhou, Y., Yang, X., Zhao, T. Y., Fan, C. F., … Wang, Y. C. (2020). 'A Mouse Model of SARS-CoV-2 Infection and Pathogenesis'. *Cell Host & Microbe*, 28(1), 124–133.e4. https://doi.org/10.1016/j.chom.2020.05.020

185 *A seminal 2008 paper by Dr Baric and Dr Mark Denison* Becker, M. M., Graham, R. L., Donaldson, E. F., Rockx, B., Sims, A. C., Sheahan, T., Pickles, R. J., Corti, D., Johnston, R. E., Baric, R. S., & Denison, M. R. (2008). 'Synthetic Recombinant Bat SARS-Like Coronavirus Is Infectious in Cultured Cells and in Mice'. *Proceedings of the National Academy of Sciences of the United States of America*, 105(50), 19944–19949. https://doi.org/10.1073/pnas.0808116105

186 *Dr Baric met Dr Shi at a scientific meeting* Rowan Jacobsen. 'Inside the Risky Bat-Virus Engineering That Links America to Wuhan'. *MIT Technology Review*. 29 June 2021. https://www.technologyreview.com/2021/06/29/1027290/gain-of-function-risky-bat-virus-engineering-links-america-to-wuhan/

186 *In the words of Dr Daszak* 'TWiV 615: Peter Daszak of EcoHealth Alliance'. Video Report. 19 May 2020. https://www.youtube.com/watch?v=IdYDL_RK--w&t=1700s

186 *Dr Vineet Menachery, then a postdoctoral researcher in Baric's lab* Menachery, V. D., Yount, B. L., Jr, Debbink, K., Agnihothram, S., Gralinski, L. E., Plante, J. A., Graham, R. L., Scobey, T., Ge, X. Y.,

Donaldson, E. F., Randell, S. H., Lanzavecchia, A., Marasco, W. A., Shi, Z. L., & Baric, R. S. (2015). 'A SARS-Like Cluster of Circulating Bat Coronaviruses Shows Potential for Human Emergence'. *Nature Medicine*, *21*(12), 1508–1513. https://doi.org/10.1038/nm.3985

187 *An article in* Nature *quoted virologists* Declan Butler. 'Engineered Bat Virus Stirs Debate Over Risky Research'. *Nature*. 12 November 2015. https://www.nature.com/articles/nature.2015.18787; see also: Jef Akst. 'Lab-Made Coronavirus Triggers Debate'. *The Scientist*. 16 November 2015. https://www.the-scientist.com/news-opinion/lab-made-coronavirus-triggers-debate-34502

188 *the EcoHealth Alliance won a five-year, multi-million-dollar grant from the US National Institute of Allergy and Infectious Diseases* 'Understanding the Risk of Bat Coronavirus Emergence'. NIH Report. https://reporter.nih.gov/project-details/8674931; see also https://taggs.hhs.gov/Detail/AwardDetail?arg_AwardNum=R01AI110964&arg_ProgOfficeCode=104

188 *One project funded by this grant resulted in a 2017 paper in the journal* PLoS Pathogens Hu, B., Zeng, L. P., Yang, X. L., Ge, X. Y., Zhang, W., Li, B., Xie, J. Z., Shen, X. R., Zhang, Y. Z., Wang, N., Luo, D. S., Zheng, X. S., Wang, M. N., Daszak, P., Wang, L. F., Cui, J., & Shi, Z. L. (2017). 'Discovery of a Rich Gene Pool of Bat SARS-Related Coronaviruses Provides New Insights into the Origin of SARS Coronavirus'. *PLoS Pathogens*, *13*(11), e1006698. https://doi.org/10.1371/journal.ppat.1006698

189 *civet farm in Kunming* Yuewei Wu. 'Interview with "Virus Hunter": All Genes of SARS Virus Were Found in a Bat Cave in Kunming'. *The Paper*. 8 December 2017. https://www.thepaper.cn/newsDetail_forward_1897724 (archived at https://archive.is/Zgw7b)

189 *It is not clear if Dr Hu's work, funded by the NIH, fell under the type of gain-of-function research for which new US federal funding had been paused* 'US Government Gain-of-Function Deliberative Process and Research Funding Pause on Selected Gain-of-Function Research Involving Influenza, MERS, and SARS Viruses'. 17 October 2014. https://www.phe.gov/s3/dualuse/documents/gain-of-function.pdf; see also Noah Y. Kim. 'No, Dr. Anthony Fauci Did Not Fund Research Tied to COVID-19 "Creation"'. PolitiFact. 8 February 2021. https://www.politifact.com/factchecks/2021/feb/08/worldnetdaily/no-dr-anthony-fauci-did-not-fund-research-tied-cov/

191 *the NIH announced that funding of gain-of-function experiments involving influenza, MERS and SARS would resume under a new framework* Department of Health and Human Services Framework for Guiding Funding Decisions about Proposed Research Involving Enhanced Potential Pandemic Pathogens. Public Health Emergency. https://www.phe.gov/s3/dualuse/Pages/p3co.aspx; see also Burki T. (2018). 'Ban on Gain-Of-Function Studies Ends'. *Lancet. Infectious Diseases, 18*(2), 148–149. https://doi.org/10.1016/S1473-3099(18)30006-9

192 *Dr Shi's group at the WIV, holding at any time numerous such grants and sharing funding with other top labs* Shi Zhengli. Curriculum Vitae. November 2017. https://www.ws-virology.org/wp-content/uploads/2017/11/Zhengli-Shi.pdf

193 *Dr Hu in Dr Shi's lab received a further 250,000 yuan grant* https://archive.vn/DdnaA#selection-1165.0-1165.96

194 *In 2018, a further large five-year grant was announced by the Chinese Academy of Sciences* https://threadreaderapp.com/thread/1363218267863539712.html; https://archive.is/spmNg

194 *Dr Shi's own group were also to receive funding* Latinne, A., Hu, B., Olival, K. J., Zhu, G., Zhang, L., Li, H., Chmura, A. A., Field, H. E., Zambrana-Torrelio, C., Epstein, J. H., Li, B., Zhang, W., Wang, L. F., Shi, Z. L., & Daszak, P. (2020). 'Origin and Cross-Species Transmission of Bat Coronaviruses in China'. *Nature Communications, 11*(1), 4235. https://doi.org/10.1038/s41467-020-17687-3

195 *sporadic cases of vaccine-derived poliovirus* 'Vaccine-derived Poliovirus'. Centers for Disease Control and Prevention. https://www.cdc.gov/vaccines/vpd/polio/hcp/vaccine-derived-poliovirus-faq.html

195 *recent African swine fever outbreaks in China had been caused by unlicensed vaccines* 'New China Swine Fever Strains Point to Unlicensed Vaccines'. Channel News Asia. 22 January 2021. https://www.channelnewsasia.com/news/asia/african-swine-fever-pork-china-disease-outbreak-vaccine-14017538

195 *Dr Baric's team was trying to get around this limitation of live vaccines* Graham, R. L., Deming, D. J., Deming, M. E., Yount, B. L., & Baric, R. S. (2018). 'Evaluation of a Recombination-Resistant Coronavirus as a Broadly Applicable, Rapidly Implementable Vaccine Platform'. *Communications Biology, 1*, 179. https://doi.org/10.1038/s42003-018-0175-7

196 *It would be other kinds of novel vaccines* 'Vaccine Types'. HHS. https://www.hhs.gov/immunization/basics/types/index.html

197 *Dr Limeng Yan had previously worked on influenza vaccines and is a co-first author on a* Nature *paper* Sia, S. F., Yan, L. M., Chin, A., Fung, K., Choy, K. T., Wong, A., Kaewpreedee, P., Perera, R., Poon, L., Nicholls, J. M., Peiris, M., & Yen, H. L. (2020). 'Pathogenesis and Transmission of SARS-CoV-2 in Golden Hamsters'. *Nature*, 583(7818), 834–838. https://doi.org/10.1038/s41586-020-2342-5

198 *According to stories on Fox News* Barnini Chakraborty and Alex Diaz. 'Exclusive: Chinese Virologist Accuses Beijing of Coronavirus Cover-Up, Flees Hong Kong: "I know How They Treat Whistleblowers"'. Fox News. 10 July 2020. https://www.foxnews.com/world/chinese-virologist-coronavirus-cover-up-flee-hong-kong-whistleblower

198 *in the* New York Times *(November 2020)* Amy Qin, Vivian Wang and Danny Hakim. 'How Steve Bannon and a Chinese Billionaire Created a Right-Wing Coronavirus Media Sensation'. *New York Times*. 20 November 2020. https://www.nytimes.com/2020/11/20/business/media/steve-bannon-china.html

199 *Dr Yan's escape story revealed a problem that one of us (Alina) pointed out* @ayjchan Twitter. 16 September 2020. https://twitter.com/Ayjchan/status/1306398565363744772; see also Alex Ward. 'The Bogus Steve Bannon-Backed Study Claiming China Created the Coronavirus, Explained'. Vox. 18 September 2020. https://www.vox.com/2020/9/18/21439865/coronavirus-china-study-bannon

199 *In a blog post at the end of June 2020, Dr Lucey had himself published questions* Daniel R. Lucey. 'COVID-19: Eight Questions for the WHO Team Going to China Next Week to Investigate Pandemic Origins'. Science Speaks: Global ID News. 30 June 2020. https://sciencespeaksblog.org/2020/06/30/covid-19-covid-eight-questions-for-the-who-team-going-to-china-next-week-to-investigate-pandemic-origins/

200 *Dr Yan published a lengthy preprint that month* Yan, Li-Meng, Kang, Shu, Guan, Jie, & Hu, Shanchang. (2020, 14 September). 'Unusual Features of the SARS-CoV-2 Genome Suggesting Sophisticated Laboratory Modification Rather Than Natural Evolution and Delineation of Its Probable Synthetic Route'. Zenodo. http://doi.org/10.5281/zenodo.4028830

200 *Within a month, Dr Yan had released another preprint* Yan, Li-Meng, Kang, Shu, Guan, Jie, & Hu, Shanchang. (2020,

8 October). 'SARS-CoV-2 Is an Unrestricted Bioweapon:
A Truth Revealed through Uncovering a Large-Scale,
Organized Scientific Fraud'. Zenodo. http://doi.org/10.5281/
zenodo.4073131

201 *Ralph Baric had warned in 2006* Baric, R. S. (2006). 'Synthetic
Viral Genomics'. In: *Working Papers for Synthetic Genomics:
Risks and Benefits for Science and Society*, pp. 35–81. Garfinkel,
M. S., Endy, D., Epstein, G. L., Friedman, R. M., editors. 2007.
https://www.jcvi.org/sites/default/files/assets/projects/synthetic-
genomics-options-for-governance/Baric-Synthetic-Viral-
Genomics.pdf

9. The furin cleavage site

202 *Furin is a vital protein doing steady work in human cells*
Benedette Cuffari. 'What Are Furin Proteases?' News Medical.
https://www.news-medical.net/health/What-are-Furin-Proteases.
aspx; see also 'Furin. The Human Protein Atlas'. https://www.
proteinatlas.org/ENSG00000140564-FURIN/tissue

203 *A team of scientists in France and Canada* Coutard, B., Valle, C.,
de Lamballerie, X., Canard, B., Seidah, N. G., & Decroly, E.
(2020). 'The Spike Glycoprotein of the New Coronavirus 2019-
nCoV Contains a Furin-Like Cleavage Site Absent in CoV of the
Same Clade'. *Antiviral Research*, 176, 104742. https://doi.
org/10.1016/j.antiviral.2020.104742

203 *a team of scientists from four universities across cities spanning
China* Zhang, W., Li, X., Chen, J., Shi, J., Duan, G., Chen, X.,
Gao, S., Ruan, J. 'A Furin Cleavage Site Was Discovered in the S
Protein of the 2019 Novel Coronavirus'. ResearchGate preprint.
January 2020. https://www.researchgate.net/publication/
338804501_A_furin_cleavage_site_was_discovered_in_the_S_
protein_of_the_2019_novel_coronavirus

205 *In April 2021, scientists from the National Institutes of Health
found long inserts* Garushyants, S. K., Rogozin, I. B., & Koonin,
E. V. (2021). 'Insertions in SARS-CoV-2 Genome Caused By
Template Switch and Duplications Give Rise to New Variants
of Potential Concern'. *bioRxiv: The Preprint Server for
Biology*, 2021.04.23.441209. https://doi.org/10.1101/2021.04.23.
441209

206 *Several groups of scientists have found that when they remove the
furin cleavage site* Hoffmann, M., Kleine-Weber, H., & Pöhlmann,
S. (2020). 'A Multibasic Cleavage Site in the Spike Protein of

SARS-CoV-2 Is Essential for Infection of Human Lung Cells'. *Molecular Cell, 78*(4), 779–784.e5. https://doi.org/10.1016/j.molcel.2020.04.022

206 *In 1992, a team at the Institute of Virology in Marburg, Germany, discovered that a type of furin* Stieneke-Gröber, A., Vey, M., Angliker, H., Shaw, E., Thomas, G., Roberts, C., Klenk, H. D., & Garten, W. (1992). 'Influenza Virus Hemagglutinin with Multibasic Cleavage Site Is Activated by Furin, a Subtilisin-Like Endoprotease'. *EMBO Journal, 11*(7), 2407–2414

206 *Just a few months later, a different group of scientists led by the same senior authors demonstrated that such a sequence motif* Hallenberger, S., Bosch, V., Angliker, H., Shaw, E., Klenk, H. D., & Garten, W. (1992). 'Inhibition of Furin-Mediated Cleavage Activation of HIV-1 Glycoprotein gp160'. *Nature, 360*(6402), 358–361. https://doi.org/10.1038/360358a0

207 *Dr Jack Nunberg, a biologist whose career has included spells at a pharmaceutical firm and two biotech firms* http://hs.umt.edu/dbs/people/faculty.php?s=Nunberg

207 *carried out the pioneering experiment in 2006* Follis, K. E., York, J., & Nunberg, J. H. (2006). 'Furin Cleavage of the SARS Coronavirus Spike Glycoprotein Enhances Cell-Cell Fusion But Does Not Affect Virion Entry'. *Virology, 350*(2), 358–369. https://doi.org/10.1016/j.virol.2006.02.003

208 *In 2009, a team at Cornell University* Belouzard, S., Chu, V. C., & Whittaker, G. R. (2009). 'Activation of the SARS Coronavirus Spike Protein Via Sequential Proteolytic Cleavage at Two Distinct Sites'. *Proceedings of the National Academy of Sciences of the United States of America, 106*(14), 5871–5876. https://doi.org/10.1073/pnas.0809524106

208 *In May 2015, a scientific collaboration between the Huazhong Agricultural University in Wuhan and Utrecht University in the Netherlands* Li, W., Wicht, O., van Kuppeveld, F. J., He, Q., Rottier, P. J., & Bosch, B. J. (2015). 'A Single Point Mutation Creating a Furin Cleavage Site in the Spike Protein Renders Porcine Epidemic Diarrhea Coronavirus Trypsin Independent for Cell Entry and Fusion'. *Journal of Virology, 89*(15), 8077–8081. https://doi.org/10.1128/JVI.00356-15

208 *Also in 2015, Drs Shi Zhengli and Ralph Baric were co-authors on a paper* Yang, Y., Liu, C., Du, L., Jiang, S., Shi, Z., Baric, R. S., & Li, F. (2015). 'Two Mutations Were Critical for Bat-to-Human Transmission of Middle East Respiratory Syndrome Coronavirus'.

Journal of Virology, 89(17), 9119–9123. https://doi.org/10.1128/
JVI.01279-15

209 *This disease of chickens was the earliest coronavirus disease to be
identified* Mahase E. (2020). 'Covid-19: Coronavirus Was First
Described in *The BMJ* in 1965'. *BMJ (Clinical Eesearch ed.), 369*,
m1547. https://doi.org/10.1136/bmj.m1547

209 *The experiment showed that putting a furin cleavage site into the
S2 domain* Cheng, J., Zhao, Y., Xu, G., Zhang, K., Jia, W., Sun, Y.,
Zhao, J., Xue, J., Hu, Y., & Zhang, G. (2019). 'The S2 Subunit of
QX-type Infectious Bronchitis Coronavirus Spike Protein Is an
Essential Determinant of Neurotropism'. *Viruses, 11*(10), 972.
https://doi.org/10.3390/v11100972

210 *On 6 May 2021, in a lengthy online essay, the veteran science
journalist Nicholas Wade* Nicholas Wade. 'Origin of Covid –
Following the Clues'. *Medium.* 2 May 2020. https://nicholaswade.
medium.com/origin-of-covid-following-the-clues-6f03564c038

211 *He also told* Caltech Weekly 'The Debate Over Origins of SARS-
CoV-2'. *Caltech Weekly.* 22 June 2021. https://www.caltech.edu/
about/news/the-debate-over-origins-of-sars-cov-2

211 *'The Proximal Origin of SARS-CoV-2' published in* Nature
Andersen, K. G., Rambaut, A., Lipkin, W. I., Holmes, E. C., &
Garry, R. F. (2020). 'The Proximal Origin of SARS-CoV-2'. *Nature
Medicine, 26*(4), 450–452. https://doi.org/10.1038/s41591-020-
0820-9

211 *Dr Andersen had been quoted in the Scripps press release* 'The
COVID-19 Coronavirus Epidemic Has a Natural Origin,
Scientists Say. Scripps Research'. 17 March 2020. https://www.
scripps.edu/news-and-events/press-room/2020/20200317-
andersen-covid-19-coronavirus.html

212 *Dr Andersen pointed out Mr Wade's 'troubled history of
misrepresenting (and/or misunderstanding) the very basics of
evolutionary biology'* 'Letters: "A Troublesome Inheritance"'.
Stanford University. https://cehg.stanford.edu/letter-from-
population-geneticists

213 *Within the genome of SARS-CoV-2, the commonest codon for
arginine is AGA and the rarest is CGG* Hou W. (2020).
'Characterization of Codon Usage Pattern in SARS-CoV-2'.
Virology Journal, 17(1), 138. https://doi.org/10.1186/s12985-020-
01395-x

213 *In 2004, a mainly Boston-based team of scientists* Moore, M. J.,
Dorfman, T., Li, W., Wong, S. K., Li, Y., Kuhn, J. H., Coderre, J.,

Vasilieva, N., Han, Z., Greenough, T. C., Farzan, M., & Choe, H. (2004). 'Retroviruses Pseudotyped with the Severe Acute Respiratory Syndrome Coronavirus Spike Protein Efficiently Infect Cells Expressing Angiotensin-Converting Enzyme 2'. *Journal of Virology*, 78(19), 10628–10635. https://doi. org/10.1128/JVI.78.19.10628-10635.2004

215 *Dr Nunberg likewise was quoted in June 2020* David Cyranoski. 'The Biggest Mystery: What It Will Take to Trace the Coronavirus Source'. *Nature*. 5 June 2020. https://www.nature.com/articles/d41586-020-01541-z

216 *a bat virus was found and given a lot of publicity* Zhou, H., Chen, X., Hu, T., Li, J., Song, H., Liu, Y., Wang, P., Liu, D., Yang, J., Holmes, E. C., Hughes, A. C., Bi, Y., & Shi, W. (2020). 'A Novel Bat Coronavirus Closely Related to SARS-CoV-2 Contains Natural Insertions at the S1/S2 Cleavage Site of the Spike Protein'. *Current Biology : CB*, *30*(11), 2196–2203.e3. https://doi. org/10.1016/j.cub.2020.05.023

216 *Rm stands for* Rhinolophus malayanus, *the Malayan horseshoe bat* Liang, J., He, X., Peng, X., Xie, H., Zhang, L. 'First Record of Existence of Rhinolophus Malayanus (Chiroptera, Rhinolophidae) in China'. *Mammalia*. 84(4). DOI:10.1515/mammalia-2019-0062. Available at: https://www.researchgate.net/publication/337145895_First_record_of_existence_of_Rhinolophus_malayanus_Chiroptera_Rhinolophidae_in_China

217 *later published in* BioEssays *in May 2021* Deigin, Y. & Segreto, R. (2021). 'SARS-CoV-2's Claimed Natural Origin Is Undermined by Issues with Genome Sequences of Its Relative Strains: Coronavirus Sequences RaTG13, MP789 and RmYN02 Raise Multiple Questions to Be Critically Addressed by the Scientific Community'. *BioEssays: News and Reviews in Molecular, Cellular and Developmental Biology*, 43(7), e2100015. https://doi.org/10.1002/bies.202100015

218 *In December 2020, scientists from the University of Tokyo published the genome of a SARS-CoV-2-like virus* Murakami, S., Kitamura, T., Suzuki, J., Sato, R., Aoi, T., Fujii, M. Horimoto, T. (2020). 'Detection and Characterization of Bat Sarbecovirus Phylogenetically Related to SARS-CoV-2, Japan'. *Emerging Infectious Diseases*, 26(12), 3025-3029. https://doi.org/10.3201/eid2612.203386

219 *In early 2021, French and Cambodian scientists found two nearly identical viruses* Vibol Hul, Deborah Delaune, Erik A. Karlsson,

Alexandre Hassanin, Putita Ou Tey, Artem Baidaliuk, Fabiana Gámbaro, Vuong Tan Tu, Lucy Keatts, Jonna Mazet, Christine Johnson, Philippe Buchy, Philippe Dussart, Tracey Goldstein, Etienne Simon-Lorière, Veasna Duong. 'A Novel SARS-CoV-2 Related Coronavirus in Bats from Cambodia'. *bioRxiv: The Preprint Server for Biology* 2021.01.26.428212; doi: https://doi.org/10.1101/2021.01.26.428212

219 *The next new virus to be reported was in Thailand* Wacharapluesadee, S., Tan, C. W., Maneeorn, P., Duengkae, P., Zhu, F., Joyjinda, Y., Kaewpom, T., Chia, W. N., Ampoot, W., Lim, B. L., Worachotsueptrakun, K., Chen, V. C., Sirichan, N., Ruchisrisarod, C., Rodpan, A., Noradechanon, K., Phaichana, T., Jantarat, N., Thongnumchaima, B., Tu, C., ... Wang, L. F. (2021). 'Evidence for SARS-CoV-2 Related Coronaviruses Circulating in Bats and Pangolins in Southeast Asia'. *Nature Communications*, 12(1), 972. https://doi.org/10.1038/s41467-021-21240-1

219 *the species* Rhinolophus acuminatus, *one not reported to be found in China* Accuminate Horseshoe Bat. IUCN Red List. 30 August 2018. https://www.iucnredlist.org/species/19520/21974227

219 *seven bat species so far have been found to harbour them* Zhou, H., Ji, J., Chen, X., Bi, Y., Li, J., Wang, Q., Hu, T., Song, H., Zhao, R., Chen, Y., Cui, M., Zhang, Y., Hughes, A. C., Holmes, E. C., & Shi, W. (2021). 'Identification of Novel Bat Coronaviruses Sheds Light on the Evolutionary Origins of SARS-CoV-2 and Related Viruses'. *Cell*, 184(17), 4380–4391.e14. Advance online publication. https://doi.org/10.1016/j.cell.2021.06.008

221 *The first published paper to analyse and discuss the newly discovered genome sequence* Zhou, P., Yang, X. L., Wang, X. G., Hu, B., Zhang, L., Zhang, W., Si, H. R., Zhu, Y., Li, B., Huang, C. L., Chen, H. D., Chen, J., Luo, Y., Guo, H., Jiang, R. D., Liu, M. Q., Chen, Y., Shen, X. R., Wang, X., Zheng, X. S., ... Shi, Z. L. (2020). 'A Pneumonia Outbreak Associated with a New Coronavirus of Probable Bat Origin'. *Nature*, 579(7798), 270–273. https://doi.org/10.1038/s41586-020-2012-7

222 *another paper was published online, on 31 January 2020, in the journal* Emerging Microbes & Infections Jiang, S., Du, L., & Shi, Z. (2020). 'An Emerging Coronavirus Causing Pneumonia Outbreak in Wuhan, China: Calling for Developing Therapeutic and Prophylactic Strategies'. *Emerging Microbes & Infections*, 9(1), 275–277. https://doi.org/10.1080/22221751.2020.1723441

223 *Some scientists have pointed out that* @trvrb Twitter. 10 June 2021. https://twitter.com/trvrb/status/1403045938009022466; see also Wu, F., Zhao, S., Yu, B., Chen, Y. M., Wang, W., Song, Z. G., Hu, Y., Tao, Z. W., Tian, J. H., Pei, Y. Y., Yuan, M. L., Zhang, Y. L., Dai, F. H., Liu, Y., Wang, Q. M., Zheng, J. J., Xu, L., Holmes, E. C., & Zhang, Y. Z. (2020). 'A New Coronavirus Associated With Human Respiratory Disease in China'. *Nature*, 579(7798), 265–269. https://doi.org/10.1038/s41586-020-2008-3

10. The other eight

225 *in the addendum to the paper published the following November* Zhou, P., Yang, X. L., Wang, X. G., Hu, B., Zhang, L., Zhang, W., Si, H. R., Zhu, Y., Li, B., Huang, C. L., Chen, H. D., Chen, J., Luo, Y., Guo, H., Jiang, R. D., Liu, M. Q., Chen, Y., Shen, X. R., Wang, X., Zheng, X. S., … Shi, Z. L. (2020). 'Addendum: A Pneumonia Outbreak Associated with a New Coronavirus of Probable Bat Origin'. *Nature*, 588(7836), E6. https://doi.org/10.1038/s41586-020-2951-z

228 *Alina confirmed that at least one US scientist had reached out* @notoriousFIL Twitter. 13 September 2020. https://twitter.com/notoriousFIL/status/1305206970937479170

228 *On 19 May 2020, the amplicon sequences were deposited into the GenBank database* https://www.ncbi.nlm.nih.gov/sra/SRX8357956

229 *Alina pointed out Dr Daszak's mistake in the* Sunday Times George Arbuthnott, Jonathan Calvert and Philip Sherwell. 'Revealed: Seven Year Coronavirus Trail from Mine Deaths to a Wuhan Lab'. *Sunday Times*. 4 July 2020. https://www.thetimes.co.uk/article/seven-year-covid-trail-revealed-l5vxt7jqp

230 *It started with a tweet on 2 July 2020 from a Spaniard called Francisco de Ribera* @franciscodeasis Twitter. 2 July 2020. https://twitter.com/franciscodeasis/status/1278698980940173314

233 *8 July, Ribera asked Dr Daszak* @franciscodeasis Twitter. 8 July 2020. https://twitter.com/franciscodeasis/status/1280980860850909189

234 *Then, on 8 July, up popped the anonymous and indefatigable internet sleuth called the Seeker with a helpful clue* @TheSeeker268 Twitter. 7 July 2020. https://twitter.com/TheSeeker268/status/1280702509640634368; https://bigd.big.ac.cn/biosample/browse/SAMC191048

234 *On 2 August, he found there was a 'cluster of consecutive samples from 7895 to 7966 (72 samples, 47 positives, 3 co-infections) suggesting a same trip'* @franciscodeasis Twitter. 1 August 2020. https://twitter.com/franciscodeasis/status/1289716856052264960

234 *Ribera gradually became all but certain that the eight new viruses came from the copper mine in Mojiang County, where they had been collected two years later than the RaTG13 sample, in the spring of 2015* @franciscodeasis Twitter. 21 May 2021. https://twitter.com/franciscodeasis/status/1395915528234012674

235 *On 3 December 2020, Dr Shi spoke at a webinar* 'Covid-19: diaporama et conférences de l'académie vétérinaire de France'. Academie Veterinaire de France. https://academie-veterinaire-defrance.org/publications/videos-de-lavf/covid-19

236 *A bioinformatics consultant, Moreno Colaiacovo, who spotted the slide* @emmecola Twitter. 3 December 2020. https://twitter.com/emmecola/status/1334505779718729729

236 *on 18 December, Dr Shi gave another online lecture at a conference of the European Scientific Working Group on Influenza* ESWI2020 Keynote Lecture Zhengli Shi: From SARS to COVID-19, Understanding The Interspecies Transmission of SARS-Related Coronaviruses. ESWI. https://eswi.org/eswi-tv/eswi2020-keynote-lecture-zhengli-shi-from-sars-to-covid-19-understanding-the-interspecies-transmission-of-sars-related-coronaviruses/

236 *On 24 February 2021, Dr Shi gave another online talk* Available at: https://www.youtube.com/watch?v=LSISkvlesnA

236 *On 23 March, Dr Shi gave yet another talk* 'From SARS, MERS to COVID-19: Understanding of Interspecies Transmission of Bat Coronaviruses'. Rutgers. 23 March 2021. https://globalhealth.rutgers.edu/event/from-sars-mers-to-covid-19-understanding-of-interspecies-transmission-of-bat-coronaviruses/

236 *seven of the eight had a different prefix: Rst* @AntGDuarte Twitter. 23 March 2021. https://twitter.com/AntGDuarte/status/1374457519838437378

237 *in May 2021, Dr Shi and colleagues did at last publish in a preprint* Hua Guo, Ben Hu, Hao-rui Si, Yan Zhu, Wei Zhang, Bei Li, Ang Li, Rong Geng, Hao-Feng Lin, Xing-Lou Yang, Peng Zhou, Zheng-Li Shi. *bioRxiv: The Preprint Server for Biology* 2021.05.21.445091; doi: https://doi.org/10.1101/2021.05.21.445091

238 *A Spanish business consultant working in his spare time painstakingly worked out* @franciscodeasis Twitter. 18 November 2020. https://twitter.com/franciscodeasis/status/1329057812815765504

240 *In September 2021, it was revealed that Dr Wang* https://www.science.org/content/article/why-many-scientists-say-unlikely-sars-cov-2-originated-lab-leak

242 *Remember that Francisco de Ribera had asked, as long ago as July 2020, 'Could it be that the 7896 was used somehow as an aid for sequencing RaTG13?'* @franciscodeasis Twitter. 7 July 2020. https://twitter.com/franciscodeasis/status/1280518834965950472

11. Popsicle origins and the World Health Organization

244 *the World Health Organization's director general Dr Tedros Adhanom Ghebreyesus said China's speed* WHO Director-General's Statement on IHR Emergency Committee on Novel Coronavirus (2019-nCoV). World Health Organization. 30 January 2020. https://www.who.int/director-general/speeches/detail/who-director-general-s-statement-on-ihr-emergency-committee-on-novel-coronavirus-(2019-ncov)

244 *on 14 January, the WHO had even issued a tweet* @WHO Twitter. 14 January 2020. https://twitter.com/who/status/1217043229427761152?lang=en

244 *the acting head of the WHO's emerging diseases unit, Dr Maria Van Kerkhove, announced that although there were clusters of cases* Stephanie Nebehay. 'WHO Says New China Coronavirus Could Spread, Warns Hospitals Worldwide'. Reuters. 14 January 2020. https://www.reuters.com/article/us-china-health-pneumonia-who-idUSKBN1ZD16J

245 *A WHO team visited Wuhan on 20 January 2020* Mission summary: WHO Field Visit to Wuhan, China 20–21 January 2020. World Health Organization. 22 January 2020. https://www.who.int/china/news/detail/22-01-2020-field-visit-wuhan-china-jan-2020

245 *WHO officials were complaining* 'Gravitas: Wuhan Virus Pandemic: Leaked Audio Featuring WHO Officials Reveal China's Lapses'. WION. Video Report. 27 January 2021. https://www.wionews.com/videos/gravitas-wuhan-virus-pandemic-leaked-audio-featuring-who-officials-reveal-chinas-lapses-359761

245 *In February Dr Lawrence Gostin of Georgetown University complained* Emily Rauhala. 'Chinese Officials Note Serious

Problems in Coronavirus Response. The World Health Organization Keeps Praising Them'. *Washington Post*. 9 February 2020. https://www.washingtonpost.com/world/asia_pacific/chinese-officials-note-serious-problems-in-coronavirus-response-the-world-health-organization-keeps-praising-them/2020/02/08/b663dd7c-4834-11ea-91ab-ce439aa5c7c1_story.html

246 *Dr Gostin to the* New York Times Selam Gebrekidan, Matt Apuzzo, Amy Qin and Javier C. Hernández. 'In Hunt for Virus Source, WHO Let China Take Charge'. *New York Times*. 2 November 2020. https://www.nytimes.com/2020/11/02/world/who-china-coronavirus.html

246 *The decision was reversed on 30 January* Statement on the Second Meeting of the International Health Regulations (2005) Emergency Committee Regarding the Outbreak of Novel Coronavirus (2019-nCoV). World Health Organization. 30 January 2020. https://www.who.int/news/item/30-01-2020-statement-on-the-second-meeting-of-the-international-health-regulations-(2005)-emergency-committee-regarding-the-outbreak-of-novel-coronavirus-(2019-ncov)

246 *Dr Gauden Galea, the WHO representative in China, complained publicly* Tom Cheshire. 'Coronavirus: WHO "Not Invited" to Join China's COVID-19 Investigations'. Sky News. 1 May 2020. https://news.sky.com/story/coronavirus-who-not-invited-to-join-chinas-covid-19-investigations-11981193

246 *'little had been done in terms of epidemiological investigations around Wuhan since January 2020'* Stephanie Kirchgaessner. 'China Did "Little" to Hunt for Covid Origins in Early Months, Says WHO Document'. *Guardian*. 23 February 2021. https://www.theguardian.com/world/2021/feb/23/china-did-little-hunt-covid-origins-early-months-says-who-document

247 *The final composition of the international team was to be agreed by both China and the WHO* Simone McCarthy. 'Coronavirus: WHO Waits for China to Approve Pandemic Origins Investigators'. *South China Morning Post*. 8 October 2020. https://www.scmp.com/news/china/diplomacy/article/3104578/coronavirus-who-waits-china-approve-pandemic-origins

248 *he wrote an opinion article in the* New York Times Peter Daszak. 'We Knew Disease X Was Coming. It's Here Now'. *New York Times*. 27 February 2020. https://www.nytimes.com/2020/02/27/opinion/coronavirus-pandemics.html

248 *Dr Daszak showed he had not changed his mind, writing an opinion piece for the* Guardian Peter Daszak. 'Ignore the Conspiracy Theories: Scientists Know Covid-19 Wasn't Created in a Lab'. *Guardian*. 9 June 2020. https://www.theguardian.com/commentisfree/2020/jun/09/conspiracies-covid-19-lab-false-pandemic

248 *later revealed under FOI requests, Dr Daszak had said* https://usrtk.org/wp-content/uploads/2021/02/Baric-Emails-2.17.21.pdf; see also @CharlesRixey Twitter. 18 May 2021. https://twitter.com/CharlesRixey/status/1394555646071160833

249 *Later the journalist Nathan Robinson asked* Nathan J. Robinson. 'The Stakes of Finding COVID-19's Origins'. *Current Affairs*. 14 May 2021. https://www.currentaffairs.org/2021/05/the-stakes-of-finding-covid-19s-origins

249 *The first phase of investigation* Drew Hinshaw and Jeremy Page. 'WHO Mission to Look for Answers to Covid-19's Origin in Wuhan'. *Wall Street Journal*. 14 January 2021. https://www.wsj.com/articles/who-mission-to-look-for-answers-to-covid-19s-origin-in-wuhan-11610566665; see also John Sudworth. 'Wuhan Marks Its Anniversary with Triumph and Denial'. BBC. 23 January 2021. https://www.bbc.com/news/world-asia-china-55765875

249 *resulting in what Bloomberg called 'a rare rebuke' from the WHO on 5 January 2021* 'WHO Issues Rare Rebuke to China for Delaying Virus Team'. Bloomberg. 6 January 2021. https://www.bloomberg.com/news/articles/2021-01-06/who-issues-rare-rebuke-to-china-for-delaying-virus-origins-trip; see also Drew Hinshaw. 'WHO Criticizes China for Stymying Investigation Into Covid-19 Origins'. *Wall Street Journal*. 6 January 2021. https://www.wsj.com/articles/world-health-organization-criticizes-china-over-delays-in-covid-19-probe-11609883140

250 *The Chinese foreign ministry spokeswoman responded* 'Covid: WHO Team Investigating Virus Origins Denied Entry to China'. BBC. 6 January 2021. https://www.bbc.com/news/world-asia-china-55555466

251 *It was 'part of a global study, not an investigation,' said a foreign ministry spokesman* Janis Mackey Frayer and Adela Suliman. 'As WHO Mission into Covid Origin Begins in China, Bereaved Son Seeks Answers'. NBC. 29 January 2021. https://www.nbcnews.com/news/world/who-mission-covid-origin-begins-china-bereaved-son-seeks-answers-n1256121

251 *a press conference in Wuhan* COVID-19 Virtual Press conference transcript – 9 February 2021. World Health Organization. 9 February 2021. https://www.who.int/publications/m/item/covid-19-virtual-press-conference-transcript---9-february-2021

251 *Dr Ben Embarek told* Science *magazine* Kai Kupferschmidt. '"Politics Was Always in the Room." WHO Mission Chief Reflects on China Trip Seeking COVID-19's Origin'. *Science.* 14 February 2021. https://www.sciencemag.org/news/2021/02/politics-was-always-room-who-mission-chief-reflects-china-trip-seeking-covid-19-s

252 *Interviewed on CNN, one of the members of the team, Dr Daszak* '"Striking piece of evidence": WHO Researcher on Products in Wuhan Market'. CNN. Video Report. https://edition.cnn.com/videos/world/2021/02/09/who-coronavirus-origin-wuhan-china-peter-daszak-interview-anderson-ctw-intl-ldn-vpx.cnn/video/playlists/coronavirus-intl/

253 *Dr Fabian Leendertz from Germany, later told the* New York Times Javier C. Hernández and James Gorman. 'On WHO Trip, China Refused to Hand Over Important Data'. *New York Times.* 12 February 2021. https://www.nytimes.com/2021/02/12/world/asia/china-world-health-organization-coronavirus.html

253 *the total amount of time the WHO-convened team had spent at the Wuhan Institute of Virology* Emily Wang Fujiyama. 'WHO Team Visits Wuhan Virus Lab at Center of Speculation'. Associated Press. 2 February 2021. https://apnews.com/article/who-visits-wuhan-virus-lab-china-8798374a3c6acc56b1e8b973309045ba; see also Gerry Shih and Emily Rauhala. 'After Wuhan Mission on Pandemic Origins, WHO Team Dismisses Lab Leak Theory'. *Washington Post.* 9 February 2021. https://www.washingtonpost.com/world/asia_pacific/coronavirus-who-china-investigation-wuhan/2021/02/09/2af3c44c-6a79-11eb-a66e-c27046e9e898_story.html

254 *the host of* 60 Minutes, *Lesley Stahl, later called it out in an interview* Lesley Stahl. 'What Happened in Wuhan? Why Questions Still Linger on the Origin of the Coronavirus'. *60 Minutes.* 28 March 2021. https://www.cbsnews.com/news/covid-19-wuhan-origins-60-minutes-2021-03-28/

254 *said one of the international experts on the team, Dr Vladimir Dedkov* Daria Litvinova and Jamey Keaten. '1 Report, 4 Theories: Where Did COVID-19 Come From?' Associated Press. 25 March

2021. https://www.king5.com/article/news/health/coronavirus/1-report-4-theories-coronavirus-origin/507-ff2bd079-b3ad-48a6-9988-0e2c574a3c99

255 *the report's map of confirmed early cases in Wuhan* Alina Chan.
'A Response to "The Origins of SARS-CoV-2: A Critical Review"'.
Medium. 12 July 2021. https://ayjchan.medium.com/a-response-to-the-origins-of-sars-cov-2-a-critical-review-5d4a644d9777

255 *the* Washington Post *reported several important errors in the
China-WHO joint report* Eva Dou and Emily Rauhala. 'WHO
Clarifies Details of Early Covid Patients in Wuhan after Errors in
Virus Report'. *Washington Post*. 15 July 2021. https://www.
washingtonpost.com/world/asia_pacific/covid-wuhan-outbreak-
who/2021/07/15/51e7e8a6-e2c6-11eb-88c5-4fd6382c47cb_story.
html

256 *a group of prominent scientists had already attempted* Holmes,
Edward C., Goldstein, Stephen A., Rasmussen, Angela L.,
Robertson, David L., Crits-Christoph, Alexander, Wertheim, Joel
O., Anthony, Simon J., Barclay, Wendy S., Boni, Maciej F.,
Doherty, Peter C., Farrar, Jeremy, Geoghegan, Jemma L, Jiang,
Xiaowei, Leibowitz, Julian L., Neil, Stuart J. D, Skern, Tim, Weiss,
Susan R., Worobey, Michael, Andersen, Kristian G., ... Rambaut,
Andrew. (2021). 'The Origins of SARS-CoV-2: A Critical Review
(1.0)'. Zenodo. https://doi.org/10.5281/zenodo.5075888

257 *Dr Marion Koopmans, a Dutch virologist on the team, told the*
Wall Street Journal ... *Dr Thea Fischer, a Danish epidemiologist,
told the* Wall Street Journal Jeremy Page and Drew Hinshaw.
'China Refuses to Give WHO Raw Data on Early Covid-19
Cases'. *Wall Street Journal*. 12 February 2021. https://www.wsj.
com/articles/china-refuses-to-give-who-raw-data-on-early-covid-
19-cases-11613150580; see also Betsy McKay, Drew Hinshaw
and Jeremy Page. 'Covid-19 Was Spreading in China Before First
Confirmed Cases, Fresh Evidence Suggests'. *Wall Street Journal*.
19 February 2021. https://www.wsj.com/articles/covid-19-was-
spreading-in-china-before-first-confirmed-cases-fresh-evidence-
suggests-11613730600

257 *The possibility of analysing Wuhan blood banks* Betsy McKay
and Amy Dockser Marcus. 'Origin of Covid-19 Pandemic Is
Sought in Old Blood Samples'. *Wall Street Journal*. 8 April 2021.
https://www.wsj.com/articles/origin-of-covid-19-pandemic-is-
sought-in-old-blood-samples-11617874200; see also James
Gorman. 'With Virus Origins Still Obscure, WHO and Critics

Look to Next Steps'. *New York Times*. 7 April 2021. https://www.
nytimes.com/2021/04/07/health/coronavirus-lab-leak-who.html

258 *Dr Rasmus Nielsen of the University of Berkeley expressed
surprise* Keld Vrå Andersen. 'China Does Not Want Corona
Theory Investigated Further'. TV2. 13 August 2021.

258 *Dr Daszak was retweeting Chinese state-owned media* Twitter. 9
February 2021. https://twitter.com/globaltimesnews/
status/1359156331769430028

259 *researchers published an original study of the wild animals sold in
Wuhan's markets* Xiao, X., Newman, C., Buesching, C. D.,
Macdonald, D. W., & Zhou, Z. M. (2021). 'Animal Sales from
Wuhan Wet Markets Immediately Prior to the COVID-19
Pandemic'. *Scientific Reports*, *11*(1), 11898. https://doi.
org/10.1038/s41598-021-91470-2

261 *Not all went as far as Dr Bruno Canard* Ian Birrell. 'World
Experts Condemn WHO Inquiry as a "Charade"'. *Mail on
Sunday*. 14 February 2021. http://www.ianbirrell.com/world-
experts-condemn-who-inquiry-as-a-charade/

261 The Australian *newspaper, which wrote* 'WHO Whitewash Leaves
the World in Darkness on Covid'. *The Australian*. 10 February
2021. https://www.theaustralian.com.au/commentary/editorials/
who-whitewash-leaves-the-world-in-darkness-on-covid/news-stor
y/5b31f4b2c3521d0b18b807729cf5d53e

262 *national security advisor, Jake Sullivan, said* Statement by
National Security Advisor Jake Sullivan. The White House. 13
February 2021. https://www.whitehouse.gov/briefing-room/
statements-releases/2021/02/13/statement-by-national-security-
advisor-jake-sullivan/

262 *Dr Daszak went another way. He tweeted* @PeterDaszak Twitter.
9 February 2021. https://twitter.com/PeterDaszak/
status/1359357862254841856

262 *participated in a public Chatham House webinar* 'COVID-19
briefing: The WHO-China mission'. Chatham House. 10 March
2021. https://www.chathamhouse.org/events/all/members-event/
covid-19-briefing-who-china-mission

263 *A* Wall Street Journal *report corroborated this story* Jeremy
Page, Betsy McKay and Drew Hinshaw. 'Search for Covid's
Origins Leads to China's Wild Animal Farms – and a Big
Problem'. *Wall Street Journal*. 30 June 2021. https://www.wsj.
com/articles/covid-origins-china-wild-animal-farms-pandemic-
source-11625060088

263 *Dr Daniel Lucey of Georgetown University in Washington, had told the BBC* John Sudworth. 'Covid: Wuhan Scientist Would "Welcome" Visit Probing Lab Leak Theory'. BBC. 21 December 2020. https://www.bbc.co.uk/news/world-asia-china-55364445

263 *Dr Lucey told the* South China Morning Post Simone McCarthy. 'Hunt for Covid-19 Origins: Questions Surface Over China's Testing of Suspect Animals'. *South China Morning Post.* 26 February 2021. https://www.scmp.com/news/china/science/article/3123147/hunt-covid-19-origins-questions-surface-over-chinas-testing

264 *the China-WHO joint team scrapped its plan to publish an interim report* Betsy McKay, Drew Hinshaw and Jeremy Page. 'WHO Investigators to Scrap Plans for Interim Report on Probe of Covid-19 Origins'. *Wall Street Journal.* 5 March 2021. https://www.wsj.com/articles/who-investigators-to-scrap-interim-report-on-probe-of-covid-19-origins-11614865067

265 *The* Global Times, *a Chinese state organ* 'Presumption of Guilt in Virus Study Not Conducive, Chinese FM Comments on "Open Letter" of "New Probe"'. *Global Times.* 5 March 2021. https://www.globaltimes.cn/page/202103/1217465.shtml

265 *This time, Dr Tedros said* WHO Director-General's Remarks at the Member State Briefing on the Report of the International Team Studying the Origins of SARS-CoV-2. World Health Organization. 30 March 2021. https://www.who.int/director-general/speeches/detail/who-director-general-s-remarks-at-the-member-state-briefing-on-the-report-of-the-international-team-studying-the-origins-of-sars-cov-2

266 *Governments from around the world weighed in on the report* Joint Statement on the WHO-Convened COVID-19 Origins Study. US Department of State. 30 March 2021. https://www.state.gov/joint-statement-on-the-who-convened-covid-19-origins-study/; see also EU Statement on the WHO-led COVID-19 Origins Study. Delegation of the European Union to the UN and Other International Organisations in Geneva. 30 March 2021. https://eeas.europa.eu/delegations/un-geneva/95960/eu-statement-who-led-covid-19-origins-study_en

266 *An unidentified Chinese scientist on the China-WHO study team was reported* Zhuang Pinghui. 'Coronavirus: Chinese Expert Rails Against WHO Chief and Wuhan Lab Leak Theory'. *South China Morning Post.* 21 April 2021. https://www.scmp.com/news/china/diplomacy/article/3130470/coronavirus-chinese-expert-rails-

against-tedros-and-wuhan-lab; see also Natalie Liu. 'Beijing Urges WHO Leader Not to Pursue "Lab Leak" Theory'. *Voice of America*. 22 April 2021. https://www.voanews.com/covid-19-pandemic/beijing-urges-who-leader-not-pursue-lab-leak-theory

267 *It was published on 14 May with eighteen authors* Bloom, J. D., Chan, Y. A., Baric, R. S., Bjorkman, P. J., Cobey, S., Deverman, B. E., Fisman, D. N., Gupta, R., Iwasaki, A., Lipsitch, M., Medzhitov, R., Neher, R. A., Nielsen, R., Patterson, N., Stearns, T., van Nimwegen, E., Worobey, M., & Relman, D. A. (2021). 'Investigate the Origins of COVID-19'. *Science (New York, N.Y.)*, 372(6543), 694. https://doi.org/10.1126/science.abj0016

267 *The editor-in-chief at* Science *wrote an accompanying blog post* H. Holden Thorp. 'Continued Discussion on the Origin of COVID-19'. *Science*. 13 May 2021. https://blogs.sciencemag.org/editors-blog/2021/05/13/continued-discussion-on-the-origin-of-covid-19/

267 *Baric had told the People's Pharmacy* Nicholson Baker. 'The Lab Leak Hypothesis'. *New York Magazine*. 4 January 2021. https://nymag.com/intelligencer/article/coronavirus-lab-escape-theory.html

268 *He told the Italian television programme* Presa Diretta *in mid-2020* Presa Diretta. Video Report. https://www.raiplay.it/video/2020/11/SARS-COV2---Identikit-di-un-Killer-f4b539cc-07ac-41c2-a1a4-1aab862f75ff.html

268 *The* New York Times *altered its description of the lab-leak theory in a tab heading* @McWLuke Twitter. 14 May 2021. https://twitter.com/McWLuke/status/1393290456717484034

269 *The fact-checking site* PolitiFact *removed a claim of 'debunked conspiracy theory'* @SohrabAhmari Twitter. 19 May 2021. https://twitter.com/SohrabAhmari/status/1395159392287342592

269 *Long articles by journalists Nathan Robinson in* Current Affairs Nathan J. Robinson. 'The Stakes of Finding COVID-19's Origins'. *Current Affairs*. 14 May 2021. https://www.currentaffairs.org/2021/05/the-stakes-of-finding-covid-19s-origins

269 *Donald McNeil Jr in* Medium Donald G. McNeil Jr. 'How I Learned to Stop Worrying and Love the Lab-Leak Theory'. *Medium*. 17 May 2021. https://donaldgmcneiljr1954.medium.com/how-i-learned-to-stop-worrying-and-love-the-lab-leak-theory-f4f88446b04d

12. Spillover

271 *The same has already been found for the 2003 SARS virus* Hu, B., Zeng, L. P., Yang, X. L., Ge, X. Y., Zhang, W., Li, B., Xie, J. Z., Shen, X. R., Zhang, Y. Z., Wang, N., Luo, D. S., Zheng, X. S., Wang, M. N., Daszak, P., Wang, L. F., Cui, J., & Shi, Z. L. (2017). 'Discovery of a Rich Gene Pool of Bat SARS-Related Coronaviruses Provides New Insights into the Origin of SARS Coronavirus'. *PLoS Pathogens*, 13(11), e1006698. https://doi.org/10.1371/journal.ppat.1006698

272 *Or take Ebola* Ebola (Ebola Virus Disease). Centers for Disease Control and Prevention. https://www.cdc.gov/vhf/ebola/history/chronology.html; see also 'Chronology of Ebola Virus Disease Outbreaks, 1976–2014'. *Public Health Intelligence*. 6 October 2014. http://publichealthintelligence.org/content/chronology-ebola-virus-disease-outbreaks-1976-2014

274 *Dr Fabian Leendertz, said they agreed to include the frozen-food theory* Javier C. Hernández and James Gorman. 'On WHO Trip, China Refused to Hand Over Important Data'. *New York Times*. 12 February 2021. https://www.nytimes.com/2021/02/12/world/asia/china-world-health-organization-coronavirus.html

274 *Even the China-WHO team leader, Dr Peter Ben Embarek, conceded* Kai Kupferschmidt. '"Politics Was Always in the Room." WHO Mission Chief Reflects on China Trip Seeking COVID-19's Origin'. *Science*. 14 February 2021. https://www.sciencemag.org/news/2021/02/politics-was-always-room-who-mission-chief-reflects-china-trip-seeking-covid-19-s

277 *an experiment involving a bovine coronavirus* Borucki, M. K., Allen, J. E., Chen-Harris, H., Zemla, A., Vanier, G., Mabery, S., Torres, C., Hullinger, P., & Slezak, T. (2013). 'The Role of Viral Population Diversity in Adaptation of Bovine Coronavirus to New Host Environments'. *PLoS One*, 8(1), e52752. https://doi.org/10.1371/journal.pone.0052752

13. Accident

288 *But if an engineered virus does escape, there might be no way to distinguish it from a natural mutant* Sarah Scoles. 'How Do We Know If a Virus Is Bioengineered?' *Future Human*. 5 August 2020. https://futurehuman.medium.com/how-do-we-know-if-a-virus-is-bioengineered-541ff6f8a48f

289 *Dr Ian Lipkin of Columbia University* Donald G. McNeil Jr.
     'How I Learned to Stop Worrying and Love the Lab-Leak
     Theory'. *Medium.* 17 May 2021. https://donaldgmcneiljr1954.
     medium.com/how-i-learned-to-stop-worrying-and-love-the-lab-
     leak-theory-f4f88446b04d

290 *Dr Bernard Roizman, a virologist at the University of Chicago*
     Jeremy Page, Betsy McKay and Drew Hinshaw. 'The Wuhan Lab
     Leak Question: A Disused Chinese Mine Takes Center Stage'.
     Wall Street Journal. 24 May 2021. https://www.wsj.com/articles/
     wuhan-lab-leak-question-chinese-mine-covid-
     pandemic-11621871125

290 *Dr Charles Calisher, a Colorado State University virologist also
     told ABC News* Kaitlyn Folmer, Sony Salzman, Sasha Pezenik,
     Mark Abdelmalek and Lucien Bruggeman. 'Nature-Based or Lab
     Leak? Unraveling the Debate Over the Origins of COVID-19'.
     ABC News. 14 June 2021. https://abcnews.go.com/US/nature-
     based-man-made-unraveling-debate-origins-covid/
     story?id=78268577

290 *Dr Peter Palese told the* New York Post Bruce Golding.
     'Prominent NYC Scientist Backtracks on COVID Origin Over
     "Disturbing Information"'. *New York Post.* 4 June 2021. https://
     nypost.com/2021/06/04/prominent-nyc-scientist-backtracks-on-
     natural-origin-of-covid/

290 *Dr Gary Whittaker, a specialist in influenza and coronaviruses*
     Whittaker, G. R. (2021). 'SARS-CoV-2 Spike and Its Adaptable
     Furin Cleavage Site'. *Lancet. Microbe*, 10.1016/S2666-
     5247(21)00174-9. Advance online publication. https://doi.
     org/10.1016/S2666-5247(21)00174-9

14. The origin of Covid-19

291 *There is a neat phrase used by the philosopher Daniel Dennett:
     burden-tennis* Review of J. Fodor, *Psychosemantics, Journal of
     Philosophy*, LXXXV, 384–389, July 1988. Jerry Fodor,
     Psychosemantics, Cambridge, MA: The MIT Press/A Bradford
     Book, 1987, 173 pp. Available at: http://cogprints.org/253/1/
     fodor.htm

294 *In the words of the journalist Rowan Jacobsen* Rowan Jacobsen.
     'Humans, Not Animals, Likely Took the COVID Virus to Wuhan,
     Contrary to China's Claims'. *Newsweek.* 25 March 2021. https://
     www.newsweek.com/humans-not-animals-likely-took-covid-virus-
     wuhan-contrary-chinas-claims-1578861

295 *Dr Baric's group published a proof-of-concept* Martinez, D. R.,
Schäfer, A., Leist, S. R., De la Cruz, G., West, A., Atochina-
Vasserman, E. N., Lindesmith, L. C., Pardi, N., Parks, R., Barr,
M., Li, D., Yount, B., Saunders, K. O., Weissman, D., Haynes, B.
F., Montgomery, S. A., & Baric, R. S. (2021). 'Chimeric Spike
mRNA Vaccines Protect Against Sarbecovirus Challenge in Mice'.
Science (New York, N.Y.), eabi4506. Advance online publication.
https://doi.org/10.1126/science.abi4506

297 *As Dr David Relman of Stanford University put it* Relman
D. A. (2020). 'Opinion: To Stop the Next Pandemic, We
Need to Unravel the Origins of COVID-19'. *Proceedings of the
National Academy of Sciences of the United States of America*,
117(47), 29246–29248. https://doi.org/10.1073/
pnas.2021133117

299 *On 26 May 2021, the US President Joe Biden issued a statement*
Statement by President Joe Biden on the Investigation into the
Origins of COVID-19. The White House. 26 May 2021. https://
www.whitehouse.gov/briefing-room/statements-
releases/2021/05/26/statement-by-president-joe-biden-
on-the-investigation-into-the-origins-of-covid-19/

299 *bipartisan hearings and Senate briefings in the US* 'PRINCIPLES
FOR OUTBREAK INVESTIGATION: COVID-19 AND FUTURE
INFECTIOUS DISEASES'. 14 July 2021. https://science.house.
gov/hearings/principles-for-outbreak-investigation-covid-19-and-
future-infectious-diseases; see also 'Sens. Marshall and Gillibrand
Hold Bipartisan COVID Origins Briefing'. 3 August 2021. https://
www.marshall.senate.gov/press-releases/sens-marshall-and-
gillibrand-hold-bipartisan-covid-origins-briefing/

Epilogue: Truth will out

303 *On 27 August 2021, the US government's intelligence agencies*
Office of the Director of National Intelligence. The unclassified
United States Intelligence Community report on the Origin of
Covid-19. 27 August 2021. https://www.dni.gov/files/ODNI/
documents/assessments/Unclassified-Summary-of-Assessment-on-
COVID-19-Origins.pdf; see also: The White House. Statement by
President Joe Biden on the Investigation into the Origins of
COVID-19. 27 August 2021. https://www.whitehouse.gov/
briefing-room/statements-releases/2021/08/27/statement-by-
president-joe-biden-on-the-investigation-into-the-origins-of-covid-
%E2%81%A019/

304 *much of the American public and the international media had already shifted* Alice Miranda Ollstein. 'POLITICO-Harvard poll: Most Americans Believe Covid Leaked from Lab'. 9 July 2021. https://www.politico.com/news/2021/07/09/poll-covid-wuhan-lab-leak-498847

305 *notable exceptions* The notable exceptions among journalists included: Ian Birrell of the *Mail on Sunday*, John Sudworth of the BBC, Josh Rogin of the *Washington Post*, Aksel Fridstrøm of *Minervanett*, Jackson Ryan of CNET, Antonio Regalado of *MIT Technology Review*, and Rowan Jacobsen who wrote for *Boston Magazine* and *Newsweek*.

305 *only Jamie Metzl publicly expressed scepticism* Jamie Metzl. Origins of SARS-CoV-2. https://jamiemetzl.com/origins-of-sars-cov-2/

305 *Yuri Deigin's essay in* Medium *in April 2020* Deigin, Y. 'Lab-Made? SARS-CoV-2 Genealogy Through the Lens of Gain-of-Function Research'. *Medium*. 22 April 2020. https://yurideigin.medium.com/lab-made-cov2-genealogy-through-the-lens-of-gain-of-function-research-f96dd7413748

305 *Milton Leitenberg's essay in June 2020* Milton Leitenberg. 'Did the SARS-CoV-2 Virus Arise from a Bat Coronavirus Research Program in a Chinese laboratory? Very Possibly'. *Bulletin of the Atomic Scientists*. 4 June 2020. https://thebulletin.org/2020/06/did-the-sars-cov-2-virus-arise-from-a-bat-coronavirus-research-program-in-a-chinese-laboratory-very-possibly/

305 *Rowan Jacobsen's profile of Alina* Rowan Jacobsen. 'Could Covid-19 Have Escaped from a Lab?' *Boston Magazine*. 9 September 2020. https://www.bostonmagazine.com/news/2020/09/09/alina-chan-broad-institute-coronavirus/

305 *Nicholson Baker's long essay in* New York Magazine *in January* Nicholson Baker. 'The Lab-Leak Hypothesis'. *New York Magazine*. 4 January 2021. https://nymag.com/intelligencer/article/coronavirus-lab-escape-theory.html

305 *Our own articles in the* Wall Street Journal *in January and the* Telegraph *in March* Alina Chan and Matt Ridley. 'The World Needs a Real Investigation into the Origins of Covid-19'. *Wall Street Journal*. 15 January 2021. https://www.wsj.com/articles/the-world-needs-a-real-investigation-into-the-origins-of-covid-19-11610728316; see also Matt Ridley and Alina Chan. 'Did the Covid-19 Virus Really Escape From a Wuhan Lab?' *Daily Telegraph*. 29 March 2021. https://www.telegraph.

co.uk/news/2021/02/06/did-covid-19-virus-really-escape-wuhan-lab/

305 *Dr Virginie Courtier-Orgogozo of the Institut Jacques Monod in Paris* Courtier V. and Decroly, E. 'Covid-19: Why the Lab Leak Theory Must Be Formally Investigated'. *The Conversation.* 2 June 2021. https://theconversation.com/covid-19-why-the-lab-leak-theory-must-be-formally-investigated-161297

306 *Steven Quay's lengthy manuscript* Quay MD PhD, Steven Carl. (2021, 29 January). 'A Bayesian Analysis Concludes Beyond a Reasonable Doubt that SARS-CoV-2 Is Not a Natural Zoonosis But Instead Is Laboratory Derived (Version 2)'. Zenodo. http://doi.org/10.5281/zenodo.4477081

306 *Angus Dalgleish and Birger Sørensen* Sørensen, B., Dalgleish, A., & Susrud, A. (2020). 'The Evidence Which Suggests That This Is No Naturally Evolved Virus A Reconstructed Historical Aetiology of the SARS-CoV-2 Spike'

306 *The* Washington Post *accused Senator Tom Cotton* Paulina Firozi. 'Tom Cotton Keeps Repeating a Coronavirus Conspiracy Theory That Was Already Debunked'. *Washington Post.* 17 February 2020. https://www.washingtonpost.com/politics/2020/02/16/tom-cotton-coronavirus-conspiracy/

306 *the* New York Times *said the Wuhan laboratory had been 'the focus of unfounded conspiracy theories promoted by the Trump administration' New York Times* on Facebook. 3 February 2021. https://www.facebook.com/nytimes/posts/10152595350679999

306 *National Public Radio reported that 'scientists debunk lab accident theory' when actually they had (barely) rebutted the genetic-manipulation theory* Geoff Brumfiel. 'Scientists Debunk Lab Accident Theory of Pandemic Emergence'. NPR, All Things Considered. 22 April 2020. https://www.npr.org/2020/04/22/841925672/scientists-debunk-lab-accident-theory-of-pandemic-emergence

307 *the letter in* Science *that eighteen scientists signed in mid-May* Bloom, J. D., Chan, Y. A., Baric, R. S., Bjorkman, P. J., Cobey, S., Deverman, B. E., Fisman, D. N., Gupta, R., Iwasaki, A., Lipsitch, M., Medzhitov, R., Neher, R. A., Nielsen, R., Patterson, N., Stearns, T., van Nimwegen, E., Worobey, M., & Relman, D. A. (2021). 'Investigate the Origins of COVID-19'. *Science (New York, N.Y.), 372*(6543), 694. https://doi.org/10.1126/science.abj0016

307 *Nicholas Wade's long essay in* Medium Nicholas Wade. 'Origin of Covid – Following the Clues'. *Medium*. 2 May 2020. https://nicholaswade.medium.com/origin-of-covid-following-the-clues-6f03564c038

307 *The* Wall Street Journal *detailed an intelligence report* Michael R. Gordon, Warren P. Strobel and Drew Hinshaw. 'Intelligence on Sick Staff at Wuhan Lab Fuels Debate on Covid-19 Origin'. *Wall Street Journal*. 23 May 2021. https://www.wsj.com/articles/intelligence-on-sick-staff-at-wuhan-lab-fuels-debate-on-covid-19-origin-11621796228

307 *Facebook reversed its policy* Cristiano Lima. 'Facebook No Longer Treating "Man-Made" Covid as a Crackpot Idea'. *Politico*. 26 May 2021. https://www.politico.com/news/2021/05/26/facebook-ban-covid-man-made-491053

307 *Ian Birrell of the* Mail on Sunday *wrote an excoriating article* Ian Birrell. 'What a Wuhan Lab Leak Would Really Mean'. Unherd. 26 May 2021. https://unherd.com/2021/05/what-if-there-was-a-lab-leak

308 *Even the* New York Times *finally caved in* Zeynep Tufekci. 'Where Did the Coronavirus Come From? What We Already Know Is Troubling'. *New York Times*. 25 June 2021. https://www.nytimes.com/2021/06/25/opinion/coronavirus-lab.html

308 *the Chinese state-controlled media appeared to have given up* Huang Lanlan. 'With Records of "Lab-Created Coronaviruses" Incidents, Supervision Loopholes and Audacious Germ Researchers, What Really Happened in US' UNC Labs?' *Global Times*. 9 August 2021. https://www.globaltimes.cn/page/202108/1230914.shtml; see also: Huang Lanlan and Zhang Hui. 'With Records of "Lab-Created Coronaviruses" Incidents, Supervision Loopholes, and Audacious Germ Researchers, What Really Happened in US' UNC and Fort Detrick Labs?' *Global Times*. 18 August 2021. https://www.globaltimes.cn/page/202108/1231868.shtml

308 *On 10 August the Swiss embassy in Beijing took to Twitter* Embassy of Switzerland in Beijing @SwissEmbChina Twitter. 10 August 2021. https://twitter.com/swissembchina/status/1425042973289504770; see also: Suranjana Tewari. 'China: Swiss Embassy Urges Media to Remove Scientist Fake News'. BBC. 11 August 2021. https://www.bbc.com/news/world-asia-china-58168588

309 *Two young activists* Gao Feng. 'Researchers Jailed Over GitHub Stash of Pandemic Content Banned in China'. Radio Free Asia. 18

August 2021. https://www.rfa.org/english/news/china/banned-08182021133234.html

309 *A citizen journalist, Zhang Zhan* Amy Chang Chien and Austin Ramzy. 'Chinese Citizen Who Documented Wuhan Outbreak Falls Ill in Prison Hunger Strike'. *New York Times*. 25 August 2021. https://www.nytimes.com/2021/08/25/world/asia/china-zhang-zhan-hunger-strike.html

309 *Chinese reporters who spoke to Peter Hessler from the* New Yorker Peter Hessler. 'Nine Days in Wuhan, the Ground Zero of the Coronavirus Pandemic'. *New Yorker*. 5 October 2021. https://www.newyorker.com/magazine/2020/10/12/nine-days-in-wuhan-the-ground-zero-of-the-coronavirus-pandemic

310 *The* New York Times *reported* Julian E. Barnes. 'Intelligence Review Yields No Firm Conclusion on Origins of Coronavirus'. *New York Times*. 27 August 2021. https://www.nytimes.com/2021/08/27/us/politics/covid-origin-lab-leak.html

310 *The* Wall Street Journal *remarked* The Editorial Board. 'The Covid Origin Muddle'. *Wall Street Journal*. 27 August 2021. https://www.wsj.com/articles/the-covid-origin-muddle-joe-biden-intelligence-community-report-11630102087

310 *The three WIV workers* Bari Weiss. 'You're Already Living in China's World Pt 1: The Lab Leak Lies'. *Honestly with Bari Weiss*. 23 August 2021. https://podcasts.apple.com/us/podcast/honestly-with-bari-weiss/id1570872415?i=1000532793045

310 *At the end of July* Francis S. Collins. Letter to Senator Grassley. 28 July 2021. https://www.grassley.senate.gov/imo/media/doc/national_institutes_of_health_to_grassley_-_covid_origins_grant_oversight.pdf

310 *On 6 September 2021* Sharon Lerner and Mara Hvistendahl. 'New Details Emerge About Coronavirus Research at Chinese Lab'. *Intercept*. 6 September 2021. https://theintercept.com/2021/09/06/new-details-emerge-about-coronavirus-research-at-chinese-lab/

312 *the WIV is known to have* Wuhan Institute of Virology. Measures for the Administration of Information Disclosure Work of Wuhan Institute of Virology, Chinese Academy of Sciences. 30 October 2017. https://web.archive.org/web/20210427041645/http://www.whiov.cas.cn/xxgk_160268/xxgkzd_160271/201911/t20191103_5420266.html

Illustrations

p. 12 Tongguan Township in Mojiang County, Yunnan. (*Zhou Lei/Xinhua/Alamy Live News Xinhua/Alamy Stock Photo*)

p. 34 June Almeida in 1963 at the Ontario Cancer Institute in Toronto. (*Norman James/Toronto Star via Getty Images*)

p. 65 The city of Wuhan in Hubei province. (*SleepingPanda/Alamy Stock Photo*)

p. 75 Masked palm civet (*Paguma larvata*) on sale in a market in Southeast Asia. (*travelib prime/Alamy Stock Photo*)

p. 91 Sunda pangolin (*Manis javanica*). (*Suzi Eszterhas/Wild Wonders of China/naturepl.com*)

p. 110 Two greater horseshoe bats (*Rhinolophus ferrumequinum*). (*Rudmer Zwerver/Alamy Stock Photo*)

p. 159 A scientist working in a biotechnology laboratory, biosafety level 2. (*imageBROKER/Alamy Stock Photo*)

p. 174 Ferrets used in influenza virus gain-of-function research. (*ITAR-TASS News Agency/Alamy Stock Photo*)

p. 204 Three spike proteins fitted together on the surface of a SARS-CoV-2 virus. (*Martin Brown*)

p. 218 An alignment of the S1/S2 region of the spike gene of SARS-CoV-2 as compared to other closely related sarbecoviruses. (*Alina Chan*)

p. 233 Twitter user @babarlelephant's tree from 7 July 2020. (*@babarlelephant*)

p. 252 WHO-China origins of Covid-19 joint study news conference in Wuhan, 9 February 2021. (*REUTERS/Aly Song/Alamy Stock Photo*)

p. 276 Wuhan's Huanan seafood market. (*Imaginechina Limited/Alamy Stock Photo*)

p. 284 Dr Peter Daszak and Dr Thea Fischer arriving at the WIV, 3 February 2021. (*REUTERS/Alamy Stock Photo*)

p. 293 Plausible paths that the virus could have taken from bats to people. (*Martin Brown*)

Index

ACTAsia, 107–8
African green monkeys, 146
African swine fever, 195
ageing science, field of, 214
Ai Fen, Dr, 50, 52, 55, 59, 250
Alibekov, Dr Kanatjan, 146–7
Almeida, June, 33–5
Andersen, Dr Kristian, 130–1, 151–2, 153, 154, 211–12, 214–16, 292, 304, 306
Angiotensin Converting Enzyme-II (ACE2), 41, 118, 125, 287, 288; in lab studies/research, 41–2, 182–3, 184–5, 186–7, 188–9, 193, 207–8, 219, 241, 287, 288; as SARS-CoV-2 RBD, 41–2, 97, 204, 204, 265, 275, 276, 287, 288; versions of in other animals, 41–2, 118, 184, 275, 276
Animal Care Trust, 107–8
animals/wildlife: Covid-19's capacity to infect wide range of species, 41, 275, 282; domestic cat meat, 70; escaped Hangzhou leopards, 141–2; high-density chicken farms, 173; safe disposal of lab animals, 165–6; sampling of for viruses, 5, 6–7, 69, 74, 75, 81, 96–7, 117–18, 124–5, 258, 268, 282, 311–13; versions of ACE2 in, 41–2, 118, 184, 185, 188–9, 275, 276 see also bats; pangolins; wildlife trade
anthrax, 145, 146–7, 178–9, 281
antibiotics, 10, 11, 36, 256–7
Asher, David, 140
Asilomar conference (1975), 298
Australia, 19, 45, 61, 64, 73, 111, 261, 266

avian influenza, 145, 172–8, 205, 206

Babarlelephant, 229, 232, 233, 236, 237, 299
bacteria, 36–7, 143–4, 146–7, 178, 298
Bahulikar, Rahul, 26
Baker, Nicholson, 305
Baltimore, Dr David, 210, 213, 215, 290
Bangladesh, 19
Bannon, Steve, 199
Baric, Dr Ralph, 153, 157, 160, 180–2, 183–7, 190, 201, 308; cooperation with Dr Shi's group, 185, 186–7, 208, 267; synthesising of novel viruses, 196–7; and vaccines, 193, 194–6, 295; on WIV, 267–8
bat coronaviruses, 7, 11, 17, 35, 65–7, 83–4, 118–27, 131–2, 182–7, 216–20, 240–3, 270–2; Babarlelephant's tree diagram, 232, 233, 236, 237; BtCoV/4991 sample from Mojiang mine, 20–1, 22, 24–5, 26–32, 83, 126, 222, 229, 233, 239–40, 241, 242; and the common cold, 43; MERS, 79, 84, 111, 209, 222, 272, 289; Ra4991 renamed, 24–5, 28–9, 30, 226, 238; RaTG13 fragment, 22–32, 90, 169, 170, 200, 205, 219, 220, 222, 225, 228–30, 232–4, 239–40, 242, 275, 286, 288, 292; reviewed in Latinne et al. (25 August 2020), 231–2, 287–8; and SARS epidemic (2002–3), 77–8; SARS-CoV-2 genome resembles, 2, 21–5, 28–32, 222; thorough/detailed nature of Chinese research, 287, 288, 292–3, 311–13; virus RmYN02,

bat coronaviruses (*cont ...*)
216–18; WIV library of genomes, 186,
188–9, 267–8, 286, 287–8; WIV work
on, 113–16, 118–19, 121–3, 125–9,
131–2, 182–3, 186, 188–90, 192–4,
221–4, 240–3, 267–8, 284–5;
WIV-EcoHealth Alliance team,
115–16, 118–23, 125–6, 127, 166,
170, 186, 188–90, 191–2, 232, 237,
305, 310–11; WIV's '7896' group
(eight other viruses), 21, 225–6,
232–9, 243, 274, 285, 288, 292;
Wuhan Institute of Virology (WIV),
311–13; Zhoushan viruses, 126, 198,
200, 222 *see also* horseshoe bats;
Mojiang copper mine, Yunnan
bats: bat guano trade, 16, 81, 111, 311;
dangers of catching and handling, 81,
114, 127–8, 311–12; eating of in
southern China, 81, 83–4, 261; fruit
bats, 73, 81; kept in WIV laboratory,
128–9; as living in large aggregations,
37–8, 110, 111, 112; as long-lived
animals, 111–12; as not traded in
Wuhan's markets (2017–19), 104–5;
not typically consumed in central
China, 105; protease system, 209;
RmYN02 virus, 216–18; sampling of
in Hubei province, 258; species of,
110, 111; unique immune systems,
112–14; Wuhan CDC's sampling of,
123–5, 127, 128, 241; as zoonotic
source of viruses, 37–8, 43, 65–7, 73,
79, 81–2, 111, 114–16, 117–29,
131–2, 216–20 *see also* horseshoe bats
'The Bats Behind the Pandemic' (Matt
Ridley essay, April 2020), 1–2, 231
Baumgarth, Dr Nicole, 176, 177
Bedford, Trevor, 153, 155
Beijerinck, Martinus, 37
Beijing Institute of Pathogen Biology,
11, 19
Biden, Joe, 299, 303, 309–10
biosecurity/biosafety: accidents as rare
events, 147–8, 281, 292; BSL-4
laboratories world-wide, 160;
cautious limits set at Asilomar (1975),
298; China's mobile BSL-3
laboratories, 160, 161; concerns over
safety at WIV's high-level BSLs,
163–4, 165; dangers of catching/
handling bats, 81, 114, 127–8,
311–12; deliberately making viruses
more dangerous to humans, 172–8;

disposal of biological waste, 165–6;
levels of, 133–4, 139, 158–60, 253,
254, 289–90, 311; mistakes in US
laboratories (2014), 178–9; no
enforceable international standards,
281; plans for high-level BSLs across
China, 164–5; safety precautions,
133–4, 139; WIV's BSL-4 laboratory,
158–9, 160–4, 253, 254
bioweapons, 197, 200–1, 278, 297, 306
Birrell, Ian, 98, 307–8
Bloom, Dr Jesse, 56–8, 266–7, 268
Bostickson, Billy (Drastic member),
27–8, 129, 170, 213, 299
Botao Xiao, Dr, 64–6, 124
brucellosis, 144
Brundtland, Gro Harlem, 71

Cai Wei, 309
Calisher, Dr Charles, 290
Cambodia, 219, 258, 286
camels, 79, 84
Canada, 64, 70, 72
Canard, Dr Bruno, 203, 261, 265
Carroll, Dennis, 130
Center for Disease Control and
Prevention (CDC), China: Chinese
Institute of Virology, 136–8;
crackdown on whistleblowers, 62,
247; doctoral thesis from (2016), 13,
16, 17, 18, 26, 31, 226; headquarters
in Beijing, 11, 60, 68, 69, 161–2; and
Huanan seafood market, 69, 85, 86,
88, 98, 251; and Mojiang copper
mine, Yunnan, 11, 13, 17, 19, 226;
and outbreak of pandemic, 51; sale of
testing kit rights, 60; and SARS
epidemic (2002–3), 70, 71, 136–8;
Wuhan regional CDC, 21, 65–6, 69,
85, 123–5, 127, 128, 158, 165, 241,
280
Chakravarti, Aravinda, 153
Chen Jinping, Dr, 93–4
Chen Mai, 309
Chen Wu, Dr, 93–4
Chen Zhu, Dr, 160
China-WHO joint study, 7, 32, 284;
and apparent failures of Chinese
scientists, 257–8, 262–4; and
ascertainment bias, 256–7; China
dictate terms of, 248, 249–50, 251,
253, 254–5, 257–8, 261–2, 264, 274,
297; composition of international
team, 157, 247–9; frozen-food/cold

chain theory, 250, 251, 252–3, 259, 261, 274, 294; investigation, 58, 249, 250–5; laboratory leak theory dismissed by, 251, 253–4, 261, 292; lack of access to samples and data, 254–5, 257–8, 264, 283; Paris group's open letter, 264–5; press conference (9 February 2021), 251, 252, 261–2, 263–4, 265, 292, 305–6; report of, 87–8, 251, 255–7, 261–2, 265–6, 310; terms of reference, 246–7; WHO team denied access to raw data, 254–5, 264, 283; widespread criticism of in west, 261–3, 266, 305–6; and wildlife sampling, 258

Chinese government/authorities: behaviour of during pandemic, 49–58, 59–62, 85–7, 94–5, 96–7, 99–102, 106–8, 167–71, 274–5, 283; Biden's comments on cover-up by, 309–10; critics and enemies of, 278; harsh treatment of whistleblowers, 309; issuing of gag orders, 58, 247; and laboratory leak theory, 283, 308; lobbying of WHO by, 245–6; Ministry of Agriculture, 209–10; Ministry of Science and Technology, 164–5; restricting of academic publications on Covid, 231; secrecy during SARS epidemic (2002–3), 71–2, 245, 275; secrecy over laboratory research, 162, 164, 167–71; secrecy over Mojiang mine, 12–15, 24–6, 28–32, 243, 286; secretive and autocratic system, 274–5; and US intelligence community, 303–4; WHO rebukes (January 2021), 249–50; WHO's praise for, 244, 245

Chinese Human Rights Defenders, 53
Chirac, Jacques, 160
Chmura, Aleksei, 127
Chu, Dr C. M., 150
Chuang Yin-ching, Professor, 53
civets, 41, 74–8, 75, 80, 84, 118, 189, 260, 272, 282, 288, 312
Cobey, Dr Sarah, 268
Colaiacovo, Moreno, 236, 299
Collins, Drs Francis, 176–7
coronaviruses: Junc Almeida identifics (1960s), 34–5; alpha-coronavirus genus, 35, 119–21, 131; beta-coronavirus genus, 18, 31, 35, 170, 205, 231, 235; bovine (BCoV), 43, 207; the common cold, 34–5, 37, 40,

43, 207; four groups of, 35, 43; HKU1, 35, 205; HKU4, 208–9; infectious bronchitis virus, 209–10; OC43 version, 35, 43–4, 205, 207; in pangolins, 94–5, 99, 102–3, 106 see also MERS (Middle East Respiratory Syndrome); SARS-like/related viruses

Cotton, Tom, 306
Covid-19 pandemic: asymptomatic but infectious patients, 54, 139; children as largely spared, 44; December 2019–January 2020 in Wuhan, 5–6, 21–3, 28–9, 49–54, 55–60, 255–7, 282–3; early criteria identifying cases in Wuhan, 256–7; first confirmed case outside China, 63; hypothesis on 1889–90 pandemic, 43–4; impact of, 4, 95; importance of finding origin of, 4, 6–7; men affected more severely than women, 44; pre-symptomatic transmitters of virus, 6, 139; SARS-CoV-2 as causative agent, 1; snake theory, 82–3; start of global spread, 63–4; symptoms of virus, 41, 44, 139; Wuhan locked down (23 January 2020), 63

Crimean-Congo haemorrhagic fever, 133–4
crocodiles, 82, 87
Cui Jie, Dr, 127, 240

Dalgleish, Angus, 306
Daszak, Dr Peter: background of, 116–17; and Drastic, 169–70, 213, 233; and frozen-food/cold chain theory, 252; and grants from US sources, 115, 117–18, 188–90, 191–2; inclusion in the WHO team, 157, 247, 248–9, 252, 254, 262, 263, 284; *Lancet* letter (19 February 2020), 156–7, 248, 290, 304; and Latinne et al., 232, 233; as long-time collaborator with WIV/Dr Shi, 25, 29, 82, 115, 118, 128, 129, 131–2, 186, 229, 238, 247, 248, 249; NASEM consults, 153–4; and pangolins, 103; and sequencing of RaTG13, 29, 170, 229, 230; and Shi-Baric collaboration, 186, 187; supports GVP, 130; supports natural spillover theory, 82, 123, 155–7, 248, 249, 306; tweet on Cambodia, 258; and vaccines, 193; and WIV's pathogen database, 25, 166, 169–70, 310

Decroly, Dr Etienne, 203, 265
Dedkov, Dr Vladimir, 254–5
Deigin, Yuri, 213–15, 217, 299, 305, 306
Demaneuf, Gilles, 136, 299
Denison, Dr Mark, 185, 193
Dennett, Daniel, 291–2
Deverman, Dr Ben, 2, 97–9
diabetes, 41
Drastic ('Decentralized Radical Autonomous Search Team Investigating COVID-19'), 27–8, 129, 169–70, 213–14, 230–5, 299–300, 305
Durrell, Gerald, 116
Dwyer, Dr Dominic, 255

Ebola, 133–4, 145, 272
Ebright, Dr Richard, 188, 264–5
EcoHealth Alliance: grants from US sources, 115, 117–18, 188–90, 191–2; and natural spillover theory, 117–19, 131, 155–7; roots/history of, 116–17; WIV collaboration, 115–16, 118–23, 125–6, 127, 163, 164, 166, 170, 186, 188–90, 191–2, 232, 237, 305, 310–11; and WIV medical records/ data, 312; WIV receives funding from, 29, 118, 156, 163, 186, 188–90, 191–2, 229, 310–11 see also Daszak, Dr Peter
Embarek, Dr Peter Ben, 246, 250, 251, 254, 257, 262, 274
Emerging Pandemic Threats (EPT) programme, USAID, 116, 130
engineered virus theory, 151–5, 210–16, 304, 306; confused with lab leak question, 154, 306; engineered with intent theory, 153, 154, 197–200, 250, 307; and furin cleavage site, 205–6, 210–11, 213, 215, 220, 277, 288–9, 290, 293; Limeng Yan's claims, 197–200, 250, 307 see also 'gain-of-function' research
European Virus Archive Global (EVAg) project, 167
Ewald, Dr Paul, 45–6, 47

Fang Li, Dr, 208–9
Farrar, Sir Jeremy, 54, 60–1, 151, 152
Fauci, Dr Anthony, 152, 176–7, 189–90, 307
Feng Zijian, Dr, 247
ferrets, 174–5, 177

Field, Dr Hume, 115
First World War, 47–8
Fischer, Dr Thea, 258, *284*
Fletcher, John, 39
Fodor, Jerry, 291
foot-and-mouth virus, 145, 281
Fort Detrick (US army biomedical facility), 144, 147
Fort Dix, New Jersey, 150
Fouchier, Dr Ron, 173–5, 177, 179
France, 44, 64, 160–2; 'Paris group', 264–5, 305
Frieden, Dr Thomas, 178
frozen-food/cold chain theory, 250, 251, 252–3, 259, 261, 274, 294
furin cleavage site: and amino acid sequences, 202, 203, 205, 206, 207, 208, 210, 213; and burden-of-proof issues, 292, 293; CGG-CGG doublet, 213, 215; Deigin's argument on, 213–15, 217, 306; discovery of (1992), 206; laboratory manipulation of, 207–10, 224, 277, 293; MERS-related viruses, 205, 208–9, 222, 223, 224, 289, 310–11; S1/S2 junction in virus spike, 207–9, 216, 217, *218*, 219, 222, 223–4, 288; of SARS-CoV-2, 42, 202–6, 207, 210–11, 213, 215, 220, 221–4, 265, 277, 288–9, 290, 293; widespread searches related to (from 2019), 216–20, 293; WIV ignores insertion, 221–4, 288–9
Furmanski, Dr Martin, 151

'gain-of-function' research: and Dr Ralph Baric, 180–2, 183–7; into bird flu, 172–8; Cambridge Working Group, 179, 191; chimeric viruses, 66, 175, 183, 185, 186–7, 192, 196–7, 288, 295, 310–11, 312; in China, 1, 66, 131–2, 177–8, 184–7, 188–90, 191–2, 287, 288, 310–11, 312–13; competition amongst scientists, 183; controversy/debate over, 27, 176–80, 187–8, 191, 305, 307; on coronaviruses, 181–4, 287, 288; danger of viral escape, 177, 178–9, 187–8, 287, 294; deliberately making viruses more dangerous to humans, 172–8; and enhanced PPPs, 191; Fauci's definition of, 189–90; funding moratorium in US (2014–17), 179–80, 188, 189–91; as ongoing issue, 313; 'passaging' practice, 173–5, 181, 287;

reverse genetics, 66, 180, 181–90, 192; synthetic virus creation, 180, 181–7, 196–7, 200–1, 287, 311–13; temporary moratorium on (2012), 177

Galea, Dr Gauden, 246

Gansu province, 143–4

Gao, Dr 'George' Fu, 11, 13, 51, 68, 86, 98, 130, 161, 251

Garry, Dr Robert, 152

GenBank (international virus sequence database), 23, 24, 29, 59, 61, 169, 226, 228, 231, 232, 237, 241–2

genetic sequences: amplicon sequencing, 227–8, 229, 230; assembly of, 227–8; Babarlelephant's tree diagram, 232, 233, 233, 236, 237; BigD database, 234; early sequencing dataset from Wuhan University, 56–8; genome of SARS virus (2003), 76, 84, 95–7, 282; genomes, 1–2, 5, 20, 38, 39, 40, 118, 120, 288; next-generation sequencing (NGS), 227, 228, 229–30; recombination, 38, 64, 112, 120–1, 125–6, 131, 132, 143, 178, 180, 188–90, 195–7, 219, 241, 242, 271, 273; Ribera's 'big sudoku', 231–5; Ribera's crucial 2 July tweet, 231–2 see also SARS-CoV-2 (causative agent of Covid-19), genome of

genetics: ambiguity of genetic code, 217–18; analysis of genomic texts, 38, 39–40; codons, 39, 82–3, 203, 205, 210, 213, 217–18, 219; four-letter codes for DNA/RNA, 38, 39; genetic manipulation, 3, 20, 24, 42; and language of proteins, 39–40; literary analogy, 38–9; pangolin immune systems, 93; reverse genetics research, 66, 180, 181–90; RNA, 23, 31, 38, 39–40, 42–3, 119, 149, 202, 205–6; scientific advances in, 5, 59, 73–4, 84–5, 288; synthetic DNAs, 180, 181–7; transcription regulatory sequence (TRS), 196; universal genetic code, 213

Germany, 44, 45, 64, 145–6

Gillibrand, Kirsten, 299

GISAID (international virus sequence database), 61

GitHub website, 309

Global Environmental Reporting Collective, 91

Global Public Health Intelligence Network, 71

Global Virome Project (GVP), 130–1, 164

Gostin, Dr Lawrence, 86, 245, 246

Gronvall, Dr Gigi Kwik, 150–1, 154

Guangdong province, 67, 69, 70–1, 74–7, 84, 92, 116, 118, 119, 120, 260, 272, 281–2; Institute of Applied Biological Resources, 93–4; Wildlife Rescue Center, 92–3, 99

Guangjian Zhu, 127

Guangxi province, 67, 92, 103

Guangzhou, 11, 70–1, 76, 89, 92–3, 94, 99, 260, 272

Guizhou province, 126

Guo, Miles, 199

Hangzhou safari park, 141–2

hantaviruses, 79, 142–3

Harriss, Joseph, 161, 162

Hendra virus, 73, 79, 111, 114, 272

herpes, 93

Hessler, Peter, 309

HIV, 73, 153, 182, 205

Hoffmann, Dr Markus, 205

Holmes, Dr Edward, 61, 130–1, 151, 203, 223–4, 250

Hong Kong, 54, 70, 72, 74, 77–8, 83–4, 120, 172, 197–8

horseshoe bats, 22–3, 109–16, 110, 118–27, 182–7, 216–20, 235–6, 270–2; in Mojiang copper mine, Yunnan, 15–16, 20–1, 83; and SARS epidemic (2002–3), 77–8, 82, 84, 116, 125–6, 131; use of in TCM, 81; warnings from scientists over, 83–4

Hoy, Cyrus, 39

Hu, Dr Ben, 115, 125, 163, 188–90, 193, 221

Huanan seafood market, Wuhan, 276; and apparant failures of Chinese scientists, 263–4; cases who had not visited, 54, 56, 63, 64, 255, 282; Chinese authorities change mind over, 68, 98, 251; Chinese authorities suggest as origin, 22, 68, 69, 85–8, 106, 256–7, 261; evidence against being source of pandemic, 2, 3, 57–8, 69, 86–8, 97, 98, 255–6; genome sequences from, 57–8; hurriedly closed and sanitised, 51, 68–9, 85, 277; initial focus on, 22, 68–9, 85–8, 256–7, 261, 282; no animals test positive at, 3, 69, 85, 97, 98, 282; samples taken at, 3, 51, 69, 85–8, 96,

Huanan seafood market (*cont* ...)
97, 98, 282; staff fall ill, 21–2, 50,
57–8; study of animals sold in,
259–61, 276–7; as 'wet market', 82,
87; WHO team visit, 250–1; wild
animals sold at, 259; Wuhan CDC's
proximity to, 65, 66, 123, 158, 165
Huazhong Agricultural University,
Wuhan, 124, 158, 208, 280, 289
Hubei province, 54, 63, 123–5, 241,
258
Hubei Provincial Hospital, 50
Hubei Wildlife Rescue Centre, 158
Husseini, Sam, 117
hydroxychloroquine treatment, 199
hypertension, 41

IgG antibodies, 18
IgM (immunoglobulin-M) antibodies,
17–18, 31
India, 64, 153
influenza, 5, 35, 37, 38, 48; 1918 flu
pandemic, 46–8; avian, 145, 172–8,
205, 206; H1N1 strain, 46–8, 148–51,
178, 308; H5N1 strain, 172–8, 179;
Russian 'flu' pandemic (1889–90),
43–5; Russian flu pandemic (1977),
148–51, 308
Inglesby, Tom, 154
intelligence community, US, 299, 303–4
interferons, 93, 113
Iran, 95
Italy, 44, 64, 95
Ivins, Bruce, 147
Iwasaki, Dr Akiko, 268

Jacobsen, Rowan, 268, 294, 305
Japan, 63, 218, 276, 286
Jersey Wildlife Preservation Trust, 116
Jiangxi province, 80
Jin Qi, Dr, 19, 21
Jinning caves (Shitou and Yanzi), 14,
114, 118–19, 122, 125, 188–9, 234,
273
Jinyintan hospital, Wuhan, 52, 250

Kawaoka, Dr Yoshihiro, 172–3, 174,
175–6, 177, 179
Kean, Jefferson, 47
Kerkhove, Dr Maria Van, 244–5
Kleine-Weber, Dr Hannah, 205
Koch's postulates, 37
Koopmans, Dr Marion, 257, 259
Kuiken, Dr Thijs, 54

Kunming Medical University Hospital,
9–11, 16, 17–18

laboratory leaks: anthrax leak in Soviet
Union (1979), 146–7; bacterial leaks,
143–4, 146–7; danger of SARS-CoV-2
leaks, 139–41; of first SARS virus,
133, 134–9, 281; and Marburg virus,
145–6; need to learn from, 6; as rare
but not infrequent, 147–8, 281, 292;
Russian flu pandemic (1977), 148–51,
308; in UK, 145; in USA, 144–5, 148;
during vaccine trials/production,
143–4, 150–1, 308; of viruses, 133,
134–9, 142–3, 281
laboratory origin theory: Kristian
Andersen's view, 151–2, 211–12, 214,
216, 304, 306; arguments in support
of, 280–90; David Baltimore's view,
210–11, 213, 215, 290; and Baric's
'no-see'm' method, 181–2; burden-of-
proof issues, 292–3, 294; change in
tone of media coverage of, 268–9,
304, 307–8; China-WHO joint study
dismisses, 251, 253–4, 261, 292; and
China-WHO joint study report,
265–6; and Chinese authorities, 283,
308; competing interests issue, 155–8;
confused with engineered virus theory,
154, 306; Daszak rejects, 155–7, 248,
249, 306; Dr Segreto's view, 24; Dr
Shi's *Scientific American* article, 66–7;
furin cleavage site debate, 220, 277,
288–9; implications if true, 295–6;
initial scientific assumptions over, 3,
292, 304–6; lack of direct evidence
for, 278; *Lancet* letter (19 February
2020), 156–8, 290, 304; motivations
of those pursuing, 278–9; NASEM
letter (6 February 2020), 153–5; as
not allegation of malfeasance, 280–1;
as not precluding natural origin, 220,
306; as not requiring genetic
modification, 288; as now strong
possibility, 312–13; online article by
the Xiaos, 64–6, 124; politicisation of,
98, 269, 306, 307–8; 'Proximal
Origin' paper (Andersen et al.), 211,
214, 216, 289–90, 292; Ribera's work
on, 231–5; *Science* letter from top
scientists (May 2020), 266–9, 307;
Segreto-Deigin paper (November
2020), 214; senior virologists' views
(January 2020), 151–2; several

versions of, 290; shift in views of scientists, 289–90, 297–8, 300–1, 304, 307; and US intelligence community, 299, 303–4; view of authors of this book, 300–1; WHO preliminary document (November 2020), 98–9; why Wuhan? question, 7, 283–4, 292; Zhan et al. paper (2 May 2020), 97–9

laboratory research: Chinese government's lack of openness, 162, 164, 167–71; downstream of virus-hunting, 131–2; gene-synthesis machines, 197; growing of novel viruses, 120–1; human 'airway' cell cultures, 183–4, 187; humanised animals, 3, 42, 184–7, 194, 287, 288, 312; 'immortal' cells, 114; 'isolating' of live viruses, 118–19, 120–1, 163, 188–9, 192–4, 311–13; manipulation of furin cleavage site sequences, 207–10, 224, 277, 293; 'passaging' practice, 173–5, 181, 287; pseudoviruses, 182–3, 207–8, 209; recombinant viruses, 120, 132, 143, 178, 180, 188, 190, 195–7; reverse genetics, 66, 180, 181–90; sampling of people, 74–5, 121–3, 240, 311; serial passaging, 120–1, 287; synthetic virus creation, 185–7, 196–7, 200–1, 287, 311–13; timelags in reporting of discoveries, 287–8; wider implications of Covid origin search, 297–8 see also 'gain-of-function' research

Lanying Du, Dr, 222
Lassa fever, 133–4
Latinne, Alice, 231–2
Leendertz, Dr Fabian, 253, 274
Lei Xiao, 64–6, 124
Leitenberg, Milton, 305
Lentzos, Dr Filippa, 265
Li Wenliang, Dr, 50, 52, 53, 59, 250
Lipkin, Dr Ian, 289–90
Lipsitch, Dr Marc, 179, 191
Liu Jianlun, Dr, 70
Liu Ping, Dr, 93–4
Liu Yingle, Dr, 21–2
lllandca (anonymous Twitter user), 99
Loman, Dr Nick, 60
'long Covid' symptoms, 41
Lu De (YouTuber), 198, 199
Lucey, Dr Daniel, 199–200, 263–4

malaria, 4, 37, 46
Malaysia, 73, 111

Marburg, Institute of Virology, 206
Marburg virus, 133–4, 145–6, 272
market origin thesis, 3–4, 247, 251–2, 255–7, 282; and 2003 SARS epidemic, 51, 69, 71–2, 74–7, 80, 84–5, 96–7, 260, 281–2; multi-market hypothesis, 256; study of animals sold in Wuhan, 259–61, 276–7 see also Huanan seafood market, Wuhan
Marshall, Roger, 299
Mazet, Dr Jonna, 116, 130
McNeil Jr, Donald, 269
measles, 205
medicine, traditional Chinese (TCM), 69, 80, 81, 83, 90, 92, 107–8
Merieux, Alain, 161
MERS (Middle East Respiratory Syndrome), 35, 40, 134, 159, 180, 181, 191, 240; and furin cleavage site, 205, 208–9, 222, 223, 224, 289, 310–11; as originating in bats, 79, 84, 209, 222, 272, 289; Dr Zaki discovers, 78–9
Metabiota (viral database firm), 116
Metzl, Jamie, 261–2, 265, 305
Military World Games (Wuhan, October 2019), 55
Miller, Dr Maureen, 263
Moderna (biotech firm), 61–2
Mojiang copper mine, Yunnan, 7, 9–12, 15–32, 225–6; Chinese secrecy over, 12–15, 24–6, 28–32, 243, 286; Ning Wang's thesis (2014), 239–40; peculiar behaviour of Wuhan scientists, 274–5, 286, 292; WIV's expeditions to (2012/13), 19–21, 22, 30–2, 114, 169, 220, 222, 239, 240, 285–6, 292; and WIV's '7896' group (eight other viruses), 21, 234–9, 242, 243, 274, 285, 288, 292
Mong La, Myanmar, 108
mouse hepatitis virus (MHV), 181, 183, 207

Nabel, Dr Gary, 176–7
National Institute of Allergy and Infectious Diseases (NIAID), 188, 189–90, 191–2
National Institutes of Health (NIH), US, 173, 176–7, 178–9, 180, 187, 189–91, 205–6, 248, 307, 310–11
natural spillover theory: and 2003 SARS epidemic, 51, 69–78, 80, 82, 84–5, 96–7, 116, 125–6, 192, 260, 272, 273,

natural spillover theory (*cont …*)
281–2; arguments in support of,
270–9; burden-of-proof issues, 292,
293, 294–5; Chan/Zhan/Deverman
preprint (2 May 2020), 2–3, 97–9;
Daszak supports, 82, 123, 155–7,
248, 249, 306; and EcoHealth
Alliance, 117–19, 131, 155–7; and
farming of wildlife for food, 6, 80–2,
83, 108, 189; implications if true,
294–5; initial scientific assumptions
over, 304–6; intermediate hosts, 84,
85, 89, 97, 102–3, 107, 108; *Lancet*
letter (19 February 2020), 156–8, 290,
304; mammals as greatest risk for
humans, 83; and Mojiang copper
mine, Yunnan, 7, 9–21, 22–3, 24–5,
27, 28–32, 83, 114, 126; NASEM
letter (6 February 2020), 153–5; no
direct evidence for, 281–2, 290, 293;
as not precluding lab leak, 220, 306;
pangolin theory, 2, 89–90, 93–5,
99–102, 105–8; recombination,
125–6, 131, 143, 189, 271, 273;
senior virologists' views (January
2020), 151–2; snake theory, 82–3; and
US intelligence community, 299,
303–4; viruses stemming from
zoonoses, 5–6, 65–7, 73–82, 84–5,
96–7, 111, 114–16, 117–19; and
WHO, 247 *see also* bats; market
origin thesis; pangolins
Nielsen, Dr Rasmus, 57, 258
Ning Wang, 239–40
Nipah virus, 73, 79, 111, 114, 117,
272
norovirus, 37
Nunberg, Dr Jack, 207–8, 215

Palese, Dr Peter, 150, 290
pangolins: biology of, 90–1, 93, 104;
confiscated from smugglers, 1, 92–5,
99, 102–3, 106; coronaviruses in,
1–2, 94–5, 99, 102–3, 106; data
behind 2019 paper released (22
January 2020), 94, 102, 106; at
Guangdong Wildlife Rescue Center,
92–3, 99; illegal trade in, 80, 90,
91–5, 99, 102–3, 104, 107, 108; Liu
et al. paper (May 2020), 94, 101;
lives in the wild, 91, 104; *Manis
javanica*, 90–4, *91*; as not traded in
Wuhan's markets (2017–19), 104–5;
political motives for studies on, 102;

proposed as source of Covid-19
pandemic, 89–90, 93–5, 99–102,
105–8; questions for Chinese
scientists, 106; and TCM, 80, 81, 90,
92, 107; as unlikely to be source of
pandemic, 2, 102–4, 261
paramyxoviruses, 19, 93–4
Peng Zhou, Dr, 113, 119, 221
Perlman, Dr Stanley, 101, 154
Petrovsky, Dr Nikolai, 265
Pickles, Dr Raymond, 184
pig coronaviruses, 119–21, 131,
158.272, 208, 289
the plague, 36, 37, 73
Pöhlmann, Dr Stefan, 205
polio, 4, 195
Pollack, Dr Marjorie, 51
Poon, Dr Leo, 198
Porcine Epidemic Diarrhea
Coronavirus, 208, 289
pork industry, Chinese, 195
Predict, USAID, 5, 116, 117–18,
129–30, 167
ProMED-mail, 51
proteins, 36, 39–40, 41, 42–3, 82–3, 93,
113–14, 202–7, 209
protozoa, 37

Quammen, David, *Spillover*, 127
Quay, Dr Steven, 212, 306

Racaniello, Dr Vincent, 23
Raffarin, Jean-Pierre, 160
Rahalkar, Dr Monali, 26, 30, 264, 299
Rambaut, Dr Andrew, 61, 130–1, 132,
152
Ranst, Dr Marc van, 43–4
Ray, Prasenjit 'Jeet' (the Seeker), 7, 17,
26–7, 30, 234, 239, 299
Reddit, 28
Reed, Brian, 15, 299
Relman, Dr David, 248, 266–7, 268,
297
Research Institute for Medicine of
Nanjing Command, 126–7, 198
retroviruses, 210
Ribera, Francisco de, 193–4, 229,
230–5, 238, 239, 242, 264, 299
Robinson, Nathan, 269
rodents, 19, 22, 35, 41–2, 43, 73,
142–3, 312
Rogin, Josh, 55, 163–4
Roizman, Dr Bernard, 290
Rosen, Gary, 2

Rozo, Dr Michelle, 150–1
Russian 'flu' pandemic (1889–90), 43–5
Ryan, Dr Mike, 245

salamanders, 82, 87
Salzberg, Dr Steven, 179
sarbecovirus subgenus *see* SARS-like/ related viruses
SARS epidemic (2002–3), 5, 11, 14, 20, 22, 43, 53, 181, 215; Chinese government's lack of openness, 71–2, 245, 275; furin cleavage site of virus, 222–4; genomes of virus, 76, 84, 95–7, 282; in Guangdong, 70–1, 74–7, 84, 116, 118, 260, 272, 281–2; infection of variety of animals, 41; laboratory leaks, 133, 134–9, 281; Metropole Hotel superspreader event, 70, 72, 74; research into origins of, 114–16, 118–19, 125–6; second emergence of (2003–4), 76–7, 85, 282; technology available to scientists, 84–5; virus compared to SARS-CoV-2, 95–7, 282–3; zoonotic source of, 51, 69–78, 80, 82, 84–5, 96–7, 116, 125–6, 192, 260, 272, 273, 281–2
SARS-CoV-2 (causative agent of Covid-19), 1; and 2003 SARS virus, 95–7, 282–3; and ACE2 receptors, 41–2, 97, 204, 204, 265, 275, 276, 287, 288; adaptation scenarios, 97–9; Alpha variant, 283; arginine codons, 203, 205, 210, 213, 215; biosecurity levels at WIV, 159–60, 253, 254, 289–90, 311; capacity to infect wide range of species, 41, 275, 282; close relatives n Southeast Asia/Japan, 276, 286; danger of laboratory leaks, 139–41; Delta variant, 283; effect on the lungs, 40; entry into cells, 41–2, 89–90; genome of published (January 2020), 61, 203, 223–4; as here to stay, 48; human-to-human transmission identified, 50–1, 53–4, 198–9, 244–5; infection as systemic, 40; Li-Meng Yan as whistleblower, 197–200, 250, 307; long incubation time, 139; pangolin theory, 2, 89–90, 93–5, 99–102, 105–8; receptor-binding domain (RBD), 41, 89–90, 94, 99, 100, 106, 118, 182–4, 221, 224, 275; re-infection of Covid survivors, 283; replication of, 42–3; size of particle, 36; as sneaky and

stealthy virus, 40, 139–40; unique characteristics, 6, 42, 202, 205–6, 222–4; as well adapted to human beings, 2, 3, 96, 97–9, 282–3; ZXC21 and ZC45 viruses, 126
SARS-CoV-2 (causative agent of Covid-19), genome of: earliest reported virus genomes, 57–8; furin cleavage site insertion, 42, 202–6, 207, 210–11, 213, 215, 220, 221–4, 265, 277, 288–9, 290, 293; Drs Holmes and Zhang publish, 61–2, 203, 223–4; length of, 42–3; sequencing of (by early January 2020), 21–2, 28, 52–3, 59, 124; similarity of early genomes, 95–7, 282–3; textual analysis of, 38; and virus from pangolin, 1–2, 89–90; Dr Zhang uploads to GenBank, 59–60
SARS-like/related viruses: anticoagulant treatments, 11; Babarlelephant's tree diagram, 232, *233*, 233, 236, 237; and Dr Ralph Baric, 153, 157, 160, 180–2, 183–7; bat virus RmYN02, 216–18; as beta-coronaviruses, 35; BSL-3 conditions required, 134, 139, 159, 160; clusters as geographic, 241; isolated and grown at WIV, 118–19, 163, 188–9, 192–4, 311–13; meaning of term, 18; Rs8561_Guangdong, 241; searching of caves in southern China, 114–15, 118–19, 122, 124–5, 127–8, 188–9, 284–5; SL-CoV-WIV1 virus, 119; spike sequences of, 205–6; WIV's four different lineages, 242–3; WIV's pathogen database, 166–71, 286, 287–8, 292, 310; ZXC21 and ZC45 viruses, 126 *see also* bat coronaviruses
Saudi Arabia, 78–9
Schaffner, Dr William, 177
scientific papers: Chan/Zhan/Deverman preprint (2 May 2020), 2–3, 97–9; doctoral thesis from CDC (2016), 13, 16, 17, 18, 26, 31, 226; Dr Shi et al. in *Emerging Microbes & Infections* (31 January 2020), 222–3; Dr Shi et al. in *Nature* (3 February 2020), 22–3, 24, 26, 28–9, 221–2, 225, 228–9, 289; Dr Yingle's paper (5 February 2020), 22, 24; Kunming medical thesis (2013), 7, 17, 26, 27, 30, 31, 226; Lam et al. paper in *Nature*, 94, 103; Latinne et al. (25 August 2020), 231–2, 233,

scientific papers (*cont ...*)
236, 237, 241–2, 243, 287–8; Ning
Wang's thesis (2014), 239; timelags in
reporting of discoveries, 287–8; Xiao
et al. paper in *Nature*, 94, 99–102; Yu
Ping's thesis (2014), 240–3; Zhan et
al. paper (2 May 2020), 97–9; Zhang
Yongzhen et al. (January 2020),
59–62, 124, 203, 223–4
the Seeker (anonymous Twitter user), 7,
17, 26–7, 30, 234, 239, 299
Segreto, Dr Rossana, 23–5, 30, 214,
217, 264, 299
Sendai virus, 93, 94
Shakespeare, William, 38–9
Shandong, 216–17
Shanghai Public Health Clinical Center,
59, 62
Shen Yongyi, Dr, 89
Shi Zhengli, Dr: addendum to paper in
Nature (17 November 2020), 30–2,
169, 225–6, 234–5, 238; admits 4991
and RaTG13 as same thing, 25; and
bat research/hunting, 19–20, 22–3,
28–32, 113, 114, 115, 118–19, 125–8,
182, 186, 193–4; and biosecurity
levels, 159–60; cooperation with Dr
Baric's group, 185, 186–7, 208, 267;
failure to mention furin cleavage site,
221–4, 288–9; 'gain-of-function'
experiments, 66, 181, 182–3, 185,
186, 188–90, 192, 310–11; interview
with *Science* magazine (July 2020),
29, 159–60, 229–30, 241; and
Kunming laboratory leak, 143; library
of bat coronavirus genomes, 186,
188–9, 267–8, 286, 287–8; and
outbreak of pandemic, 28, 66–7, 275;
paper in *Nature* (3 February 2020),
22–3, 24, 26, 28–9, 221–2, 225,
228–9, 289; reacts to *Science* letter,
268; rejects engineering accusations,
153; SADS episode, 120–1, 131; and
sample history of RaTG13, 229–30;
supports natural spillover theory, 82;
synthesising of novel viruses, 196–7;
and timing of RaTG13 sequencing,
29, 169; and vaccines, 194; and WIV's
pathogen database, 166–7, 169, 286;
and WIV's '7896' group (eight other
viruses), 235–6
Shibo Jiang, Dr, 222
Shitou cave, Jinning County, Yunnan,
14, 188–9, 234, 273

Singapore, 70, 72, 133, 134–6, 138
Small, Charles, 128–9, 167–8, 299
smallpox, 4, 36–7, 145, 178–9, 281
snakes, 74, 75, 82–3, 87
'So Where Did This Virus Come
From?' (Matt Ridley essay, May
2020), 3
social media: censorship of virus origin
discussion, 28, 307; Chinese
censorship of Covid-19 'rumours', 53;
Facebook reverses censorship policy,
307; and Wuhan Health Commission,
52
Sorensen, Birger, 306
South China Agricultural University,
Guangzhou, 89, 94
South Korea, 79
Soviet Union, 146–7, 148
Spain, 44–5
Spedding, James, 39
spike protein gene, 89–90, 94, 185–9,
212, 213, 241, 273, 287, 289; RBD
of, 41, 89–90, 99, 100, 106, 182–4,
219, 221, 241, 275; S1/S2 junction,
207–9, 216, 217, *218*, 219, 222,
223–4, 288; use of by virus, 41–2, 43,
182–3, 185, 187, 203–6, 207–10, 241
see also furin cleavage site
Stahl, Lesley, 254
Sudworth, John, 13–14
Sullivan, Jake, 262
Sweden, 44, 64
swine acute diarrhoea syndrome
(SADS), 119–21, 131, 272
swine flu, 150, 175, 176–7

Taiwan, 53, 72, 133, 135–6, 138, 160
Taiyi cave, Xianning, 113
Tedros Adhanom Ghebreyesus, Dr, 244,
249–50, 262, 265–6
testing kits, 60, 61
Thailand, 63, 219, 286
Third Military Medical University,
Chongqing, 126–7, 198, 200
Tian Junhua, Dr, 123–5, 127, 128
Tong Yigang, Dr, 247
Trump, Donald, 199, 307–8
Tufekci, Zeynep, 308
Twitter, 7, 13, 17, 25, 26, 28, 98, 99,
299–300, 305
Tyrrell, Dr David, 34–5

Uganda, 146
United Kingdom, 44, 45, 64, 145

United States: anthrax letters
(September 2001), 147; biosecurity
lapses in, 144–5, 148; Centers for
Disease Control and Prevention, 86,
116–17, 144–5, 178–9; concerns over
WIV's BSL-4 laboratory, 161, 162,
163–4; Federal Select Agent Program,
145; first confirmed case in, 64;
funding of EcoHealth Alliance, 115,
117–18, 188–90, 191–2; impact of
Covid-19 pandemic, 95; intelligence
community, 299, 303–4
United States Agency for International
Development (USAID), 116, 117–18,
167
UPMC Center for Health Security,
Baltimore, 150–1
Urbani, Carlo, 183
US Right to Know organisation, 101,
140–1, 154, 156, 157
USA Today, 144–5, 148

vaccines: Chinese development of, 52–3,
194; dead versions, 195; development
and distribution infrastructure, 5;
development of in 2020, 296; and Dr
Baric's group, 193, 194–6, 295; early
designs for SARS virus, 187; how they
work, 194–5; live attenuated, 195–6;
messenger-RNA (mRNA) vaccine,
61–2, 151, 196; mutatation back into
pathogenic form, 195; release of the
novel coronavirus genome, 61–2;
rewriting of TRS, 196; against SARS-
like viruses, 192–3, 194–5, 295
Vallance, Sir Patrick, 152
Vero cells (monkey kidney), 99, 120
Vietnam, 70, 72, 92
Virological.org website, 61
viruses, 36–48, 119; Global Virome
Project (GVP), 130–1, 164; laboratory
leaks, 133, 134–9, 142–3, 281;
paramyxoviruses, 19, 93–4 see also
coronaviruses; SARS-like/related
viruses

Wade, Nicholas, 210, 212, 307
Wain-Hobson, Dr Simon, 187–8
Wang, Dr Linfa, 102, 112–13, 114, 115,
119, 157, 240
Wang Yanyi, Dr, 62, 163
West Nile virus, 134, 138
whistleblowers, 50, 52, 53, 197–200,
250, 267, 300, 308–9

Whittaker, Dr Gary, 290
Wildlife Conservation Society, 107, 116
wildlife trade, 74–5, 84–5, 104–5; and
2003 SARS epidemic, 51, 69, 71–2,
74–7, 80, 84–5, 96–7, 260, 281–2;
apparant lack of checks of supply
chains, 263–4; breeding/farming of
exotic species in China, 6, 80–2, 83,
108, 189; Chinese regulations on,
80–1, 83, 104–5, 107–8; eating of
exotic species in southern China, 69,
70, 74, 76, 80, 81, 83–4, 189, 261;
frozen-food/cold chain theory, 250,
251, 252–3, 259, 261, 274, 294;
illegal trade in pangolins, 80, 90,
91–5, 99, 102–3, 104, 107, 108; live
markets, 51, 69, 74–7, 80, 81, 82, 84,
96–7; as ongoing problem, 313;
sampling of people in, 74–5, 122–3,
311; seasonal factors, 260; tick-borne
disease in Wuhan's markets, 104–5 see
also Huanan seafood market, Wuhan
Wolinetz, Dr Carrie, 191
World Health Organization (WHO), 7,
32, 53–4, 58, 86; assumptions over
origin of virus, 305; and Chinese lack
of transparency, 245, 246; declares a
pandemic (11 March 2020), 246;
delegation to China (February 2020),
246; emerging diseases unit, 244–5;
Health Emergencies Programme, 245;
and human-to-human transmission of
virus, 198–9, 244–5; international
experts visit Wuhan (early 2021), 87;
and laboratory leak theory, 98–9; and
laboratory leaks, 137–8, 139;
lobbying of by Chinese government,
245–6; rebukes China (January 2021),
249–50; SARS epidemic (2002–3),
71–2, 114–15, 245; TCM as
recognised form of medicine, 81; team
visits Wuhan (January 2020), 245 see
also China-WHO joint study
World Organisation for Animal Health
(OIE), 69, 85
Worobey, Dr Michael, 268
Wuhan, 7, 65; institutions with
coronavirus laboratories, 158;
Military World Games (October
2019), 55; rapid spread of Covid-19
virus, 6
Wuhan Center for Disease Control and
Prevention (CDC), 65–6, 69, 85,
123–5, 127, 128, 158, 165, 241, 280

Wuhan Central Hospital, 49, 50, 52, 59, 250

Wuhan Institute of Biological Products, 158

Wuhan Institute of Virology (WIV): '7896' group (eight other viruses), 21, 225–6, 232–9, 242, 243, 274, 285, 288, 292; addendum to Dr Shi's paper in *Nature* (17 November 2020), 30–2, 169, 225–6, 234–5, 238; and alpha-coronaviruses in pigs, 119–21, 131; barred from publishing Covid-19 data, 62; bats kept in laboratory, 128–9; behaviour over '7896' group, 225–6, 231–9, 241–2, 243, 274, 285; behaviour over RaTG13 and 4991 sample, 22–3, 24–6, 28–32, 169, 170, 222, 226, 228–30, 239–40, 242, 292; biosecurity levels at, 159–60, 253, 254, 289–90, 311; BSL-4 laboratory in, 159, 160–4, 253, 254; Center for Emerging Infectious Diseases, 20, 66; concerns over safety at high-level BSLs, 163–4, 165; EcoHealth Alliance collaboration, 115–16, 118–23, 125–6, 127, 163, 164, 166, 170, 186, 188–90, 191–2, 232, 237, 305, 310–11; and EcoHealth Alliance funding, 29, 118, 156, 163, 186, 188–90, 191–2, 229, 237, 310–11; expeditions to Mojiang mine (2012/13), 19–21, 22, 30–2, 114, 169, 220, 222, 239, 240, 285–6, 292; experiments on bat cells, 113–14; 'gain-of-function' experiments, 131–2, 181, 182–3, 185, 186, 188–90, 192–4, 287, 288, 311, 312–13; geographical location of, 66, 253–4; ignores furin cleavage site insertion, 221–4, 288–9; isolating/growing of bat SARS viruses, 118–19, 163, 188–9, 192–4, 311–13; and laboratory leak theory, 280–1; as leading lab for bat research/testing, 18, 20; low-key release of vital information, 235–6, 238; and Mojiang mine patients, 7, 11, 17–18, 28–32, 226, 239, 292; pathogen database of, 166–71, 286, 287–8, 292, 310; pathogen database taken offline, 167–71, 286, 292; Ribera's 'big sudoku', 231–5; rumours of ill workers at, 307, 310; secret projects at, 162, 201, 287; sequence for eight viruses belatedly published, 237–9; thorough/detailed nature of research at, 287, 288, 292–3, 311–13; timing of RaTG13 sequencing, 228–30, 288; two campuses of, 158–9, 253–4; US State Department fact sheet on (January 2021), 105; and vaccines, 194; WHO team visit, 253–4; work on SARS viruses in bats, 113–16, 118–19, 121–3, 125–9, 131–2, 182–3, 186, 188–90, 192–4, 221–4, 240–3, 267–8, 284–5, 311–13 *see also* Shi Zhengli, Dr

Wuhan University, 21–2, 24, 25, 56–7, 58, 158, 312

Xi Jinping, 62, 81, 246

Xiao Lihua, Dr, 89

Xie Canmao, Dr, 10

Xishuangbanna Tropical Botanical Garden, 216

Yan, Dr Li-Meng, 197–200, 250, 307

Yang Jian, 127

yellow breasted buntings, 80–1

yellow fever, 37

Young, Alison, 144–5, 148

Yu Ping, 240–3

Yuan Zhiming, Dr, 164–5

Yugoslavia, 146

Yunnan province, 67, 110–11, 216–17, 240–1, 284–5; WIV-EcoHealth Alliance team in, 118–22, 123, 125–6, 127 *see also* Mojiang copper mine, Yunnan

Zaki, Dr Ali Mohamed, 78–9

Zhan, Dr Shing Hei, 2–3, 95–9, 100

Zhang Jixian, Dr, 50, 51, 55, 56

Zhang Yongzhen, Dr, 59–61, 62, 124, 126, 203, 223–4

Zhang Zhan, 309

Zhao Lijian, 251

Zhejiang province, 126–7, 198, 200

Zhong Nanshan, Dr, 10–11, 17, 18

Zhou Xianwang, 63

zoonosis: bats as source of, 35, 37–8, 43, 65–7, 73, 79, 81–2, 111, 114–16, 117–29, 131, 216–20; source of SARS epidemic (2002–3), 51, 69, 70–1, 73–8, 82, 84–5, 96–7, 116, 125–6, 192, 260, 272, 273, 281–2; viruses stemming from, 5–6, 65–7, 73–82, 84–5, 96–7, 111, 114–16, 117–19